W9-CNJ-205

Name Reactions and Reagents in Organic Synthesis

Name Reactions and Reagents in Organic Synthesis

Bradford P. Mundy

Michael G. Ellerd

Department of Chemistry
Montana State University
Bozeman, Montana

A Wiley-Interscience Publication

JOHN WILEY & SONS

New York · Chichester · Brisbane · Toronto · Singapore

Library of Congress Cataloging in Publication Data:

Mundy, Bradford P., 1938–
Name reactions and reagents in organic synthesis/Bradford P. Mundy, Michael G. Ellerd.
p. cm.
"A Wiley-Interscience publication."
Bibliography: p.
Includes index.
ISBN 0-471-83626-5
1. Chemical reactions. 2. Chemical tests and reagents. 3. Chemistry, Organic—Synthesis. I. Ellerd, Michael G. II. Title.
QD291.M86 1988
547′ .2—dc19 88-14915
CIP

Printed in the United States of America

10 9 8 7 6 5 4 3 2

To Margaret and Mary
. . . for putting up with more than anyone should ever have to!

Preface

New Year's Day, 1988, and the book is finished. This has been an enjoyable and interesting project—we have both learned a lot. As I (B.P.M.) sit at the word processor and reflect on the last two years of effort, I know what I would do differently, and how the task could be made more simple. But this did not start as a planned project.

A number of years ago, one of us (B.P.M.) was involved in teaching a course about how to carry out simple one-step transformations, as a starting point for a second course in design of complex syntheses. It was clear that many students did not have a useful vocabulary of reactions, reagents, and knowledge of the "key players" of organic chemistry. I started by preparing a number of reaction and reagent cards. Each class period there would be 3–5 new cards added to the growing list, and each class period I would shuffle the cards and give a random quiz.

As this developed, I needed to add more examples, mechanisms, and so on. The cards developed into full sheets of paper and students were now putting these into ring binders. The format was simple to follow, however; I really didn't take a lot of time to prepare them. A senior chemistry major (M.G.E.) was in my course one year, and after the quarter was over I happened to have the opportunity to see his class notes—they were beautiful! I copied the notes for my own use.

During this developmental period, a number of visiting seminar speakers would see these notes and reaction formats as we discussed educational aspects of the graduate program—and a number of these well-known chemists encouraged me to provide them with copies. During the National Organic Symposium held in Bozeman, Dr. Ted Hoffman, of Wiley-Interscience, encouraged me to organize the handouts into a book. I agreed, and immediately asked Mike Ellerd to participate in the project. This book is the result of this collaboration.

I must give special thanks to Mike as a coauthor—he did all of the artwork and was a tireless library worker. He could find references to reactions and reagents in a spectacular way. This is more difficult these days when the indexing schemes do not list reactions and reagents as commonly as they used to.

This book is written for the students of organic chemistry—be they young or old. We have not included every reaction and reagent—this would be impossible for one text. The Fieser's *Reagents* series is an indispensable compilation for practicing organic chemists, and we must give credit to this work for many of the leads we have incorporated into the Reagent portion of this book. The reactions and reagents picked for inclusion express our own natural prejudices, but we have tried to include those materials usually listed in the most common graduate texts.

Many thanks are required. A number of classes of students have used various levels of handouts and have given useful insight to teaching effectiveness. Dr. Andrew G. Williams and Mr. Scott R. Harring read the entire manuscript and gave valuable suggestions for us and found a number of "typos" and minor errors in artwork. Kathy Deter, of Kwik-Kopy Printing, spent countless hours making copies of all the drawings, one page at a time. We have tried to check each yield, reaction condition, and reference for proper citation (it is absolutely amazing how many citation errors exist in the literature and other monographs). However, there will be errors in this text also—and for these we apologize to those whose work we have incorrectly represented. We would appreciate receiving comments from users of this book. Are there reactions or reagents that should be included in a next edition? Are there better examples? We encourage anyone to send in a two-page format (the same as in this text) for future inclusion. The authors will be properly cited for the effort.

Bradford P. Mundy
Michael G. Ellerd

Bozeman, Montana
January 1988

Contents

Name Reactions and Reagents in Organic Synthesis

In the following section are a number of common named reactions for organic chemistry. The format for each reaction is the same: The left-facing page has a general scheme for the reaction, a mechanism, some notes and references. The right-facing page will have a number of examples.

==

Left-facing page:

==

REACTION: REACTION NAME

==

March's <u>Advanced Organic Chemistry</u> : References to March's text.

GENERAL SCHEME:

A general reaction scheme

MECHANISM:

A possible mechanism

NOTES:

Where applicable, notes are keyed by △.

REFERENCES:

References to the reaction examples.

==

Right-facing page:

==

EXAMPLES:

REACTION: ACETOACETIC ESTER SYNTHESIS

March's Advanced Organic Chemistry : 413

GENERAL SCHEME:

$$CH_3C(=O)CH_2CO_2Et \xrightarrow[2.\ RX]{1.\ :B} CH_3C(=O)CH(R)CO_2Et \xrightarrow[\Delta]{Hyd.} [CH_3C(=O)CH(R)CO_2H] \xrightarrow{-CO_2} CH_3C(=O)CH_2R$$

MECHANISM:

$$CH_3C(=O)-CH(H)-CO_2Et \xrightarrow{:B} CH_3-C(O)\overset{\ominus}{-}CH-C(O)-OEt \xrightarrow[2.\ H^{\oplus}]{1.\ RX} CH_3C(=O)-CH(R)-CO_2Et$$

$$\xrightarrow{Hyd.} CH_3C(=O)-CH(R)-CO_2H \longrightarrow [CH_3C(=O)CH(R)C(=O)O-H] \longrightarrow CH_3C(OH)=CHR \rightleftharpoons CH_3C(=O)CH_2R$$

NOTES:

1. The acidic proton is readily removed by base.

2. By using very strong bases, such as **LDA** and **t-BuLi**, a dianion can be formed that will preferentially alkylate at the methyl group.

Via:

$$H_2C=C(O^{\ominus})-CH=C(O^{\ominus})-OEt$$

REFERENCES:

1. S.N. Huckin and L. Weiler, **J. Am. Chem. Soc.**, (1974), 96, 1082.

2. K. Mori, **Tetrahedron,** (1974), 30, 4223.

3. J. Kennedy, N.J. McCorkindale and R.A. Raphael, **J. Chem. Soc.**, (1961), 3813.

4. R. Nishida, T. Sato, Y. Kuwahara, H. Fukami and S. Ishii, **Agric. Biol. Chem.**, (1976), 40, 1407.

5. W.L. Meyer, M.J. Brannon, C. da G. Burgos, T.E. Goodwin and R.W. Howard, **J. Org. Chem.**, (1985), 50, 438.

EXAMPLES:

$CH_3C(=O)CH_2CO_2Et$ + $CH_2{=}CHCH_2Br$ —(NaH, n-BuLi; THF, 30 min)→ $CH_2C(=O)CH_2CO_2Et$ / $CH_2CH{=}CH_2$ (83%) ① ②

Br ... MeO H ... MeO H —($MeC(O)CH_2CO_2Et$; NaH)→ O ... EtO_2C ... MeO H ... MeO H (32%) ②

Br ... CO_2H + O ... CO_2Et —(1.) NaOEt; 2.) OH^-; 3.) H^+)→ O ... CO_2H (46%) ③

Br ... O O + O ... CO_2Et —(1.) NaOEt / Reflux; 2.) KOH; 3.) Hyd.)→ O ... O O (10%) ④

O CH ... O —(i-Bu$\overset{O}{\text{C}}\overset{\ominus}{\text{C}}HCO_2t$-Bu)→ t-BuO ... O ... O ... OH ... O (90%) —(TsOH; HOAc)→ O ... O (75%) ⑤

REACTION: ACYLOIN CONDENSATION

March's Advanced Organic Chemistry : 1113-1116

GENERAL SCHEME:

$$2\ RCOR' \xrightarrow{Na} R-C(O^{\ominus}Na^{\oplus})=C(O^{\ominus}Na^{\oplus})-R \xrightarrow{H_2O} RCH(OH)C(=O)R$$

MECHANISM:

1

$$RC(=O)OR' \xrightarrow{e^{\ominus}} R\dot{C}(O^{\ominus})OR' \xrightarrow{RCO_2R'} R\dot{C}(OR')-O-C(O^{\ominus})(R)-OR' \xrightarrow{e^{\ominus}} R\overset{\ominus}{C}(OR')-O-C(O^{\ominus})(R)-OR' \longrightarrow$$

$$R\overset{\ominus}{C}(OR')-O-C(=O)R \longrightarrow RC(OR')(-O-)CR(O^{\ominus}) \longrightarrow RC(OR')(O^{\ominus})-C(=O)R \longrightarrow RC(=O)-C(=O)R \xrightarrow{e^{\ominus}}$$

$$RC(O^{\ominus})=C(\dot{O})R \xrightarrow{e^{\ominus}} RC(O^{\ominus})=C(O^{\ominus})R$$

NOTES:

1. J.J. Bloomfield and D.C. Owsley, **J. Org. Chem.**, (1975), 40, 393. For a complete explanation of mechanism, see also, **NOTE** 2.

2. **TMSCl** = Trimethylchlorosilane, Me_3SiCl. It is very important to use freshly distilled TMSCl in this reaction. The TMSCl should be distilled from CaH_2 under nitrogen to remove any other silane impurities. See: **Organic Reactions**, 23, 306, (1976).

REFERENCES:

1. P.G. Gassman, J. Seter and F.J. Williams, **J. Am. Chem. Soc.**, (1971), 93, 1673.

2. G.D. Gutsche, I.Y.C. Tao, and J. Kozma, **J. Org. Chem.**, (1967), 32, 1782.

3. A. Krebs, **Tetrahedron Lett.**, (1968), 4511.

4. R.C. Cookson and S.A. Smith, **J. Chem. Soc., Perkin I**, (1979), 2447.

5. J.J. Bloomfield and J.R.S. Irelan, **J. Org. Chem.**, (1966), 31, 2017.

EXAMPLES:

$(CH_2)_2CO_2CH_3$ / $(CH_2)_2CO_2CH_3$ —Na / Xylene→ (76%) ①

CO_2Et —$NaC_{10}H_8$ / THF→ (11%) ②

EtO_2C … CO_2Et —Na / Xylene→ (70%) ③

CO_2Me / CO_2Me / NMe_2 —1.) Na/Toluene 2.) TMSCl→ OTMS / OTMS / NMe_2 (78%) —SiO_2→ O / OH / Me (70%) ④ △2

CO_2Me / CO_2Me —Na / C_6H_6→ OH / O (70%) ⑤

REACTION: ALDOL CONDENSATION (DIRECTED)

March's **Advanced Organic Chemistry :** 8298- 34, 844.

GENERAL SCHEME:

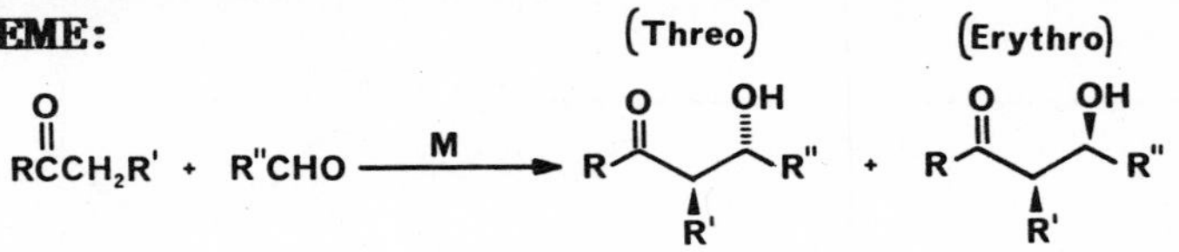

1

$RCOCH_2R'$ + R''CHO —M→ (Threo) + (Erythro)

MECHANISM:

2

$RCOCH_2R'$ + M → $RC(OM)$=CHR' + R''CHO

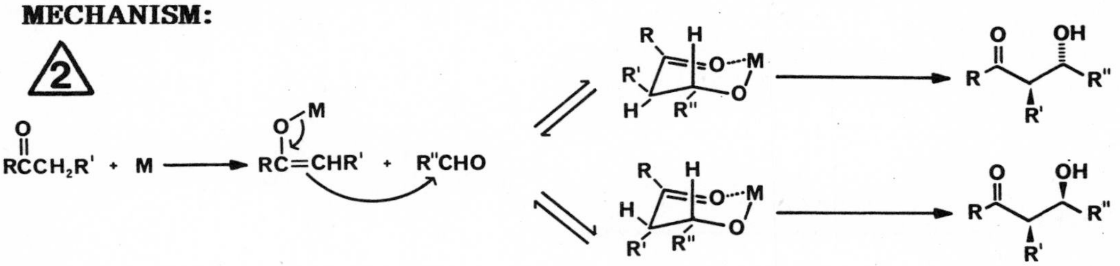

NOTES:

1. M = Metal = B, Si, Al, Zn, Mg, or Li

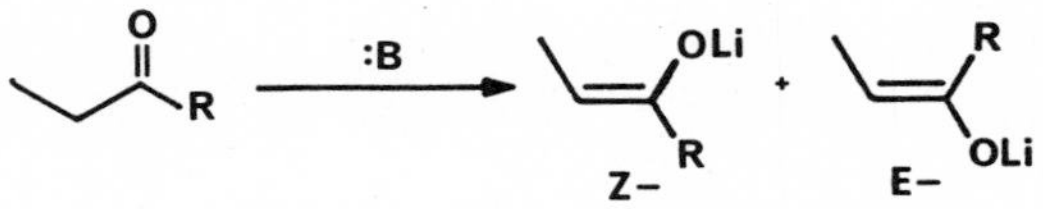

:B → Z- + E-

Base	Z / E ratio
i-Pr_2NLi	61:39
$Me_3Si)_2$NLi	85:15
$Me_2ArSi)_2$NLi	99:1
$Et_3Si)_2$NLI	94:6

S. Masamune and W. Choy, **Aldrichimica**, (1982), 15, 47.

It has been well established that (Z)-enolates give rise to erythro aldol products and that (E)- enolates provide threo aldols. Dialkylboryltriflates have been shown to be effective in the stereoselective formation of (Z)-enolates.

2. C.H. Heathcock, in Current Trends in Organic Synthesis, Proceedings of the Fourth International Conference on Organic Synthesis, Pergamon Press, New York, 1983.

REFERENCES:

1. D.A. Evans and L.R. McGee, **Tetrahedron Lett.**, (1980), 3975.

2. Y. Yamamoto, H. Yatagai and K. Maruyama, **Tetrahedron Lett.**,(1982), 2387.

3. Y. Tamura, T. Harada, S. Nishi, M. Mizutani, T. Hioki and Z. Yoshida, **J. Am. Chem. Soc.**, (1980), 102, 7806.

4. C. Siegel and E.R. Thornton, **Tetrahedron Lett.**, (1986), 457.

5. G.R. Clark, J. Lin and M. Nikaido, **Tetrahedron Lett.**, (1984), 2645.

EXAMPLES:

1.) $(i\text{-}Pr)_2NLi$ 2.) Et_2ZrCl_2 PhCHO ①

LDA + BEt_3 PhCHO ②

(90%, 97% threo)

1.) PhCHO, -78°, THF 2.) H_3O^+ ③

(85-100%, 93% erythro)

1.) LDA 2.) $ClTi(i\text{-}Pr)_3$ 3.) CHO ④

(74-85%, 99% syn)

Me_2CuLi N_2 ⑤

(70%)

REACTION: ALDOL CONDENSATION (GENERAL)

March's Advanced Organic Chemistry : 829-34, 844

GENERAL SCHEME:

$$2\ RC(=O)\!-\!CH \longrightarrow RC(=O)\!-\!C\!-\!C(R)(OH)\!-\!CH$$

MECHANISM:

$$RC(=O)\!-\!CH(H)\!-\!R' \xrightarrow{^{\ominus}B} {}^{\ominus}O\!-\!RC\!=\!CH(R') \xrightarrow{R''C(=O)CH_2R'''} RC(=O)\!-\!C(H)(R')\!-\!C(OH)(R'')\!-\!CH_2\!-\!R''' \dashrightarrow RC(=O)\!-\!C(R')\!=\!C(R'')\!-\!CH_2\!-\!R'''$$

NOTES:

1. Although both the acid (via the enol) and the base (via the enolate) conditions can be used for the **aldol condensation**, base-catalyzed reactions are more frequently used.

2. This is an example of a "**mixed aldol**" condensation. In order for this to be a useful reaction, the second carbonyl should not have acidic hydrogens next to the carbonyl. Both formaldehyde and benzaldehyde are useful for this condensation.

3. This is an example of an "**intramolecular aldol condensation**".

4. The reaction of a nitroalkane with an aldehyde in the presence of base is called the **Henry Reaction**.

$$RCHO + R'\overset{\ominus}{C}HNO_2 \longrightarrow R\!-\!CH(OH)\!-\!CH(R')\!-\!NO_2 \longrightarrow R\!-\!CH\!=\!C(R')\!-\!NO_2$$

5. The preformed enol-silyl ether is a disguised enolate that is released with fluoride ion.

REFERENCES:

1. R. Noyori, K. Yokoyama, J. Sakata, I. Kuwajima, E. Nakamura and S. Shimizu, **J. Am. Chem. Soc.**, (1977), 99, 1265.

2. W. Hoffmann and H. Siegel, **Tetrahedron Lett.**, (1975), 533.

3. B.M. Trost, C.D. Shuey, F.Dininno, Jr., and S.S. McElvain, **J. Am. Chem. Soc.**, (1979), 101, 1284.

4. S.C. Welch, J.-M. Assercq and J.-P. Loh, **Tetrahedron Lett.**, (1986), 1115.

5. S.D. Burke, C.W. Murtiashaw, J.O. Saunders and M.S. Dike, **J. Am. Chem. Soc.**, (1982), 104, 872.

EXAMPLES:

OSiMe$_3$ + PhCHO → 1.) 10% TBAF, THF, Argon, -78°; 2.) H_3O^+ → (68%) ① △2 △5

CHO ... O → ZnO_2, Decalin, Δ → (83%) ② △3

MeO_2C ... H ... O ... H → $K^+t\text{-}BuO^-$ → MeO_2C ... H ... O ... H (79%) ③ △3

O ... H ... O → 5% KOH, H_2O, Reflux → O= (88%) ④ △3

CHO ... O ... O → $TiCl_4$ → CHO ... O (50%) ⑤ △3

REACTION: ARBUZOV REACTION (MICHAELIS - ARBUZOV)

March's Advanced Organic Chemistry : 848

GENERAL SCHEME:

2

$$(CH_3CH_2O)_3P + R\text{-}CH_2\text{-}X \longrightarrow R\text{-}CH_2\text{-}\overset{O}{\overset{\|}{P}}(OCH_2CH_3)_2$$

MECHANISM:

$$(CH_3CH_2O)_3P\!: + \overset{R}{\overset{|}{C}}H_2\text{-}X \longrightarrow RCH_2\overset{OEt}{\overset{|\oplus}{\underset{OEt}{\underset{|}{P}}}}\text{-}O\text{-}CH_2\text{-}CH_3 + \overset{\ominus}{X} \longrightarrow RCH_2\overset{O}{\overset{\|}{P}}(OEt)_2 + CH_3CH_2X$$

NOTES:

1. The reaction can also be carried out with trimethoxyphosphine

2. When the reaction is carried out with an alpha-bromo ketone it is called the **Perkow Reaction**.

REFERENCES:

1. D.J. Burton and R.M. Flynn, **Synthesis**, (1979), 615.

2. A. Bhattacharya and G. Thyagarajan, **Chem. Rev.**, (1981), 81, 415.

3. S. Landon and T. Brill, **Inorg. Chem.**, (1984), 23, 4177.

4. M. Nikaido, R. Aslanian, F. Scavo, P. Helquist, B. Aakermark and J. Baeckvall, **J. Org. Chem.**, (1984), 49, 4738.

5. L.M. Harwood and M. Julia, **Synthesis**, (1980), 456.

EXAMPLES:

$(EtO)_3P + CF_2Br_2 \xrightarrow[120°,\ 12\ hrs]{NaI} (EtO)_2P(O)-CF_2Br$ (70%) ①

2-(bromomethyl)-1,4-dimethoxybenzene (CH_2Br, MeO, OMe) $\xrightarrow{P(OEt)_3}$ $CH_2P(O)(OEt)_2$ product (NYA) ②

$[CpCo(P\frown P)I]^{\oplus} + P(OCH_3)_3 \longrightarrow [CpCo(P\frown P)P(O)(OCH_3)_2]^{\oplus}$ (11%) ③

AcO–$CH_2CH{=}CHCH_2$–Cl $\xrightarrow[85°]{P(OEt)_3}$ AcO–$CH_2CH{=}CHCH_2$–$P(O)(OEt)_2$ (90%) ④

CH_2Br (bicyclic alkene) $+ P(OEt)_3 \xrightarrow{Ether}$ $CH_2P(O)(OEt)_2$ (bicyclic alkene) (95%) ⑤

REACTION: ARNDT-EISTERT SYNTHESIS

March's <u>Advanced Organic Chemistry</u> : 974-76

GENERAL SCHEME:

5

$$R\text{-}\overset{O}{\overset{\|}{\underset{*}{C}}}\text{-}Cl + CH_2N_2 \longrightarrow R\text{-}\overset{O}{\overset{\|}{\underset{*}{C}}}\text{-}CHN_2 \begin{cases} \xrightarrow[Ag^{\oplus}]{R'OH} R\text{-}CH_2\text{-}\overset{O}{\overset{\|}{\underset{*}{C}}}\text{-}O\text{-}R' \\ \xrightarrow[Ag^{\oplus}]{H_2O} R\text{-}CH_2\text{-}\overset{O}{\overset{\|}{\underset{*}{C}}}\text{-}OH \\ \xrightarrow[Ag^{\oplus}]{R'NH_2} R\text{-}CH_2\text{-}\overset{O}{\overset{\|}{\underset{*}{C}}}\text{-}NHR' \end{cases}$$

MECHANISM:

$$R\overset{O}{\overset{\|}{C}}\text{-}\overset{\ominus}{C}H\text{-}\overset{\oplus}{N}{\equiv}N: \xrightarrow{Ag^{\oplus}} \underbrace{R\overset{O}{\overset{\|}{C}}\text{-}\ddot{C}H \equiv R\text{-}\overset{O^{\ominus}}{\overset{\|}{C}}\text{-}\overset{\oplus}{C}H}_{\text{(carbene)}} \longrightarrow O{=}C{=}\overset{H}{\overset{|}{C}}R \text{ (ketene)}$$

NOTES:

1a.

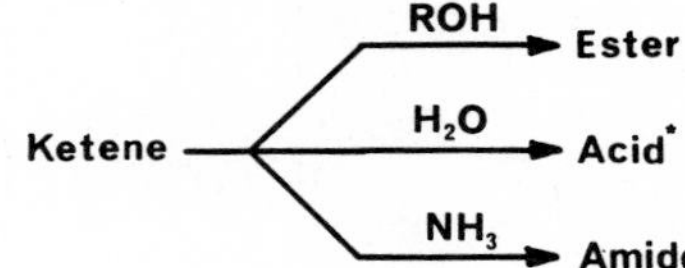

*This step is called the **Wolff Rearrangement**

b. In the absence of the catalyst, the diazoketone hydrolyzes to a keto alcohol:

$$R\underset{O}{\underset{\|}{C}}CH_2OH$$

c. Either two equivalents of diazomethane must be used or a base should be available to remove the HCl formed.

2. A photochemical **Arndt-Eistert** reaction.

3. An alternative sequence for homologation includes the steps:

Acid -----------> Alcohol --------> Tosylate ----------> Nitrile -----------> Acid

4. Trimethylsilyldiazomethane can be used for this homologation:

$$R\underset{O}{\underset{\|}{C}}Cl + Me_3SiCHN_2 \longrightarrow R\underset{O}{\underset{\|}{C}}\overset{N_2}{\overset{\|}{C}}SiMe_3 \longrightarrow R\underset{O}{\underset{\|}{C}}CHN_2$$

T. Aoyama and T. Shioiri, **Chem. Pharm. Bull.**, (1981), <u>29</u>, 3249.

5. For an equivalent approach, see: C.J. Kowalski and K.W. Fields. **J. Am. Chem. Soc.**, (1982), <u>104</u>, 321.

REFERENCES:

1. A.B. Smith, III, **Chem. Commun.**, (1974), 695.
2. A. Grussner, E. Jaeyer, J. Hellerbach and O. Schnider, **Helv.**, (1959), 42, 2431.
3. E.J. Walsh, Jr., and G.B. Stone, **Tetrahedron Lett.**, (1986), 1127.
4. A.B. Smith, III, B.D. Dorsey, M. Visnick, T. Maeda and M.S. Malamas, **J. Am. Chem. Soc.**, (1986), 108, 3110.

EXAMPLES:

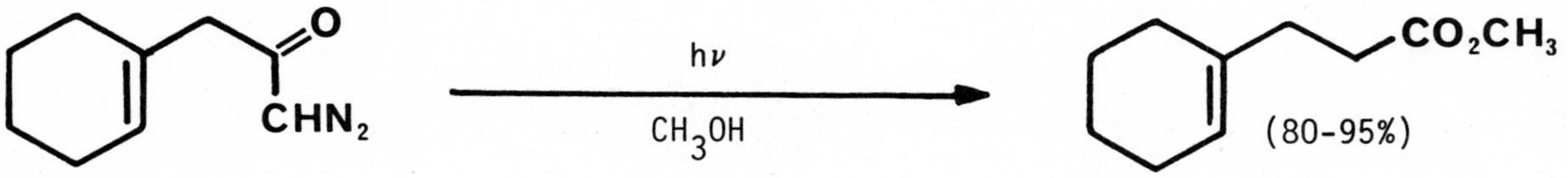

CHN$_2$ — $h\nu$, CH_3OH → CO_2CH_3 (80-95%) ① △2

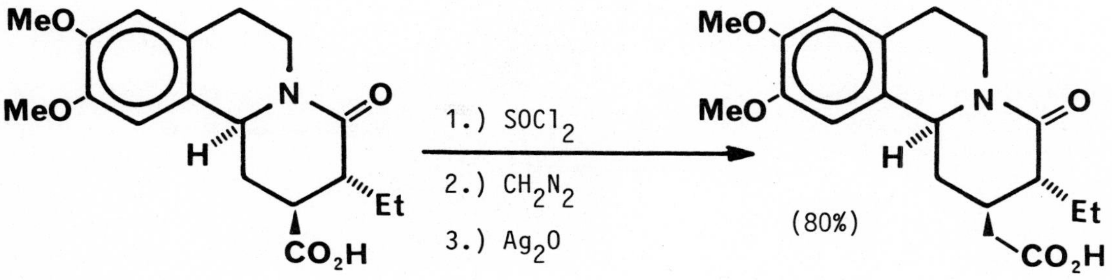

MeO, MeO, N, O, H, Et, CO_2H — 1.) $SOCl_2$ 2.) CH_2N_2 3.) Ag_2O → MeO, MeO, N, O, H, Et, CO_2H (80%) ②

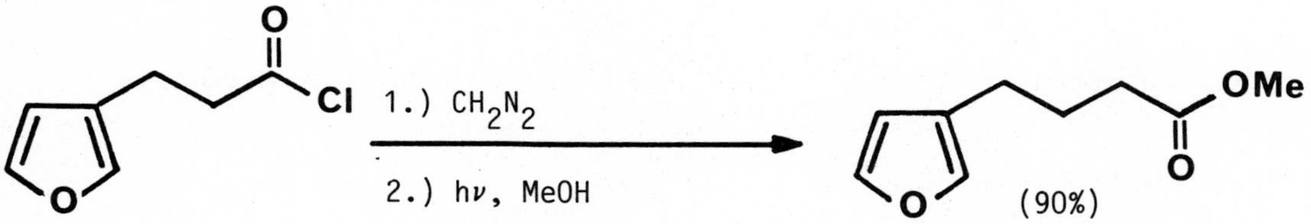

O, Cl, O — 1.) CH_2N_2 2.) $h\nu$, MeOH → OMe, O, O (90%) ③

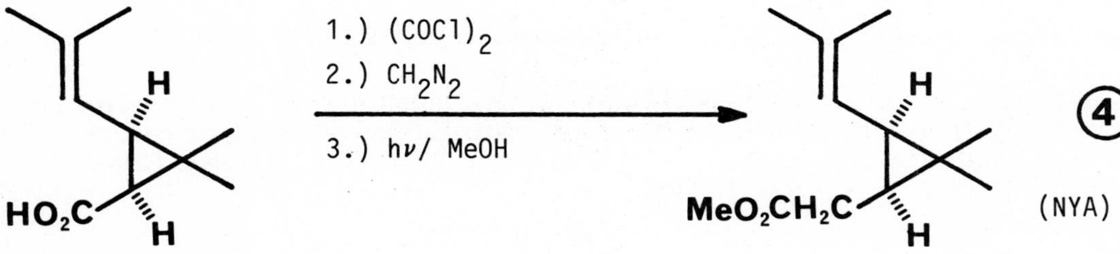

H, HO_2C, H — 1.) $(COCl)_2$ 2.) CH_2N_2 3.) $h\nu$/ MeOH → H, MeO_2CH_2C, H (NYA) ④

REACTION: BAEYER- VILLIGER REACTION

===

March's Advanced Organic Chemistry : 990-991

GENERAL SCHEME:

$R(R')C{=}O \xrightarrow{RCO_3H} R(R'{-}O)C{=}O$

MECHANISM:

$R{-}C(={}^{\oplus}OH){-}R' \xrightarrow{RCOOH} R{-}C(OH)(R')(O{-}O{-}C(=O){-}R) \longrightarrow R{-}C(=O){-}O{-}R'$

Other functional groups that will compete for the peroxyacid.

$>C{=}C< \longrightarrow$ epoxide ($-C-C-$ bridged by O)

$-S- \longrightarrow -S(=O)-$

$>N{-}R \longrightarrow -N(R){\rightarrow}O$

NOTES:

1. Typical Peroxyacids include: Peroxybenzoic acid, m-chloroperoxybenzoic acid (**MCPBA**), peroxyacetic acid, and trifluroperoxyacetic acid.

$C_6H_5CO_3H$, $m\text{-}ClC_6H_4CO_3H$, CH_3CO_3H, CF_3CO_3H

2. Migratory aptitudes:
   3° > 2°) cyclohexyl > benzyl > phenyl > 1° > methyl

Both the stereochemistry and chirality of a migrating group are retained.

$\text{C}^*{-}C(=O){-}R \longrightarrow \text{C}^*{-}O{-}C(=O){-}R$

3. This is an unique example of a ketal undergoing **Baeyer-Villiger reaction.**

4. Since alkenes react readily with peroxyacids to form epoxides, it is of interest that bis[trimethylsilyl]peroxide can be used to carry out the Baeyer-Villiger reaction in the presence of an alkene. M. Suzuki, H. Takada and R. Noyori, **J. Org. Chem.**, (1982), 47, 902.

REFERENCES:

1. G.A. Krafft and J.A. Katzenellenbogen, **J. Am. Chem. Soc.**,(1981), 103, 5459.

2. W.F. Bailey and M.-J. Shih, **J. Am. Chem. Soc.**, (1982), 104, 1769.

3. G. Magnusson, **Tetrahedron Lett.**, (1977), 2713.

4. H. Suginome and S. Yamada, **J. Org. Chem.**, (1985), 50, 2489.

EXAMPLES:

CF_3CO_3H

Na_2HPO_4

(68%) ①

MCPBA

5 hrs

(25%) ② △3

MCPBA

(69%) ③

MCPBA

(83%) ④

REACTION: BAMBERGER REACTION

March's Advanced Organic Chemistry: 606

GENERAL SCHEME:

NHOH → ($H^{\oplus}$, H_2O) → NH_2 / OH

MECHANISM:

$-H_2O$ ... H_2O

NOTES:

1. The reaction seems to resemble the **Hoffmann- Martius Reaction:**

HCl, Δ

however, while the **Hoffman- Martius Reaction** follows an electrophilic mechanism, the Bamberger reaction seems to follow a nucleophilic mechanism.

2. Via:

HCl

3. The intermediate amino-phenol derivative readily oxidizes to the quinone.

REFERENCES:

1. A.I. Krylov and B.V. Stolyarov, **J. Org. Chem.,USSR,** (1981), 17, 710.

2. T.B. Patrick, J.A. Schield and D.G. Kirchner, **J. Org. Chem.**, (1974), 39, 1758.

3. J.C. Hardy and M. Venet, **Tetrahedron Lett.**, (1982), 1255.

4. Reported in **Reagents**, Vol. 1, 1123.

EXAMPLES:

H_2SO_4 / MeOH — N-ethyl-N-phenylhydroxylamine → 4-methoxy-N-ethylaniline (66%) ①

HF / -60° — N-(m-tolyl)hydroxylamine → fluoro-methylaniline (61%) ②

1.) HCl/EtOH
2.) H_2O (79%) ③ △2

H_2SO_4, $H_2SO_4 \cdot SO_3$ / △ — 1,5-dinitronaphthalene → [HNOH / HNOH] → HO, NH_2 / H_2N, OH → HO, O / HO, O (60%) ④

REACTION: BAMFORD- STEVENS

March's Advanced Organic Chemistry : 911

GENERAL SCHEME:

$$\text{HC-C=N-NH-Ts} \xrightarrow{\text{Base}} \text{>C=C<}_{\text{H}}$$

MECHANISM:

Dianion Mechanism:

$$\text{HC-C=}\ddot{\text{N}}\text{-}\underset{\cdot\cdot}{\text{N}}\text{H-Ts} \xrightarrow{\text{2RLi}} {}^{\ominus}\text{C-C=}\ddot{\text{N}}\text{-}\overset{\ominus}{\underset{\cdot\cdot}{\text{N}}}\text{-Ts} \longrightarrow \text{>C=C<}_{\ddot{\text{N}}=\text{N}^{\ominus}}\ \text{Li}^{\oplus}$$

$$\longrightarrow \text{>C=C<}_{\text{Li}} \xrightarrow{H_2O} \text{>C=C<}_{\text{H}}$$

Carbene Mechanism:

$$\text{HC-C=}\ddot{\text{N}}\text{-}\underset{\cdot\cdot}{\overset{\text{H}}{\text{N}}}\text{-Ts} \xrightarrow{:\text{B}} \text{HC-C=}\ddot{\text{N}}\text{-}\overset{\ominus}{\underset{\cdot\cdot}{\text{N}}}\text{-Ts} \xrightarrow{-\text{Ts}^{\ominus}} \text{HC-C=}\ddot{\text{N}}\text{=}\ddot{\text{N}} \xrightarrow{-N_2}$$

$$\text{H-C-}\overset{\ominus}{\underset{\oplus}{\ddot{\text{C}}}}\text{-} \longrightarrow \text{>C=C<}_{\text{H}}$$

NOTES:

1. Tosylhydrazone

2. The major difference between the **Shapiro** and the **Bamford-Stevens** reaction is in the base used. In the Shapiro reaction two equivalents of base (**RLi**) are used, and the alkene formed is generally the <u>less-substituted</u>. With other bases, the alkene is often the <u>more</u>-substituted. When Na-ethylene glycol are used as the base, a carbene mechanism is probable; whereas with other bases, the anion or carbene mechanisms may be invoked.

3. A modification of the reaction may allow for the synthesis of aryldiazomethanes:

$$\text{Ar(R)C=N-NH-SO}_2\text{-C}_6\text{H}_2(i\text{-Pr})_3 \xrightarrow[\text{MeOH}]{\text{KOH}} \text{Ar(R)C=N}_2$$

C.D. Dirdonan and C.B. Resse, **Synthesis**, (1982), 419.

REFERENCES:

1. M.A. Gianturco, P. Friedel and V. Flanagan, **Tetrahedron Lett.**, (1965), 1847.

2. G.A. Hiegel and P. Burk, **J. Org. Chem.**, (1973), 38, 3637.

3. W.L. Scott and D.A. Evans, **J. Am. Chem. Soc.**, (1972), 94, 4779.

4. P.A. Grieco, T. Oguri, C.-L. J. Wang and E. Williams, **J. Org. Chem.**, (1977), 42, 4113.

EXAMPLES:

NNHTs — Na, Δ → (75%) + (5%) ①

NNHTs — K_2CO_3/H_2O, Δ → (53%) ②

NNHTs — MeLi → (Excess of 80%) ③

OPh, NNHTs — LDA → OPh (Excess of 98%) ④

REACTION: BARBIER-WIELAND PROCEDURE

March's Advanced Organic Chemistry : 1071

GENERAL SCHEME:

$$RCH_2\overset{O}{\overset{\|}{C}}OR'(H) \xrightarrow[2.\,[O]]{1.\ PhMgX} R\overset{O}{\overset{\|}{C}}OH$$

MECHANISM:

$$RCH_2\overset{O}{\overset{\|}{C}}OR'(H) \xrightarrow{PhMgX} RCH_2-C(Ph)_2-OH \xrightarrow{-H_2O} RCH=C(Ph)_2$$

$$\xrightarrow{CrO_3} R\overset{O}{\overset{\|}{C}}OH$$

NOTES:

1. The reaction is the oxidative cleavage of a 1,1-diphenyl alkene, with the result being the removal of one carbon atom.

2. The oxidation also converts secondary alcohols to ketones.

3. An interesting modification for a three-carbon removal has been reported: C. Djerassi, **Chem. Rev.**, (1946), 38, 526.

$$RCH_2CH_2CH=CPh_2 \xrightarrow[2.\,-HBr]{1.\ NBS} RCH=CHCH=CPh_2 \xrightarrow{CrO_3} RCO_2H$$

REFERENCES:

1. B. Riegel, R. Moffett and A. McIntosh, **Org. Synth.**, Coll. Vol. 3, 234, 237 (1955).

2. C.S. Subramaniam, P.J. Thomas, V.R. Mamdapur ad M.S. Chadha, **Synthesis**, (1978), 468.

3. F.L.M. Patterson and R.L. Buchanan, **Biochemical Journal**, (1964), 92, 100.

4. H.E. Fierz-David and W. Kuster, **Helv.**, (1939), 22, 82.

EXAMPLES:

$$\text{(3,12-dihydroxy steroid methyl ester, side chain } CH_3CHCH_2CH_2CO_2CH_3\text{)} \xrightarrow[\text{EtOH, } C_6H_6]{\text{PhMgBr}} \xrightarrow[CH_3CO_2H\text{, Reflux}]{(CH_3CO)_2O}$$

$$\text{(3,12-diacetoxy steroid, side chain } CH_3CHCH_2CH{=}CPh_2\text{)} \xrightarrow[H_2O,\ CH_3CO_2H,\ 50°]{CrO_3} \text{(3,12-diacetoxy steroid, side chain } CH_3CHCH_2CO_2H\text{)}$$

① △1

(57-68% as diol)

$$MeCH(OH)(CH_2)_8CO_2Me \xrightarrow[\text{2.) } H_2SO_4 \text{ 3.) } CrO_3,\ H_2O]{\text{1.) PhMgBr}} MeC(=O)(CH_2)_7CO_2H$$

(48%) ② △2

$$F(CH_2)_6-C(Me)_2-CH_2CO_2Me \xrightarrow[\text{2.) } Ac_2O \text{ 3.) } CrO_3]{\text{1.) PhMgBr}} F(CH_2)_6-C(Me)_2-CO_2H$$

(51%) ③

$$C_{14}H_{29}CH_2CO_2H \xrightarrow[\text{2.) } \Delta \text{ 3.) } CrO_3]{\text{1.) PhMgBr}} C_{14}H_{29}CO_2H$$

(58%) ④

REACTION: BARTON REACTION

March's Advanced Organic Chemistry: 1044

GENERAL SCHEME:

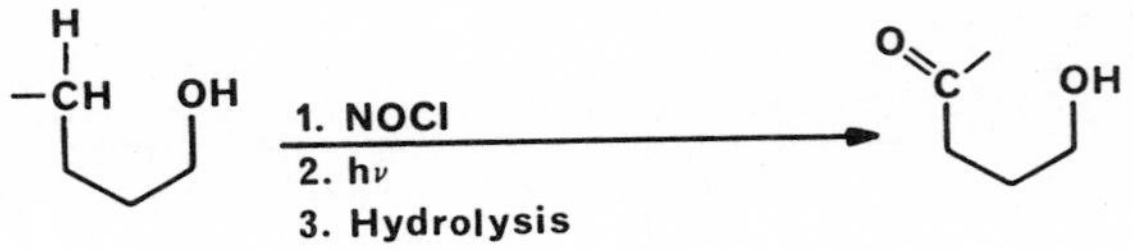

1. NOCl
2. hν
3. Hydrolysis

MECHANISM:

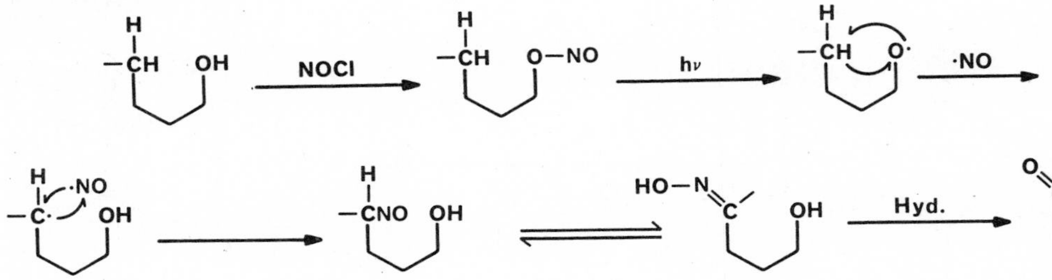

NOCl

hν

·NO

Hyd.

NOTES:

1. This method provides a way to oxidize a carbon atom separated from an -OH group by three other carbon atoms.

2. This is an example of a **Barton Fragmentation**.

3. Via:

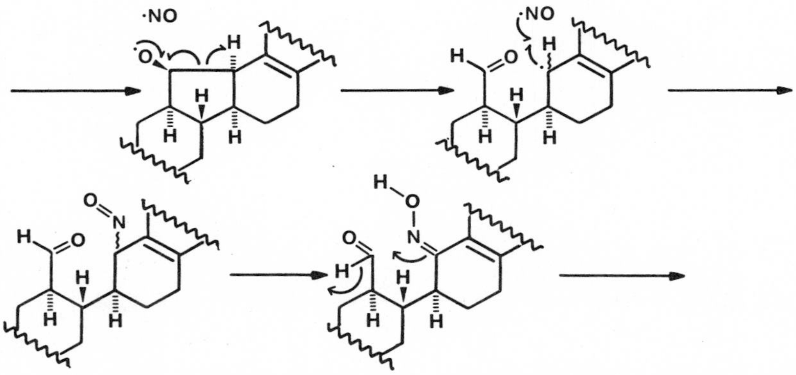

REFERENCES:

1. D.H.R. Barton and I.M. Beaton, **J. Am. Chem. Soc.**, (1961), 83, 4083.

2. D.H.R. Barton, I.M. Beaton, L.E. Geller and M.M. Pechet, **J. Am. Chem. Soc.**, (1960), 82, 2640.

3. P.D. Hobbs and P.D. Magnus, **J. Am. Chem. Soc.**, (1976), 98, 4594.

4. S.W. Baldwin and H.R. Blomquist, **J. Am. Chem. Soc.**, (1982), 104, 4990.

5. H. Suginome, N. Sato, ad T. Masamune, **Tetrahedron**, (1971), 27, 4863.

EXAMPLES:

1.) NOCl
2.) hν
3.) Hydrolysis

(Apx. 15%) ①

hν

(51%) ②

1.) NOCl
2.) hν
3.) Δ
4.) H^+

(NYA) ③

NOCl

hν

(100%) 2 ④

1.) NOCl-Pyr
2.) hν

(58%) ⑤ 3

REACTION: BECKMANN REARRANGEMENT

March's Advanced Organic Chemistry : 946, 949, 987-89

GENERAL SCHEME:

R(R')C=N-OH —H⊕→ R'-C(=O)-N(R)H

MECHANISM:

R(R')C=N-OH —H⊕→ R(R')C=N-⊕OH₂ → R'-C⊕=NR —H₂O→ R'-C(OH)=NR ⇌ R'-C(=O)-N(H)R

NOTES:

1. An **oxime**

2. Acids often include: Phosphoric, sulfuric, $POCl_3$, PCl_5, P_2O_5

3. The migrating group is anti to the oxime -OH group.

Oxime → R(R)C=N-O-PCl₃ (Cl) → R-C(=O)-N(H)R

4. These conditions result in rearrangement and reduction. The trialkyaluminum reacts with the oxime tosylate to form an imine, that is reduced to an amine with **DIBAH**.

5. **Beckmann fragmentation** is a likely competing process if one of the centers adjacent to the oxime is quaternary.

AcO, Me, Me, =NOH —1.) TsCl, Pyridine 2.) TsOH→ CN, Me, Me, OAc

P.T. Lansbury, D.J. Mazur and J.P. Springer, **J. Org. Chem.**, (1985), 50, 1632.

REFERENCES:

1. P.W. Jeffs, G. Molina, M.W. Cass and N.A. Cortese. **J. Org. Chem.**, (1982), 47, 3871.

2. S. Fujita, K. Koyona and Y. Inagaki, **Synthesis**, (1982), 68.

3. K. Hattori, Y. Matsumura, T. Miyazaki, K. Marouka and H. Yamamoto, **J. Am.**

Chem. Soc., (1981), 103, 7368.

4. O. Meth-Cohen and B. Narina, **Synthesis**, (1980), 133.

EXAMPLES:

P_2O_5, CH_3SO_3H (94%) ①

$POCl_3$ / $CH_3CON(CH_3)_2$ (82%) ②

1.) $(CH_3CH_2CH_2)_3Al$
2.) DIBAH
(48%) ③ ④

PCl_5 / Ether, 0°, 2-3 Hrs., then 15° (87%) ④

REACTION: BENZIDINE REARRANGEMENT

March's Advanced Organic Chemistry : 1034-36

GENERAL SCHEME:

$$C_6H_5-NH-NH-C_6H_5 \xrightarrow{Acid} H_2N-C_6H_4-C_6H_4-NH_2$$

MECHANISM:

2

$$C_6H_5-NH-NH-C_6H_5 \xrightarrow{2H^{\oplus}} C_6H_5-\overset{\oplus}{N}H_2-\overset{\oplus}{N}H_2-C_6H_5 \equiv$$

$$\longrightarrow H_2\overset{\oplus}{N}=C_6H_4=C_6H_4=\overset{\oplus}{N}H_2 \longrightarrow H_2N-C_6H_4-C_6H_4-NH_2$$

NOTES:

1. This is an oxygenated analog of the benzidine rearrangement.
2. The reaction can be considered by orbital symmetry.

REFERENCES:

1. M. Nojima, T. Ando and N. Tokura, **J. Chem. Soc., Perkin I**, (1976), 14, 1504.
2. T. Sheradsky and S. Auramovki-Grisaru, **J. Heterocyclic Chemistry**, (1980), 17, 189.
3. G.A. Olah, K. Dunne, D.P. Kelly and Y.K. Mo, **J. Am. Chem. Soc.**, (1972), 94, 7438.
4. O.I. Andreevskaya, G.G. Furin and G.G. Yakobson, **J. Org. Chem.,U.S.S.R.**, (1978), 13, 1558.
5. H.R. Snyder, C. Weaver and C.D. Marshall, **J. Am. Chem. Soc.**, (1949), 71, 289.

EXAMPLES:

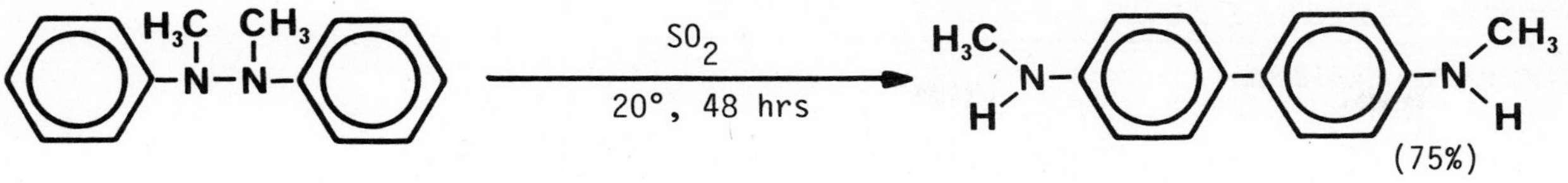

(1) SO_2, 20°, 48 hrs (75%)

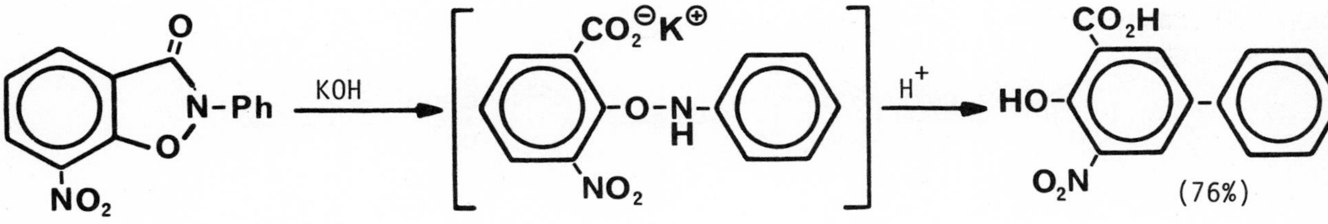

(2) KOH; H^+ (76%) [1]

(3) SbF_5-HF-SO_2(SO_2ClF), -78° (92%)

(4) SbF_5-HF (1:1), -20°, 20 min (75%)

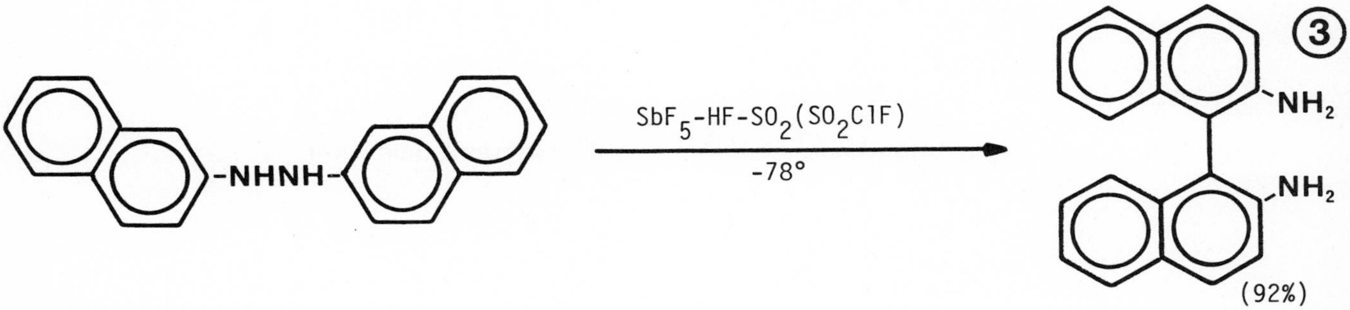

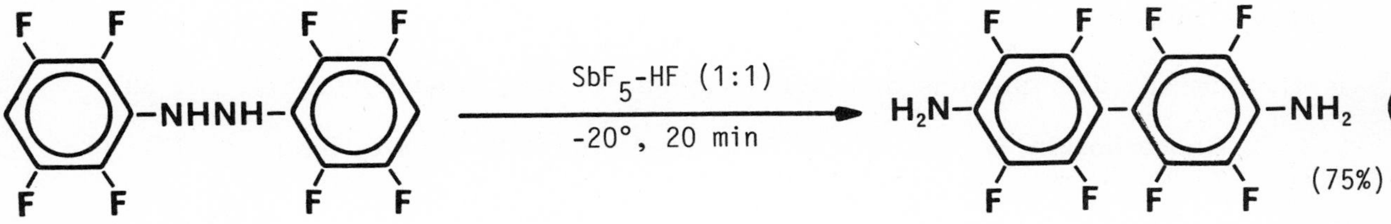

(5) HCl, Δ (25%)

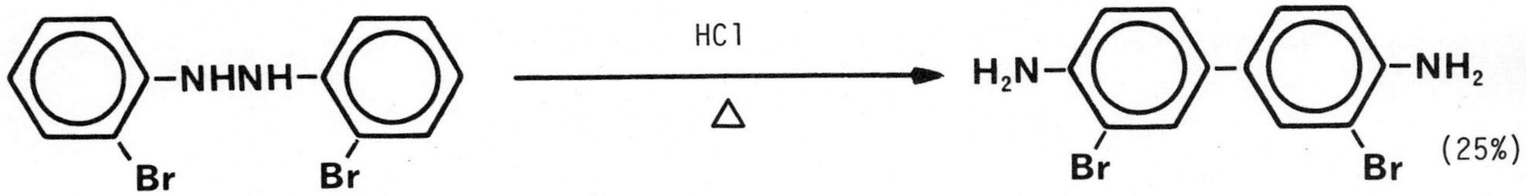

REACTION: BENZOIN CONDENSATION

March's Advanced Organic Chemistry : 859- 860

GENERAL SCHEME:

$$R-C_6H_4-CHO \xrightarrow{HCN} R-C_6H_4-CH(OH)-C(=O)-C_6H_4-R$$

MECHANISM:

$$C_6H_5CHO \underset{}{\overset{CN^{\ominus}}{\rightleftharpoons}} C_6H_5-CH(CN)-O^{\ominus} \rightleftharpoons C_6H_5-C^{\ominus}(CN)-OH \xrightarrow{C_6H_5CHO}$$

$$C_6H_5-C(OH)(CN)-CH(O^{\ominus})-C_6H_5 \longrightarrow C_6H_5-C(=O)-CH(OH)-C_6H_5$$

NOTES:

1. For details regarding the mechanism of this reaction, see: Ref. 4

REFERENCES:

1. G. Sumrell, J.I. Stevens and G.E. Goheen, **J. Org. Chem.**, (1957), 22, 39.

2. G.H. Gholamhosein, H. Hakimelahi, C.B. Boyle and H.S. Kasmai, **Helv.**, (1977), 60, 342.

3. J. Solodar, **Tetrahedron Lett.**, (1971), 287.

4. Y. Yano, Y. Tamura and W. Tagaki, **Bull. Chem. Soc. Japan**, (1980), 53, 740.

5. D.P. Macaione and S.E. Wentworth, **Synthesis**, (1974), 716.

EXAMPLES:

4-MeO-C_6H_4-CHO $\xrightarrow[H_2O]{KCN,\ EtOH}$ MeO-C_6H_4-C(=O)-CH(OH)-C_6H_4-OMe (44%) ①

4-Me-C_6H_4-CHO $\xrightarrow[RT,\ 20\ hrs]{KCN,\ DMF}$ Me-C_6H_4-C(=O)-CH(OH)-C_6H_4-Me (90%) ②

C_6H_5-CHO $\xrightarrow[H_2O]{Bu_4N^+CN^-}$ C_6H_5-C(=O)-CH(OH)-C_6H_5 (66%) ③

C_6H_5-CHO $\xrightarrow[MeOH]{\text{3-benzyl-4-methylthiazolium chloride } (PhCH_2,\ N^{\oplus},\ Cl^{\ominus},\ Me,\ S),\ Et_3N}$ C_6H_5-C(=O)-CH(OH)-C_6H_5 (79%) ④ △1

$H_2C=CH$-C_6H_4-CHO $\xrightarrow[H_2O]{KCN,\ EtOH}$ $H_2C=CH$-C_6H_4-CH(OH)-C(=O)-C_6H_4-$CH=CH_2$ (62%) ⑤

REACTION: BIRCH REDUCTION

March's Advanced Organic Chemistry : 700

GENERAL SCHEME:

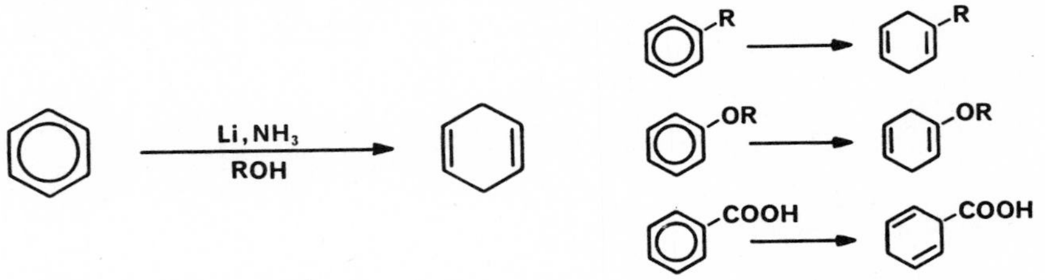

Li, NH_3

ROH

R

OR

COOH

MECHANISM:

1

2

$e^{\ominus}$ ROH $e^{\ominus}$ ROH

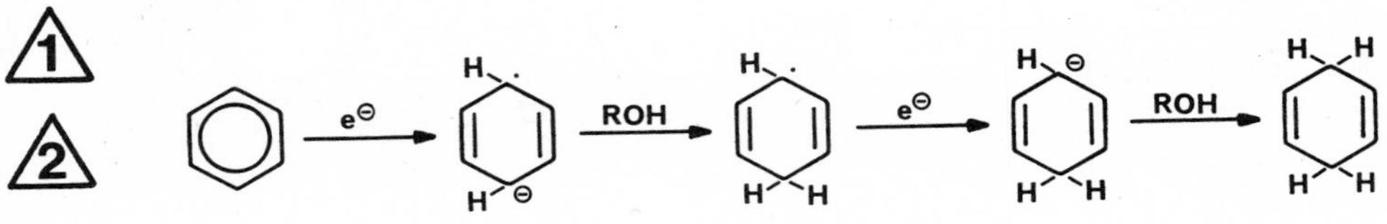

NOTES:

1. Active metal: Na, Li, K in liquid ammonia as a source of solvated electrons.
2. Other functional groups may react under Birch conditions:

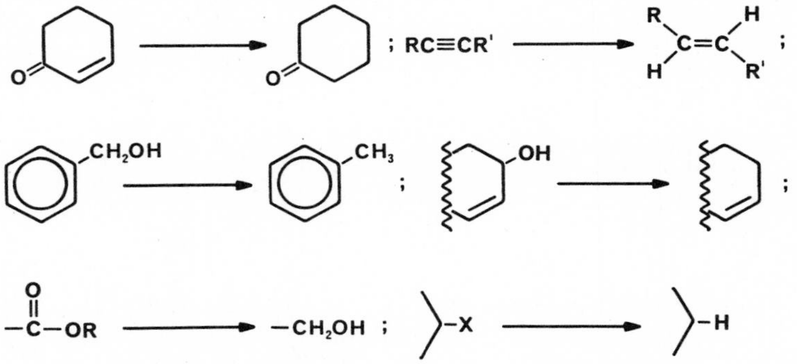

; RC≡CR' → R(H)C=C(H)R' ;

CH_2OH → CH_3 ; OH → ;

$-\overset{O}{\overset{\|}{C}}-OR \longrightarrow -CH_2OH$; $>-X \longrightarrow >-H$

REFERENCES:

1. R. Kannan, P. Geetha and S. Swaminathan, **Tetrahedron Lett.**, (1984), 1601.

2. E.J. Corey and N.W. Boaz, **Tetrahedron Lett.**, (1985), 6015.

3. E. Kariv-Miller, K.E. Swenson and D. Zemach, **J. Org. Chem.**, (1983). 48, 4210.

4. W.A. Ayer, W.R. Bowman, T.C. Joseph and P. Smith, **J. Am. Chem. Soc.**, (1968). 90, 1648.

5. A. Gopalan and P. Magnus, **J. Am. Chem. Soc.**, (1980), 102, 1756.

EXAMPLES:

Li / NH_3 — (NYA) ①

1.) Li/NH_3
2.) $HO_2C\text{-}CO_2H$, H_2O, t-BuOH — (Excess of 70%) ②

TBA/OH^-, H_2O, 60° — (92%) ③

Li/NH_3, t-BuOH — (NYA) ④

1.) Li / NH_3
2.) MeI — (87%) ⑤

REACTION: BISCHLER- NAPIERALSKI REACTION

March's Advanced Organic Chemistry : 495

GENERAL SCHEME:

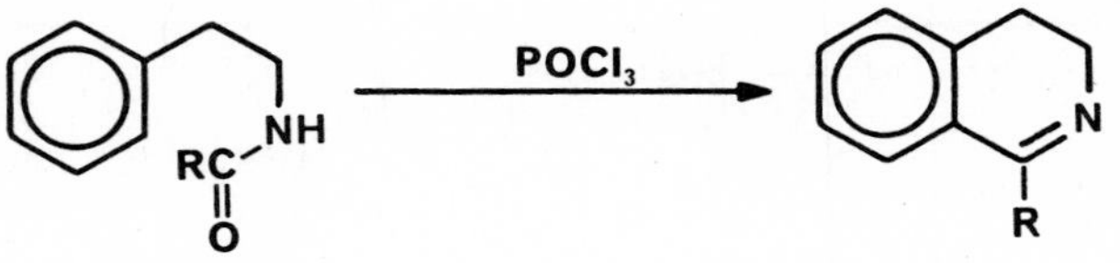

POCl$_3$

MECHANISM:

POCl$_3$

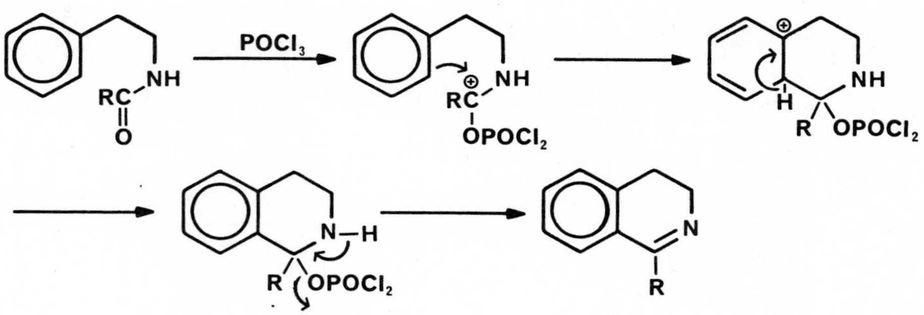

NOTES:

1. **PPE** = Polyphosphate ester, See: **Reagents**, Vol.1, 892-94

REFERENCES:

1. I. Ribas, J. Saa and L. Castedo, **Tetrahedron Lett.**, (1973), 3617.

2. S. Kessar, P. Jit, K. Mundra and A. Lumb, **J. Chem. Soc. Part C**, (1971), 266.

3. E.E. Van Tamelen and I.G. Wright, **Tetrahedron Lett.**, (1964), 295.

4. E.E. Van Tamelen, C. Placeway, G.P. Schiemenz and I.G. Wright, **J. Am. Chem. Soc.** , (1969), 91, 7359.

5. C.S. Hilger, B. Fugmann and W. Steglich, **Tetrahedron Lett.**, (1985), 5975.

EXAMPLES:

MeO, EtO, NO_2, NH, O, MeO, OMe — PPE → MeO, EtO, N, NO_2, MeO, OMe (70%) ① ⚠1

NHCCH$_3$, O — Phosphorous oxychloride / Reflux, 8 hrs → N, CH_3 (82%) ②

N H, O, N, CO_2CH_3, O, CH_3O_2C — 1.) $POCl_3$ 2.) H_2/Pd → N H, N, H, H, CO_2CH_3, O, CH_3O_2C (NYA) ③

MeO, N H, O, N, MeO_2C, OAc, OMe — $POCl_3$ → MeO, N H, $N^{\oplus}$, MeO_2C, OAc, OMe (87%) ④

MeO, MeO, NH, O, O_2N, OMe — $POCl_3$ / CH_3CN → MeO, MeO, N, O_2N, OMe (85-93%) ⑤

REACTION: BOUVEAULT-BLANC REDUCTION

==

March's Advanced Organic Chemistry : 811, 1101

GENERAL SCHEME:

$$RC(=O)OR' \xrightarrow[\text{EtOH} \ (\text{source of } e^{\ominus})]{\text{Na}} RCH_2OH + R'OH$$

MECHANISM:

$$RC(=O)OR' \xrightarrow{e^{\ominus}} R\dot{C}(O^{\ominus})OR' \xrightarrow{\text{HOEt}} R\dot{C}(OH)OR' \xrightarrow{e^{\ominus}} R\overset{\ominus}{C}(OH)OR' \xrightarrow{\text{HOEt}}$$

$$R-CH(O-H)-OR' \longrightarrow RC(=O)H \xrightarrow{e^{\ominus}} R\dot{C}H(O^{\ominus}) \xrightarrow[2.\ e^{\ominus}]{1.\ \text{HOEt}} RCH_2OH$$

NOTES:

1. Although a once popular reaction, it has largely been replaced by modern hydride reducing agents. Sometimes the reduction of an aldehyde or ketone under these conditions is called a Bouveault-Blanc reduction. Here, the thermodynamically more stable alcohol will result from the reduction. The identity of the metal used in the procedure has little influence on the composition of the product (B. Willholm, W. Thommen and U. Burger, **Helv.**, (1981), 64, 2109).

2. It is appropriate here to mention that $LiAlH_4$ would reduce both the acid and the ester; diborane would preferentially reduce the acid; and **DIBAH** would convert the ester to an aldehyde.

3. Water seems to inhibit the **Bouveault-Blanc** reduction while allowing the **Birch reduction** to take place.

REFERENCES:

1. G. Haberland, **Ber.**, (1936), 69, 1380.

2. L.A. Paquette and N.A. Nelson. **J. Org. Chem.**, (1968), 27, 2272.

3. P.W. Rabideau, D.L. Huser and S.J. Nyikos, **Tetrahedron Lett.**, (1980), 1401.

4. E.M Kaiser, **Synthesis**, (1972), 391.

5. S.G. Ford and C.S. Marvel, **Org. Synth.**, (1943), Coll. Vol. 2, 372.

EXAMPLES:

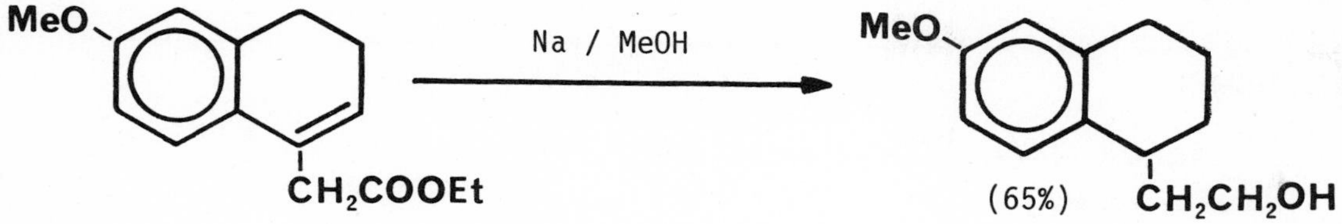

MeO — CH_2COOEt —— Na / MeOH ——→ MeO — (65%) CH_2CH_2OH ①

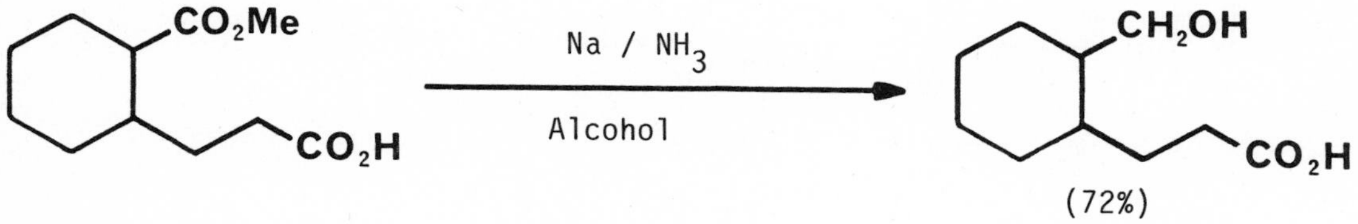

CO_2Me, CO_2H —— Na / NH_3, Alcohol ——→ CH_2OH, CO_2H (72%) ② △2

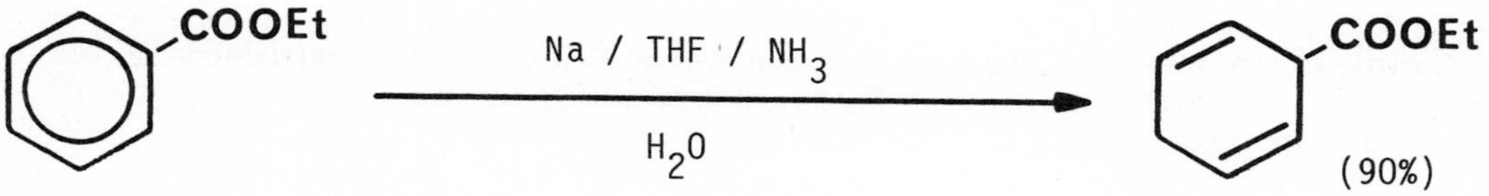

COOEt —— Na / THF / NH_3, H_2O ——→ COOEt (90%) ③

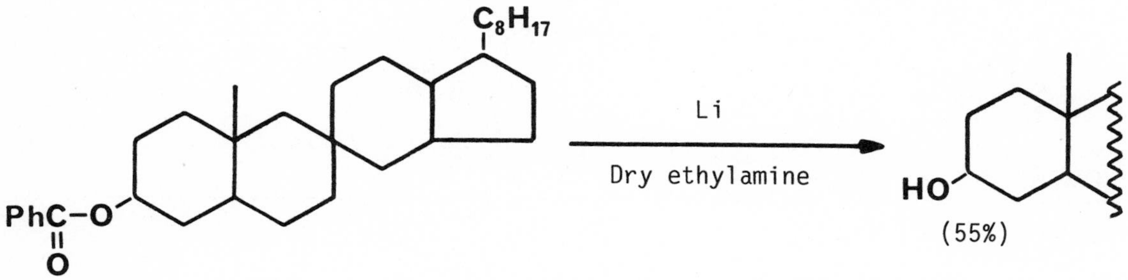

C_8H_{17}, PhC(=O)–O —— Li, Dry ethylamine ——→ HO (55%) ④

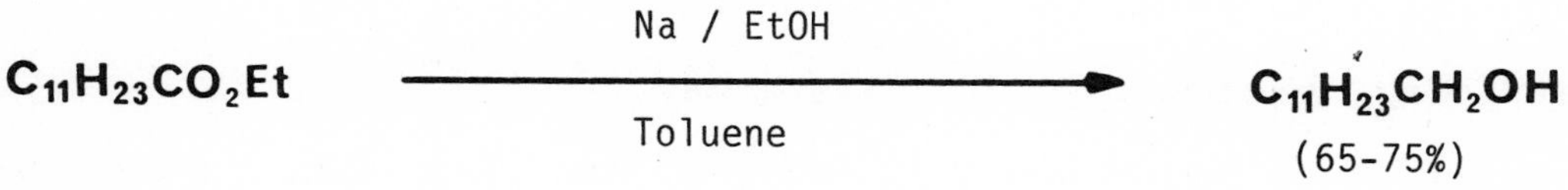

$C_{11}H_{23}CO_2Et$ —— Na / EtOH, Toluene ——→ $C_{11}H_{23}CH_2OH$ (65-75%) ⑤

REACTION: CANNIZZARRO REACTION

==

March's Advanced Organic Chemistry : 1117- 1119

GENERAL SCHEME:

1

$$\underset{\text{(No }\alpha\text{-hydrogens)}}{RCHO} \xrightarrow{\text{Base}} RCH_2OH + RCOO^{\ominus}$$

MECHANISM:

$$RCHO \quad OH^{\ominus} \longrightarrow R-\underset{OH}{\overset{O^{\ominus}}{C}}-H \xrightarrow[\text{slow}]{H(R)C=O} RCH_2-O^{\ominus} + RCOOH$$

$$\xrightarrow[\text{fast}]{} RCH_2OH + RCOO^{\ominus}$$

NOTES:

1. Strong base, about 50% alkali solution, is often used for these reactions. There is often no need to heat the reaction.

2. A **crossed-Cannizzarro reaction** utilizes a scavenger aldehyde (formaldehyde) as one of the reagents. Until the advent of hydride reducing agents, this was one of the common methods for reduction.

3. Via:

OMs, OMe, $O^{\ominus}$, H, NH, CH=O → OMs, OMe, O, NH, CH_2–$O^{\ominus}$

REFERENCES:

1. T.A. Geissman, **Org. Reactions**, Vol II., (1944), 94.

2. C.G. Swain, A.L. Powell, W.A. Sheppard and C.R. Morgan, **J. Am. Chem. Soc.**, (1979), 101, 3576.

3. R.S. McDonald and C.E. Sibley, **Can. Jour. Chem.**, (1981), 59, 1061.

4. T.A. Geissman, **Org. Reactions**, Vol II., (1944), 94.

5. J.B. Henderickson, T.L. Bogard and M.E. Fisch, **J. Am. Chem. Soc.** , (1970), 92, 5538.

EXAMPLES:

2 (2,5-dichlorobenzaldehyde) —KOH, Δ→ 2,5-dichlorobenzyl alcohol (90%) + 2,5-dichlorobenzoic acid (84%) ①

Cl, CHO, Cl; KOH; Δ; Cl, CH_2OH, Cl, +, (90%); Cl, COOH, Cl, (84%)

2 (benzaldehyde) —KOH, MeOH→ benzyl alcohol (55%) + benzoic acid (47%) ②

CHO; KOH; MeOH; CH_2OH, +, (55%); COOH, (47%)

CHO, CHO —aq NaOH, Dioxane→ CH_2OH, COOH (NYA) ③

OCH_3, CHO —CH_2O, OH^-, Δ→ OCH_3, CH_2OH (85%) ④ △2

MsO, OMe, O, O, O, N, OH —$NaBH_4$, i-PrOH, Δ→ OMe, MsO, NH, O, O, O, OH (NYA) ⑤ △3

REACTION: CARROLL REARRANGEMENT

March's Advanced Organic Chemistry : Not indexed.

GENERAL SCHEME:

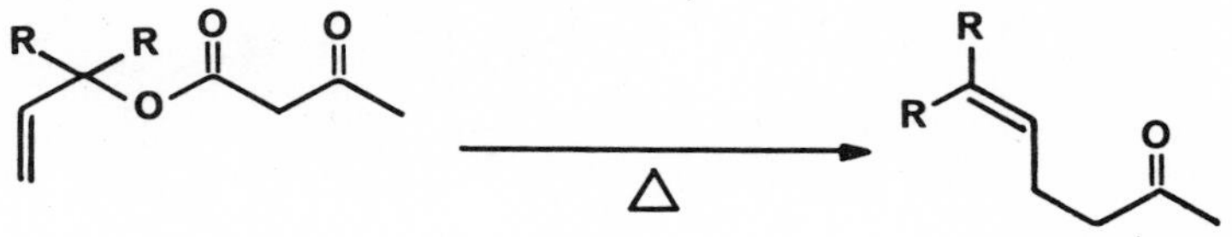

R R O O R R O

Δ

MECHANISM:

O O OH O OH O O O H O O O H O O C O O

Δ

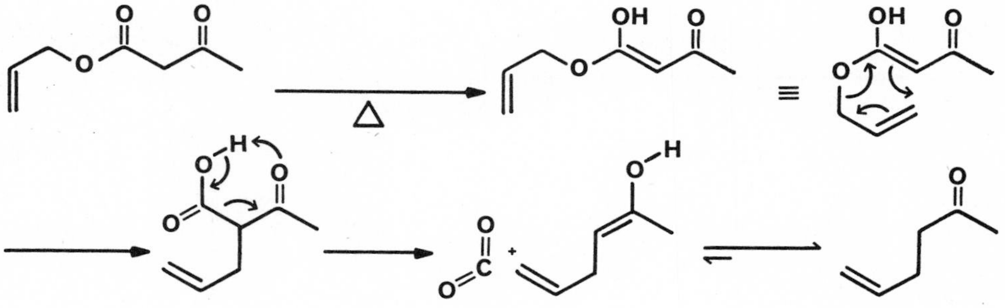

NOTES:

REFERENCES:

1. M. Tanabe and K. Hayashi, **J. Am. Chem. Soc.**, (1980), 102, 862.

2.,3. N. Wakabayashi, R.M. Waters, and J.P. Church, **Tetrahedron Lett.**, (1969), 3253

4. W. Kimel, N.W. Sax, S. Kaiser, G.G. Eichmann, G.O. Chase and A. Ofner, **J. Org. Chem.**, (1958), 23, 153.

5. C. Pinazzi and D. Reyx, **Bull. Soc. Chim. Fr.**, (1972), 3930.

EXAMPLES:

H_3C H C O Xylene Δ H H_3C O (90%) ①

O O O Δ O + O (76%, 97:3) ②

O O O Δ O + O (67%, 54:46) ③

O O O 170°-190° O (55%) ④

O O O Et n Δ Et n (n=1, 75%; 2, 77%; 3, 80%) O ⑤

REACTION: CHUGAEV REACTION (TSCHUGAEFF)

March's Advanced Organic Chemistry : 905

GENERAL SCHEME:

$$H-\overset{|}{\underset{|}{C}}-\overset{|}{\underset{|}{C}}-O-\overset{S}{\overset{\|}{C}}-S-R \xrightarrow{\Delta} \rangle C=C\langle + COS + RSH$$

MECHANISM:

RS–C(=S)–O–C–C–H (cyclic transition state) ⟶ >C=C< + RS–C(=O)–S ⟶ RSH + S=C=O

NOTES:

1. The **xanthate** intermediate is prepared from an alcohol. This reaction has the advantage of needing lower temperatures for the elimination to occur; however, this benefit is outweighed by the difficulty in purifying the products from the reaction. Often there are sulfur-containing by-products.

2. The thermal elimination of a sulfoximine.

3. The p-tolylthionocarbonate modification seems to have advantages for sterically-hindered alcohols.

4. A similar elimination of the methyl sulfite (-O-C(=S)-O-Me) gave about equal amounts of the 1- and 3-isomers.

REFERENCES:

1. G. Cernigliano and P. Kocienski, **J. Org. Chem.**, (1977), 42, 3622.

2. M. Kim and J.D. White, **J. Am. Chem. Soc.**, (1975), 97, 451.

3. H. Gerlach and W. Muller, **Helv.**, (1972), 55, 2277.

4. D.J. Cram and F.A.A. Elhafez, **J. Am. Chem. Soc.**, (1952), 74, 5828.

5. L.S. McNamara and C.C. Price, **J. Org. Chem.**,(1962), 27, 1230.

EXAMPLES:

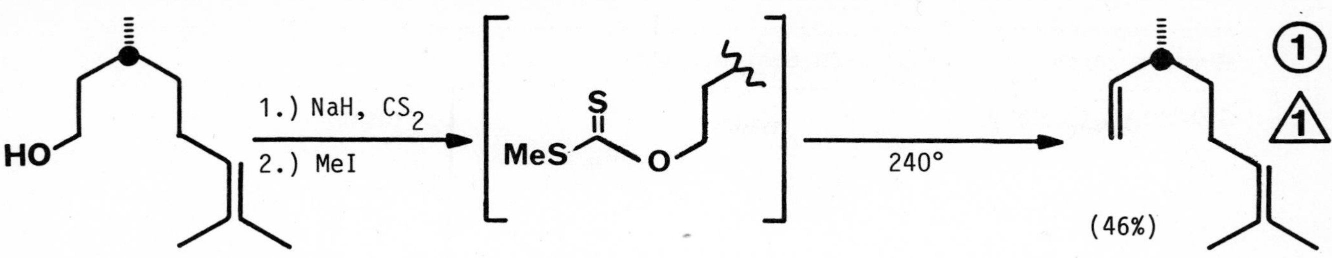

Me_2S O⊖ N=N⊕ O O

Δ

⊖N ⊕N O O

$-N_2$
$-CO_2$

(50%)

2 2

S
OCO-PhMe-4

170°

OCO-PhMe-4
S

(39%)

3 3

Me Ph
Ph C—C H
H OCSMe
S

Δ

Me Ph
C=C
Ph H

(77%)

4

CH_2Ph
OCSMe
S

140°

CH_2Ph

(90%)

5 4

REACTION: CLAISEN CONDENSATION

March's Advanced Organic Chemistry: 437- 39, 835.

GENERAL SCHEME:

$$2\ RCH_2\overset{\overset{O}{\|}}{C}OEt \xrightarrow{\ominus B} RCH(C(=O)CH_2R)COEt\ (C=O)$$

MECHANISM:

$$R{-}\underset{\ominus}{C}H{-}COOEt + R{-}CH_2{-}\underset{\underset{O}{\|}}{C}{-}OEt \longrightarrow R{-}CH_2{-}C(=O){-}CH(R){-}COOEt$$

NOTES:

1. The Claisen Condensation is the reaction of an ester (often ethyl acetate) containing an alpha hydrogen that can be removed by base with a second molecule of ester. The **Claisen-Schmidt Condensation** is a variation where the ester enolate anion reacts with an aldehyde. This resembles an **Aldol condensation.**

2. The resonance form

$$CH_3CH_2C(=O)SR \longleftrightarrow CH_3CH_2C(O^{\ominus})=\overset{\oplus}{S}{-}R$$

is less-important for thioesters; thus, the acidity of the alpha proton is greater and **Claisen condensations** are easier to carry out with thioesters.

3. A "**mixed Claisen condensation**" is preparatively useful only when one of the two esters has no alpha protons.

REFERENCES:

1. B.E. Hudson, Jr. and C.R. Hauser, **J. Am. Chem. Soc.**, (1941), 63, 3156.

2. C. Gennari, L. Colombo, S. Cardani and C. Scolastico, **Tetrahedron Lett.**, (1984), 2283.

3. F.W. Swamer and C.R. Hauser, **J. Am. Chem. Soc.**, (1950), 72, 1352.

4. T. Kubota and T. Matsuura, **Chem. and Industry**, (1956), 521.

5. C.S. Marvel and W.O. King, **Org. Synth.**, Coll. Vol. I, (1941), 252.

EXAMPLES:

$$CH_3CH_2CH(Me)COOEt \xrightarrow{Ph_3C^-Na^+} CH_3CH_2CH(Me)C(=O)\text{-}C(CH_2CH_3)(Me)\text{-}CO_2Et \quad (63\%)$$ ①

$$CH_3CH_2C(=O)S\text{-}Ph \xrightarrow[i\text{-}Pr_2NEt]{Cl\text{-}B(OCH_2CH_2O)} CH_3CH{=}C(OB(OCH_2CH_2O))S\text{-}Ph \xrightarrow[2.)\ H_2O]{1.)\ PhCHO} PhCH(OH)CH(CH_3)C(=O)SPh \quad (63\%)$$ ② △2

$$\text{3-pyridyl-}CO_2Me + Me(CH_2)_2CO_2Et \xrightarrow{NaH} \text{3-pyridyl-}C(=O)CH(CH_2Me)CO_2Et \quad (68\%)$$ △3 ③

$$\text{3-furyl-}CO_2Et + MeCO_2Et \xrightarrow[Na]{CH_3COOEt} \text{3-furyl-}C(=O)CH_2CO_2Et \quad (40\text{-}50\%)$$ △3 ④

$$C_6H_5CHO + CH_3COOEt \xrightarrow{EtO^-} C_6H_5CH{=}CHCOOEt \quad (70\%)$$ ⑤

REACTION: CLAISEN REACTION

===

March's Advanced Organic Chemistry : 832, 835

GENERAL SCHEME:

△1 △3

$$R\text{-}CH_2\text{-}C(=O)\text{-}O\text{-}R' + R''\text{-}C(=O)\text{-}R''' \xrightarrow{:B} R\text{-}CH(COOR')\text{-}C(OH)(R''')\text{-}R''$$

MECHANISM:

$$R\text{-}CH_2\text{-}C(=O)\text{-}O\text{-}R' \;(:B) \longrightarrow R\text{-}CH{=}C(O^{\ominus})\text{-}OR' + R''\text{-}C(=O)\text{-}R''' \longrightarrow$$

$$R\text{-}CH(COOR')\text{-}C(OH)(R''')\text{-}R'' \xrightarrow{\text{Elimination}}$$

NOTES:

1. This reaction will take place with any active methylene compound.

2. **PPA** = Polyphosphoric acid

3. This is very similar to the **Claisen- Schmidt Reaction:**

Reaction:

$$RC(=O)H + R'CH_2C(=O)R'' \xrightarrow{:B} RCH(OH)CH(R')C(=O)R''$$

Mechanism:

$$R'CH{=}C(O^{\ominus})R'' \quad RC(=O)H \longrightarrow RCH(OH)CH(R')C(=O)R'' \xrightarrow{\text{Elimination}}$$

REFERENCES:

1. C. Ainsworth, **Org. Synth.**, (1963), Coll. Vol. 4, 536.

2. E. Horning and A. Finelli, **Org. Synth.**, (1963), Coll. Vol. 4, 461.

3. H. Gerlach and W. Muller, **Angew. Chem. Int. Ed.**, (1972), 11, 1030.

4. D. Meuche, H. Strauss and E. Heilbronner, **Helv.**, (1958), 41, 2220.

5. R.P. Woodbury and M.W. Rathke, **J. Org. Chem.**, (1977), 42, 1688.

EXAMPLES:

+ HCOEt

1.) NaH, EtOH (Cat.), Et_2O

2.) H^+

CHO ①

(70-74%)

+ $(EtO)_2CO$

CH_2CN

1.) NaOEt / PhMe

2.) $MeCO_2H$ / H_2O

CHCN ② △1

CO_2Et

(70-78%)

COOH

HOAc / PPA

③ △2

(81%)

CHO

CHO

+ EtCEt (C=O)

NaOEt

=O ④

(60%)

H

CH_3C-CCH_3

O CH_3

1.) LDA

2.) cyclohexanone (=O)

OH

$CH_2CCH(CH_3)_2$

O ⑤

(88%)

REACTION: CLAISEN REARRANGEMENT

March's <u>Advanced Organic Chemistry</u> : 1028- 1032

GENERAL SCHEME:

MECHANISM:

NOTES:

1. The "classic" Claisen rearrangement involves with aromatic allyl ethers.

2. If an "ortho" position is blocked, the rearrangement will go to the para position.

3. This type of reaction is known as the **ester enolate Claisen rearrangement.**

4. The ester enolate reaction is very similar to the **Carroll reaction**:

S.R. Wilson and R.S. Myers, **J. Org. Chem.**, (1975), <u>40</u>, 3309.

REFERENCES:

1. J.W.S. Stevenson and T.A. Bryson, **Tetrahedron Lett.**, (1982), 3143.

2. M.R. Saidi, **Heterocycles**, (1982), <u>19</u>,1473.

3. S. Danishefsky and T. Tsuzuki, **J. Am. Chem. Soc.**, (1980), <u>102</u>, 6891.

4. E.J. Corey, R.L. Danheiser, S. Chandrasekaran, P. Siret, G.E. Keck, and J.-L. Gras, **J. Am. Chem. Soc.**, (1978), <u>100</u>, 8031.

5. S.D. Burke and G.J. Pacofsky, **Tetrahedron Lett.**, (1986),445.

EXAMPLES:

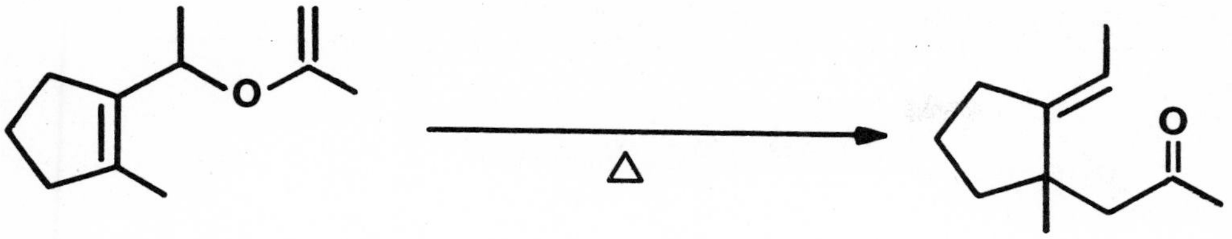

Δ

① (50-80%)

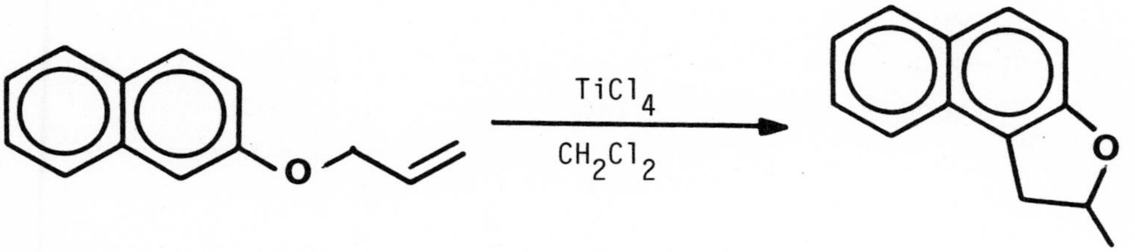

$TiCl_4$

CH_2Cl_2

② (84%)

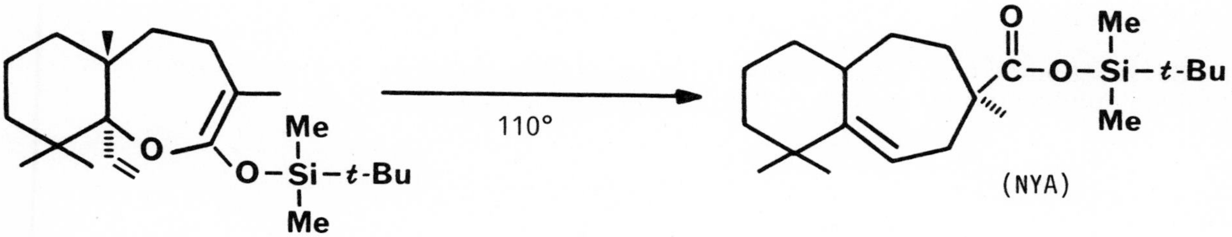

Me

O–Si–*t*-Bu

Me

110°

O

Me

C–O–Si–*t*-Bu

Me

③ (NYA)

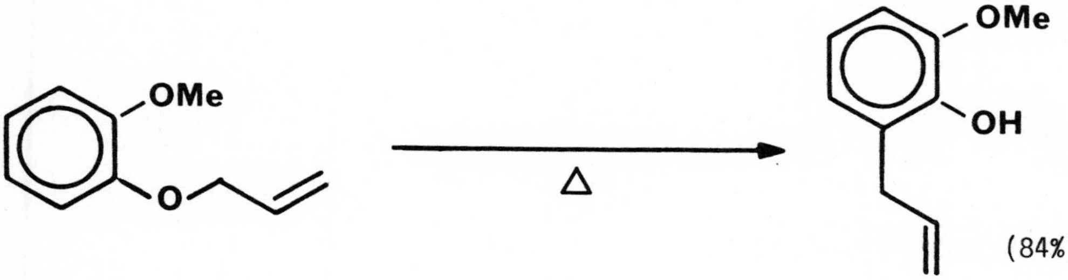

OMe

Δ

OMe

OH

④ (84%)

O

O

O

TMS

Et

O

O

Lithium hexamethyldisilazid

THF, TMS-Cl

-100°, then to 28°

O

HO

O

TMS

H

Et

O

O

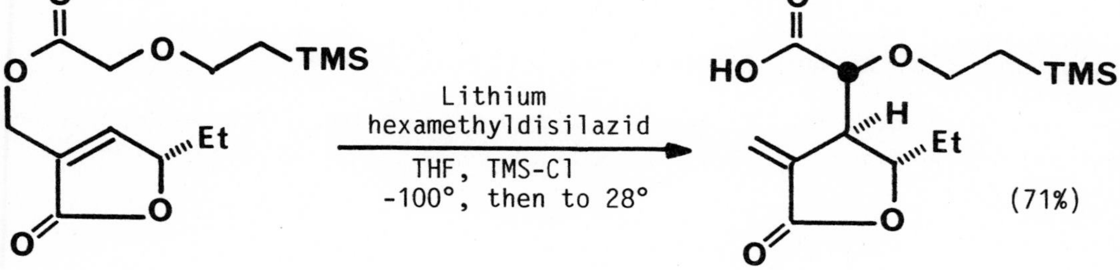

(71%)

⑤ △3 △4

REACTION: CLEMMENSEN REDUCTION

March's Advanced Organic Chemistry : 1096- 98, 1114

GENERAL SCHEME:

$$\mathrm{RR'C{=}O} \xrightarrow[\mathrm{HCl}]{[\mathrm{Zn\text{-}Hg}]} \mathrm{R{-}C(R')(H){-}H}$$

MECHANISM:

$$\mathrm{RC(O)R'} \xrightarrow[\mathrm{Zn}]{\mathrm{HCl}} \mathrm{R{-}C(OH)(ZnCl){-}R'} \xrightarrow{\mathrm{HCl}} \mathrm{R{-}C(Cl)(ZnCl){-}R'} \xrightarrow{\mathrm{Zn}} \mathrm{R{-}C(ZnCl)(ZnCl){-}R'}$$

$$\xrightarrow{\mathrm{H}^{\oplus}} \mathrm{RCH_2R'}$$

NOTES:

1. For a Review, see: E. Vedejs, **Org. Reactions**, (1975), 22, 401.

2. The reaction is subject to steric effects. The hindered ketone shown below would not undergo reduction.

REFERENCES:

1. Reported in Review by Vedejs (Note 1).

2. W.T. Borden and T. Ravindranathan, **J. Org. Chem.**, (1971), 36, 4125.

3. R. Mayer, S. Schleithauer and D. Kunz, **Ber.**, (1966), 99, 1393.

4. B. Bannister and B.B. Elsner, **J. Chem. Soc.**, (1951), 1055.

5. A.K. Banerjee, J. Alvarez G., M. Santan, and M.C. Carrasco, **Tetrahedron**, (1986), 42, 6615.

EXAMPLES:

HCl, Et_2O

Zinc dust

Ph Ph → Ph Ph (60%) ①

Me, O=, =O, Me

Zn, HCl

Ac_2O

[O=, Me, H, ZnCl, Me] → AcO (50%) Me Me ②

$PhCH_2\overset{S}{\overset{\|}{C}}SCH_3$ —[Zn-Hg], HCl→ $PhCH_2CH_2SCH_3$ (54%) ③

O, C_7H_{15} —[Zn-Hg], HCl→ C_8H_{17} (58%) ④

O —[Zn-Hg], HCl→ (80%) ⑤

REACTION: COPE ELIMINATION (REACTION)

March's Advanced Organic Chemistry : 909

GENERAL SCHEME:

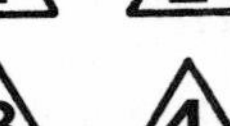

MECHANISM:

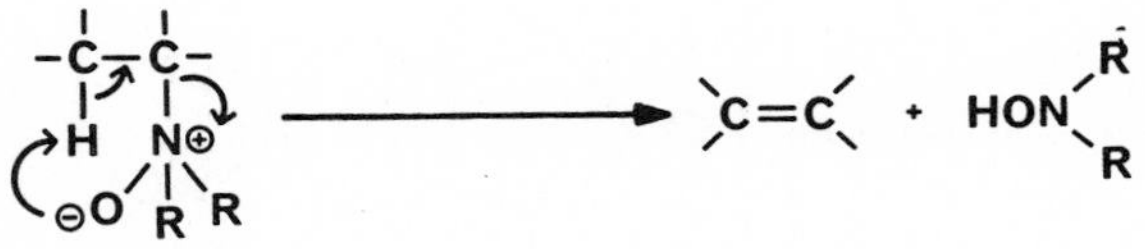

NOTES:

1. The reaction is a syn, thermal elimination reaction of an amine oxide.

2. Tertiary amines are readily converted to N-oxides by hydrogen peroxide or peroxyacids.

3. Dehydrosulfenylation is a very similar reaction:

B.M. Trost and T.N. Salzmann, **J. Org. Chem.**, (1975), 40, 148.

4. Selenoxides undergo elimination in much the same way.

REFERENCES:

1. J.-J. Barieux and J. Gore, **Bull Soc. Chim. Fr.**, (1971), 1649, 3978.

2. E.J. Corey and M.C. Desai, **Tetrahedron Lett.**, (1985), 5747.

3. W. Sucrow, **Angew. Chem. Int. Ed.**, (1968), 7, 629.

4. K. Mori, **Tetrahedron**, (1977), 33, 289.

5. L.D. Quin, J. Leimert, E.D. Middlemas, R.W. Miller and A.T. McPhail, **J. Org. Chem.**, (1979), 44, 3496.

EXAMPLES:

1.) BH_3
2.) H_2O_2 OH^-

Δ

OH

(70%) ①

$N \rightarrow O$

CO_2Me

50°

CO_2Me

(80%) ②

Me

CH_2N

Me

1.) H_2O_2

2.) 150°

(57%) ③

N

O

Δ

(69%) ④

$N \rightarrow O$

$N \rightarrow O$

Δ

(73%) ⑤

REACTION: COPE REARRANGEMENT

March's Advanced Organic Chemistry : 1021- 1027

GENERAL SCHEME: 

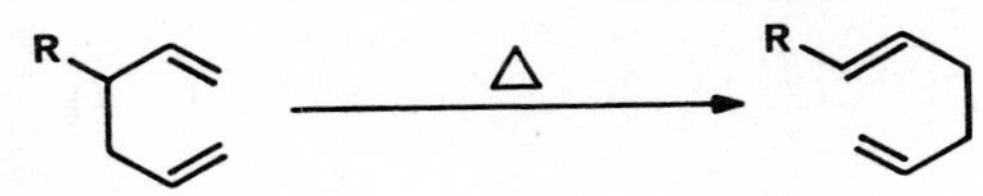

A [3,3]-sigmatropic shift is the equivalent of two [1,3]-sigmatropic shifts.

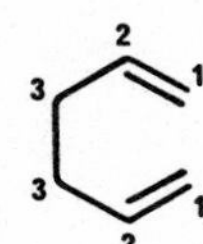

MECHANISM:

By **FMO** theory, this is preferred geometry for the concerted reaction.

NOTES:

1. R-groups in this reaction can be H-, alkyl, or aryl.

2. Catalytic amounts of Pd(II) promote the rearrangement to easily take place at around room temperature.

3. This example can be classified as a "**tandem Cope-Claisen rearrangement**". First the 1,5-diene undergoes **Cope** rearrangement, leaving an alkene bond in an appropriate place for a subsequent **Claisen rearrangement**.

4. **TFA** = Trifluroacetic acid.

REFERENCES:

1. L.E. Overman and A.F. Renaldo, **Tetrahedron Lett.**, (1983), 3757.

2. S. Nozoe, J. Furukawa, U. Sankawa and S. Shibata, **Tetrahedron Lett.**, (1976), 195.

3. J. Bruhn, H. Heimgartner and H. Schmid, **Helv.**, (1979), 62, 2630.

4. F.E. Ziegler and J.J. Piwinski, **J. Am. Chem. Soc.**, (1979), 101, 1611.

5. W.G. Dauben and A. Chollet, **Tetrahedron Lett.**, (1981), 1583.

EXAMPLES:

EtO_2C, Me — $PdCl_2$, 5 hrs → EtO_2C, Me ① (83%, E-2:Z-2=65:35)

H, H — 240° → H, H ② (Excess of 83%)

O, CO_2Et, SPh, $(CH_2)_5$ — 210°, 10^{-3} torr → O, CO_2Et, $(CH_2)_5$ — Redistill → O, CO_2Et, $(CH_2)_5$ ③ (82%)

O, H, H, OH — 1.) Et-O-CH=CH_2, Hg^{++} 2.) $Ph_3P=CH_2$ → O — 200° → (55%) CHO △3 ④

O — TFA, C_6H_6, △ → O (Quant.) ⑤ △4

REACTION: COREY- KIM OXIDATION

March's Advanced Organic Chemistry : Not indexed

GENERAL SCHEME:

$$Cl_2 + CH_3SCH_3 \xrightarrow{CCl_4} \xrightarrow{RCH(OH)R'} \xrightarrow{Et_3N} RC(=O)R'$$

MECHANISM:

$$CH_3SCH_3 + Cl_2 \longrightarrow CH_3\overset{\oplus}{S}(Cl)CH_3 \xrightarrow{RCH(OH)R'} CH_3\overset{\oplus}{S}(CH_3)\text{—O—}C(R)(R')\text{—H} \xrightarrow{Et_3N:} RC(=O)R'$$

NOTES:

1. See: K.Omura and D. Swern, **Tetrahedron**,(1978), 34, 1651; A.J. Mancuso, S.-L. Huang and D. Swern, **J. Org. Chem.**, (1978), 43, 2480 for an extensive examination of these and related oxidation reactions.

2. The reaction is cleaner when carried out using **N-chlorosuccinimide, NCS:**

$$\text{(succinimide)NCl} + CH_3SCH_3 \longrightarrow \text{(succinimide)N—}\overset{\oplus}{S}(CH_3)_2\ Cl^{\ominus}$$

3. When the base is not present, the reaction provides a technique for converting an alcohol to an alkyl chloride.

4. Assumes an intermediate similar to the sulfonium salt in the normal **Corey- Kim oxidation**.

REFERENCES:

1. E.J. Corey and C.U. Kim, **J. Am. Chem. Soc.**,(1972), 94, 7586.

2. R. Baudat and M. Petrzilka, **Helv.**, (1979), 62, 1406.

3. E.J. Corey, C.U. Kim, and M. Takeda, **Tetrahedron Lett.**, (1972), 4339.

4. Y. Tamura, L.C. Chen, M. Fujita, H. Kiyokawa and Y. Kita, **Chem. Ind.**, (1979), 668.

5. N.K.A. Dalgard, K.E. Larsen and K.B.G. Torssel, **Acta. Chem. Scand.**, (1984), 38B, 423.

EXAMPLES:

(N-chlorosuccinimide) NCl + CH_3SCH_3 —Argon / Toluene→ —(4-tert-butylcyclohexanol, OH, $C(CH_3)_3$) / -25°, 2 hrs→ —Et_3N / 5 min→ 4-tert-butylcyclohexanone ($C(CH_3)_3$) (97%) ①

OH ... SePh —Me-S-Me / NCS→ —Et_3N→ O ... SePh (74%) ②

Ph-C(OH)(H)-Ph + (succinimido) $\overset{\oplus}{N}SMe_2Cl^{\ominus}$ —CH_2Cl_2→ —-25°, 4 hrs→ Ph-C(Cl)(H)-Ph (95%) ③ △2

(N-benzyl piperidine-3,5-dione) N, Ph, O, O —1.) NCS, Me-S-Me; 2.) CF_3COOH; 3.) $MeNH_2$→ (O, N, Ph, N-H, Me) (23%) ④

(benzaldoxime) CH=NOH —NCS, Me-S-Me→ (benzonitrile) C≡N (85%) ⑤

REACTION: COREY- WINTER OLEFINATION REACTION

March's Advanced Organic Chemistry : 919

GENERAL SCHEME:

1

HO OH

−C−C− —TCDI→ (cyclic thionocarbonate, C=S) —R_3P→ >C=C<

MECHANISM:

(diol) + $G_2C=S$ → (cyclic thionocarbonate)

(thionocarbonate) —R_3P→ ($S-PR_3^{\oplus}$ adduct) —$-R_3P=S$→ (carbene) —R_3P→ ($PR_3^{\oplus}$ adduct) → $R_2C=CR_2$

NOTES:

1. Thiocarbonyldiimidazol (**TCDI**) was originally suggested by Corey, but other methods can be used. See **Ref.2.**

2. **DMAP** = 4-Dimethylaminopyridine

3. **DMPD** = 1,3-Dimethyl-2-phenyl-1,2,3-diazaphospholidine.

REFERENCES:

1. J. Davis, V. Trantz and U. Erhardt, **Tetrahedron Lett.**, (1972), 4435.

2. H. Prinzbach and H. Babsh, **Angew Chem. Internat Ed.**, (1975), 14, 753.

3. M.A. Hashem and P. Weyerstahl, **Tetrahedron**, (1981), 37, 2473.

4. E.J. Corey and P.B. Hopkins, **Tetrahedron Lett.**, (1982), 1979.

5. W.J. McGahren, G.A. Ellestad, G.O. Morton, M.P. Kunstmann and P. Mullen, **J. Org. Chem.**, (1973), 38, 3542.

EXAMPLES:

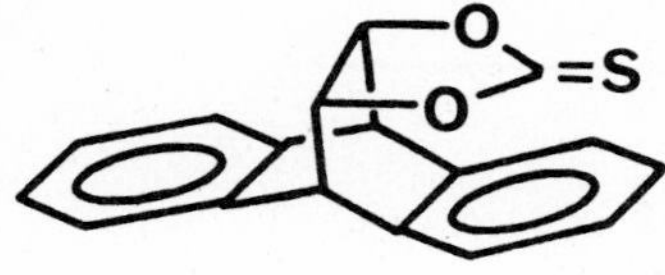

1.) TCDI, PhMe, Reflux, 1 hr

2.) $Fe(CO)_5$, PhMe, 100°, 24 hrs

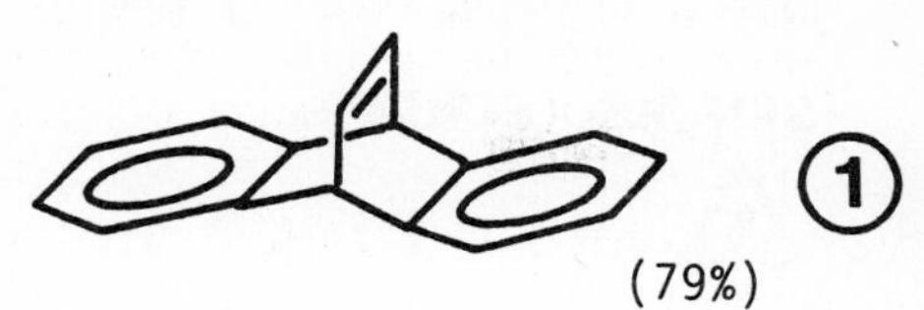

(79%) ① △1

1.) TCDI

2.) $(EtO)_3P$, 110°, 120 hrs

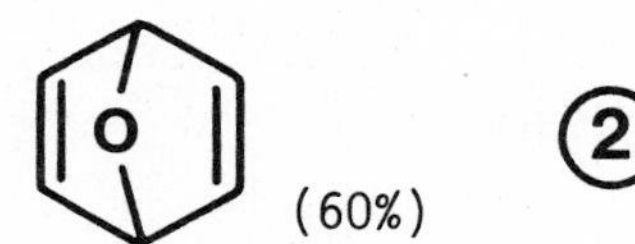

(60%) ②

1.) TCDI, PhMe, Reflux, 2 hrs

2.) $(MeO)_3P$, Reflux, 80 hrs

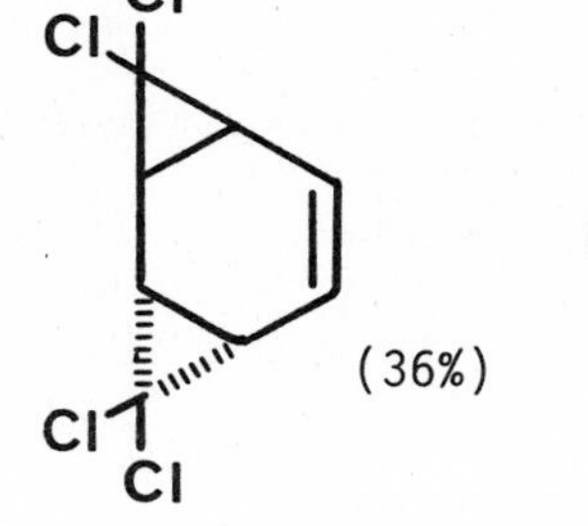

(36%) ③

1.) $CSCl_2$, CH_2Cl_2, DMAP, 0°, 1 hr

2.) DMAP, 40°, 2-24 hrs

(75%) ④ △2

1.) TCDI, PhMe, Reflux, 1 hr

2.) $(MeO)_3P$, Reflux, 90 hrs

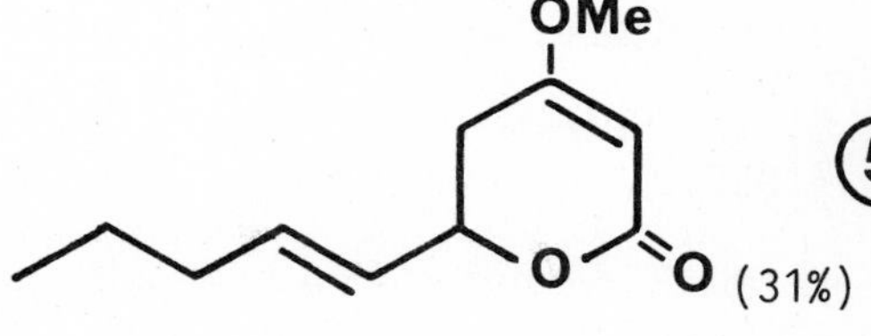

(31%) ⑤ △3

REACTION: CRIEGEE GLYCOL CLEAVAGE

==

March's <u>Advanced Organic Chemistry</u> : 1063- 1065

GENERAL SCHEME:

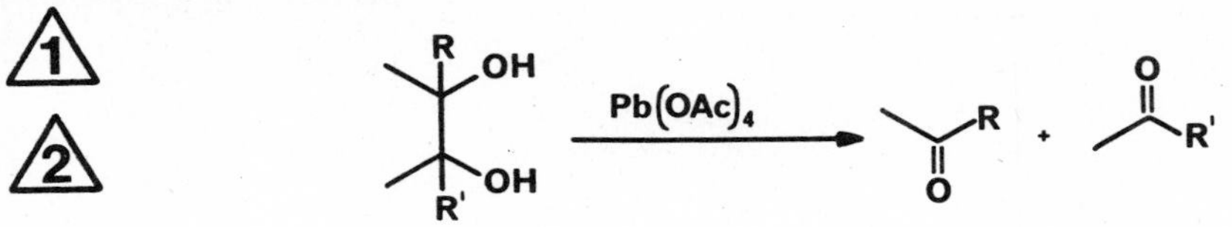

MECHANISM:

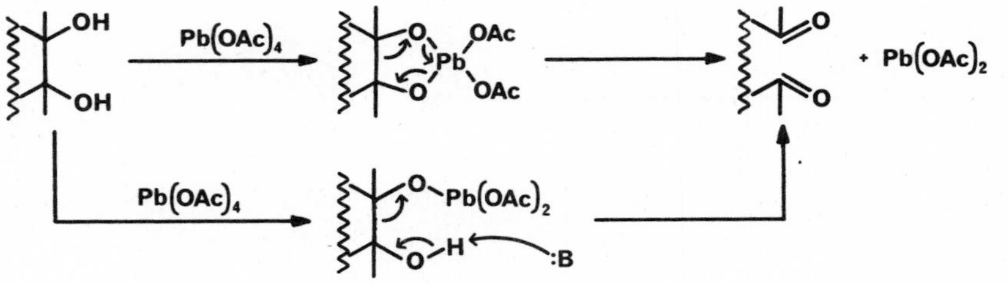

NOTES:

1. The periodic acid cleavage of diols is called the **Malprade Reaction**.

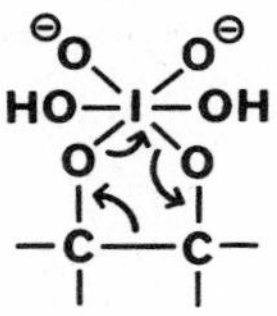

2. The oxidative cleavage using $NaBiO_3$ and H_3PO_4 is called the **Rigby Oxidation**.

3. Alkenes can be cleaved by unique processes involving <u>in situ</u> hydroxylation followed by glycol cleavage. The **Lemieux- Johnson Oxidation** uses a reaction mixture of sodium periodate and osmium tetroxide; the osmium tetroxide forms a glycol and the periodate cleaves the glycol as well as oxidizes the osmium dioxide back to the tetroxide.

REFERENCES:

1. B.H. Braun, M. Jacobson, M. Schwarz, P.E. Sonnett, N. Wakabayashi, and R.M. Waters, **J. Econom. Entomology**, (1968), <u>61</u>, 861.

2. H. Ohrui and S. Emoto, **Agric. Biol. Chem.**, (1976), <u>40</u>, 2267.

3. I.J. Borowitz, G. Gonis, R. Kelsey, R. Rapp and G.J. Williams, **J. Org. Chem.**, (1966), <u>31</u>, 3032.

4. W.L. Roelofs, M.J. Gieselmann, A.M. Carde, H. Tashiro, D.S. Moreno, C.A. Henrick and R.J. Anderson, **J. Chem. Ecol.**, (1978), <u>4</u>, 211.

5. R.J. Anderson and C.A. Henrick, **J. Am. Chem. Soc.**, (1975), 97, 4327.

EXAMPLES:

$Pb(OAc)_4$ → OHC ... COOMe (55%) ①

1.) HIO_4 2.) $NaBH_4$ → $HOCH_2$ (87%) ②

$Pb(OAc)_4$, AcOH → (83%) ③

$Pb(OAc)_4$, EtOH, C_6H_6 → OEt (NYA) ④

$Pb(OAc)_4$ → OHC ... COOEt (91%) ⑤

REACTION: CURTIUS REARRANGEMENT

==

March's **Advanced Organic Chemistry :** 945, 949, 984.

GENERAL SCHEME:

$$RCCl(=O) + NaN_3 \longrightarrow RNH_2$$

MECHANISM:

$$RC(=O)Cl \quad :\overset{\ominus}{N}=\overset{\oplus}{N}=\overset{\ominus}{N}: \longrightarrow \left[RC(=O)-N=\overset{\oplus}{N}=\overset{\ominus}{N} \leftrightarrow RC(-O^{\ominus})=N-\overset{\oplus}{N}\equiv N \;\; RC(=O)-\overset{\ominus}{N}-\overset{\oplus}{N}\equiv N \right] \longrightarrow$$

$$R-N=C=O \xrightarrow{H_2O} RNHCO_2H \xrightarrow{-CO_2} RNH_2$$

$$R-N=C=O \xrightarrow{R'NH_2} RNHCONHR' \text{ (Ureas)}$$

$$R-N=C=O \xrightarrow{R'OH} RNHCO_2R' \text{ (Urethanes)}$$

NOTES:

1. This reaction is very similar to the Schmidt and Hoffman reactions (see each under appropriate titles).

2. **Boc** = the amino protecting group, t-butoxycarbonyl

3. $(C_6H_5O_2)_2P(=O)-N_3$ = **DPPA** = diphenylphosphorazidate

---

REFERENCES:

1. W. Haefliger and E. Kloppner, **Helv.**, (1982), 65, 1837.

2. R. Bonjouklian and R.A. Ruden, **J. Org. Chem.**, (1977), 42, 4095.

3. B. Chantegrel and S,. Gelin, **Synthesis**, (1981), 315.

4. J.R. Pfister and W.E. Wymann, **Synthesis**, (1983), 39.

5. C. Kaiser and J. Weinstock, **Org. Synth.**, (1971), 51, 48.

EXAMPLES:

COOH

1.) $(PhO)_2P(O)N_3$, K_2CO_3, DMF

2.) Boc-Cl

MeO H O

NHBoc

MeO H O

(76%)

① ⚠2 ⚠3

COOH COOH

1.) $(COCl)_2$

2.) NaN_3, MeCN, 25°

[NCO NCO]

(50-70%)

②

$HOCH_2$ O—N O N$_3$

Dimethylmethoxymethane

Δ

O—N NH O O

(64%)

③

O Cl

1.) $Bu_4N^+Br^-$

2.) NaN_3

O N$_3$

1.) F_3CCO_2H

2.) K_2CO_3, H_2O

NH_2

(86%)

④

Ph COOH

1.) EtOCOCl

2.) NaN_3

3.) Δ

4.) H^+, H_2O

Ph $NH_3^{\oplus}$

(76-81%)

⑤

REACTION: DAKIN REACTION

==

March's **Advanced Organic Chemistry :** 1073

GENERAL SCHEME:

$$\text{HO-C}_6\text{H}_4\text{-C(=O)R} \xrightarrow[\ominus OH]{H_2O_2} \xrightarrow{\text{Hyd}} \text{HO-C}_6\text{H}_4\text{-OH}$$

MECHANISM:

NOTES:

1. The phenol group must be ortho or para to the carbonyl substituent.

2. This mechanism is essentially the same as that of the **Baeyer- Villiger Reaction**. See: M.B. Hocking and J.H. Ong, **Can. J. Chem.**, (1977), 55, 102 for a study of the kinetics of the Dakin Reaction.

REFERENCES:

1. Y. Agasimundin and S. Siddappa, **J. Chem. Soc.**, Perkin I., (1973), 503.

2. a. M.B. Hocking and J.H. Ong, **Can. J. Chem.**, (1977), 55, 102.
   b. M.B. Hocking, M. Ko, and T.A. Smyth, **Can. J. Chem.**, (1978), 56, 2646.

3. C.A. Bunton in Peroxide Reaction Mechanisms, edited by J.O. Edwards, Interscience.

4. H. Bretschneider, K. Hohenlowe-Oehringen, A. Kaiser and U. Wolcke, **Helv.**, (1973), 56, 2857.

5. H.D. Dakin, **Org. Synth.**, Coll. Vol. I, (1941

EXAMPLES:

H_2O_2, NaOH (1) (60%)

H_2O_2, OH^- (2) (NYA)

H_2O_2, OH^- (3) + $HCO^{\ominus}$ O_{18} (NYA)

HO-C₆H₃(COMe)-CH_2CHCO_2H (NH_2) → H_2O_2, NaOH → HO-C₆H₃(OH)-CH_2CHCO_2H (NH_2) (4) (75%)

CHO, $O^{\ominus}Na^{\oplus}$ → H_2O_2 / NaOH → OH, OH + $HCO_2^{\ominus}Na^{\oplus}$ (5) (64%)

REACTION: DARZEN'S CONDENSATION

===

March's **Advanced Organic Chemistry :** 843

GENERAL SCHEME:

$$\mathrm{RC(=O)R(H)} + \mathrm{XC(R')(H)\!-\!C(=O)OEt} \xrightarrow{\text{Base}} \mathrm{RC(R(H))\!-\!O\!-\!C(R')C(=O)OEt}$$

MECHANISM:

$$\mathrm{XC(R')(H)\!-\!C(=O)OEt} \xrightarrow{:B} \mathrm{X\overset{\ominus}{C}(R')\!-\!C(=O)OEt} + \mathrm{RC(=O)R(H)} \longrightarrow \mathrm{RC(O^{\ominus})(R(H))\!-\!C(R')(X)C(=O)OEt}$$

$$\longrightarrow \mathrm{RC(R(H))\!-\!O\!-\!C(R')C(=O)OEt}$$

NOTES:

1. The complete name for this reaction is **Darzen's Glycidic Ester Condensation,** because of the alpha-beta epoxy ester (glycidic ester) formed in the reaction.

REFERENCES:

1. R. Hunt, L. Chinn and W. Johnson, **Org. Synth.**, (1963), Coll. Vol. 4, 459.

2. R. Borch, **Tetrahedron Lett.**, (1972), 3761.

3. D. White, **Chem. Commun.**, (1975), 95.

4. H. Achenbach and J. Witzke, **Tetrahedron Lett.**, (1979), 1579.

5. A. Knorr, E. Laage and A. Weissenborn, **C.A.**, (1939), 28, 2367.

EXAMPLES:

Cyclohexanone + $ClCH_2CO_2Et$ —[$KOC(Me)_3$ / $(Me)_3COH$, 10-15°]→ ethyl 1-oxaspiro[2.5]octane-2-carboxylate (CO₂Et, HC, O) (83-95%) ①

$CH_3CH_2CH(Br)C(O)OEt$ —[1.) $LiN[Si(Me)_3]_2$ / 2.) MeCOMe]→ epoxide (O, CO_2Et, Et) (81%) ②

$CH_3C(O)CH_2CH(CO_2Et)_2$ + $NC-C(Cl)=CH_2$ —[$KOC(Me)_3$ / C_6H_6]→ bicyclic epoxide (H_3C, O, CN, EtO_2C, CO_2Et) (61%) ③

Ph–CH=CH–CHO —[$ClCH_2CO_2Et$ / EtO^-]→ Ph–CH=CH–epoxide–CO_2Et (NYA) ④

4-methylcyclohexanecarboxaldehyde (CHO) —[$ClCH_2CO_2Et$ / EtO^-]→ 4-methylcyclohexyl epoxide (O, CO_2Et) (NYA) ⑤

REACTION: DIECKMANN CONDENSATION

March's **Advanced Organic Chemistry :** 438, 1114

GENERAL SCHEME:

$$\left(\begin{array}{l} COOEt \\ CH_2COOEt \end{array}\right. \xrightarrow{:B} \left(\begin{array}{l} C{=}O \\ | \\ CHCOOEt \end{array}\right.$$

MECHANISM:

$$\left(\begin{array}{l} \overset{O}{\overset{\|}{C}}-O-Et \\ \overset{\ominus}{C}H-\underset{\underset{O}{\|}}{C}-O-Et \end{array}\right. \longrightarrow \left(\begin{array}{l} C{=}O \\ | \\ CHCOOEt \end{array}\right.$$

NOTES:

1. This is an internal **Claisen Condensation.**
2. Ideal ring sizes for this reaction are 5- 7.
3. **DME** = dimethoxyethane

REFERENCES:

1. G. Nee and B. Tchboubar, **Tetrahedron Lett.**, (1979), 3717.
2. H.-J. Liu and H.K. Lai, **Tetrahedron Lett.**, (1979), 1193.
3. C.A. Brown, **Synthesis,** (1975), 5, 326.
4. R.K. Boeckman, Jr., D.M. Blum, and S.D. Arthur, **J. Am. Chem. Soc.**, (1979), 101, 5060.
5. J. Davies and J.B. Jones, **J. Am. Chem. Soc.**, (1979), 101, 5405.

EXAMPLES:

$CH_3-CH(CO_2CH_3)\!\!\frown\!\!\frown\!\!CH(CO_2CH_3)-CH_3$ → ($Ph_3C^-K^+$, DME) → (80%) ① △1

NaH, DME / EtSH, 20° → (91%) → XS Raney Ni / EtOH, RT, 15 min → (Quant.) ② △3

$EtO_2C(CH_2)_4CO_2Et$ → (KH, THF) → CO_2Et (95%) ③

CO_2Me CO_2Me → ⊕Li ⊖N → CO_2Me (60-70%) ④

MeO_2C S CO_2Me → (1.) NaH 2.) H^+, △) → S (41%) ⑤

REACTION: DIELS- ALDER REACTION

==

March's Advanced Organic Chemistry : 738, 745- 58, 930

GENERAL SCHEME:

MECHANISM:

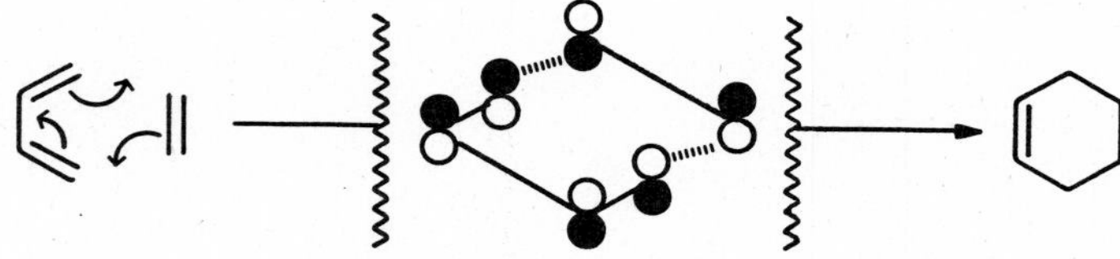

NOTES:

1. This is a [2 + 4] cycloaddition reaction between a **diene** and a **dieneophile**. In general, the diene reactivity is enhanced by electron donating groups while the dieneophile reactivity is enhanced by electron-withdrawing groups.

2.

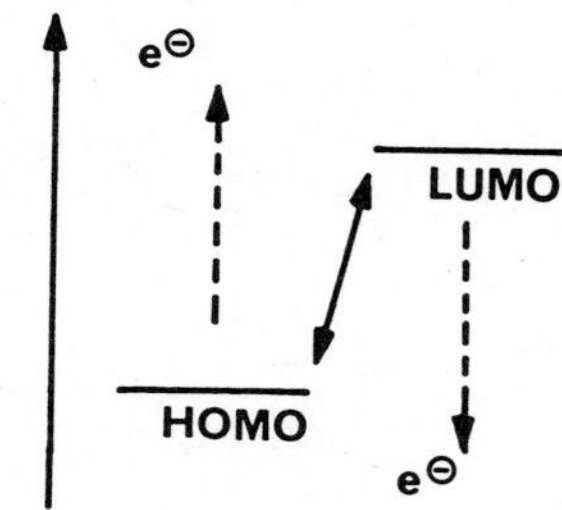

3. A **intramolecular Diels-Alder** reaction.

4. The ΔG favors endo addition by 1.2 kcal/ mol; the minimum intrinsic energy advantage associated with electronic explanation of the Alder "endo rule."

REFERENCES:

1. L.A. Van Royan, R. Mijngheer and P.J. DeClercq, **Tetrahedron Lett.**, (1982), 3283.

2. L.M. Stephenson, D.E. Smith and S.P. Current, **J. Org. Chem.**, (1982), 47, 4170.

3. R.K. Boeckman, Jr., and S.S. Kao, **J. Am. Chem. Soc.**, (1980), 102, 7149.

4. A. Ichihara, R. Kimura, S. Yamada and S. Sakamura, **J. Am. Chem. Soc.** , (1980), 102, 6353.

5. A.B. Smith,III, N.J. Liverton, N.J. Hrib, H. Sivaramakrishnan and K. Winzenberg, **J. Am. Chem. Soc.**, (1986), 108, 3040.

EXAMPLES:

CH_2Cl_2, Florisil

Room Temp

$(CH_2)_2$ (93%) ① △3

Δ

(15%) (85%) ② △4

200°

Isomerize

AcO CO$_2$Me MeO$_2$C OAc

(92%) ③ △3

50-60°

Benzene

(66%) ④

OMe

High Pressure

(80%) ⑤

REACTION: ENE REACTION

==

March's **Advanced Organic Chemistry :** 628, 647, 711, 858, 935.

GENERAL SCHEME:

MECHANISM:

NOTES:

1. This is an orbital-symmetry allowed reaction, using 4-electrons from the two pi-bonds and 2 electrons from the sigma bond (6 electron system). It is a concerted sigmatropic reaction, and can take place either intra or intermolecularly.

REFERENCES:

1. A.D. Batcho, D.E. Berger, S.G. Davoust, P.M. Wovkulich and M.R. Uskokovic, **Helv.**, (1981), 64, 1682.

2. B.B. Snider and E.A. Deutsch, **J. Org. Chem.**, (1982), 47, 745.

3. W. Oppolzer, C. Robbiani and K. Battig, **Helv.**, (1980), 63, 2015.

4. W. Oppolzer, K.K. Mahalanabis and K. Battig, **Helv.**, (1977), 60, 2388.

5. W. Oppolzer and K. Battig, **Tetrahedron Lett.**, (1982), 4669.

EXAMPLES:

CH_3, H, H_3C + $CH_2=CHCO_2CH_3$ —($EtAlCl_2$ / CH_2Cl_2)→ CH_3, CO_2CH_3, H_3C (85%) ①

CH_2 + O —($(Me)_2AlCl$ / -20°)→ C=O, CH_3 (39%) —($(Me)_2AlCl$ / 25°)→ H, OH, CH_3 (49%) ②

$COCF_3$, N, CO_2Et, CO_2Et, COO, Me_2C, Ph —(Et_2AlCl, CH_2Cl_2 / -35°)→ $COCF_3$, N, CO_2Et, CO_2Et, CH_2, COO, Me_2C, Ph (40%) ③

CO_2Et —(280°)→ EtO_2C (68%) ④

Me, H, Cl —(1.) Mg powder, Et_2O / 2.) RT, 20 hrs / 3.) O_2)→ Me, H, OH (70%) ⑤

REACTION: ESCHENMOSER FRAGMENTATION

==

March's Advanced Organic Chemistry : 929

GENERAL SCHEME:

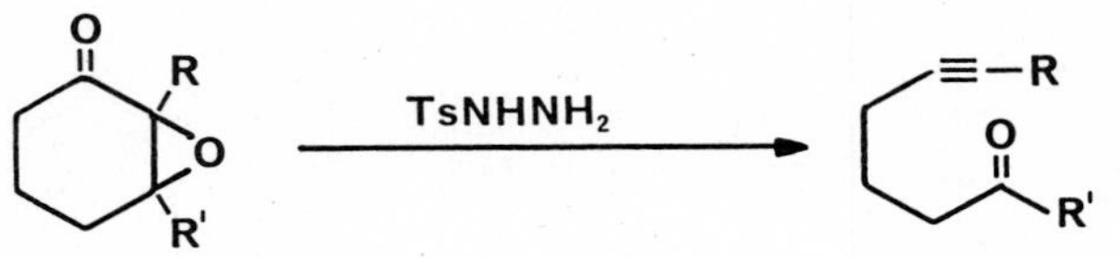

TsNHNH$_2$

MECHANISM:

NH–SO$_2$–Ph–Me-4

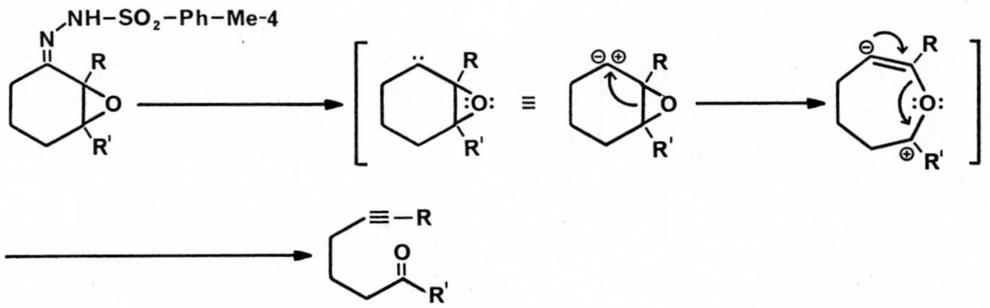

NOTES:

REFERENCES:

1. P.J. Kocienski and G.J. Cernigliaro, **J. Org. Chem.**, (1976), 41, 2927.
2. K. Mori, M. Uchida and M. Matsui, **Tetrahedron**, (1977), 33, 385.
3. J. Knolle and H.J. Shafer, **Angew. Chem. Int. Ed.**, (1975), 14, 758.
4. C.B. Reese and H.P. Sanders, **Synthesis**, (1981), 276.
5. M. Tanabe, D.F. Crowe, R.L. Dehn and G. Detre, **Tetrahedron Lett.**, (1967), 3739.

EXAMPLES:

TsNHNH$_2$ / CH$_2$Cl$_2$-HOAc → $Me(CH_2)_9\overset{O}{\overset{\|}{C}}(CH_2)_3C{\equiv}C(CH_2)_4Me$ (71%) ①

TsNHNH$_2$ / CH$_2$Cl$_2$-HOAc → $Me(CH_2)_9\overset{O}{\overset{\|}{C}}(CH_2)_3C{\equiv}CH$ (70%) ②

TsNHNH$_2$ / CH$_2$Cl$_2$-HOAc → $EtC{\equiv}C(CH_2)_3\overset{O}{\overset{\|}{C}}Me$ (91%) ③

2,4,6-(Me)$_3$Ph-SO$_2$NHNH$_2$ → (95%) ④

TsNHNH$_2$ / OH$^-$, Δ → (85%) ⑤

REACTION: ESCHWEILER- CLARKE METHYLATION

March's Advanced Organic Chemistry : 799- 800

GENERAL SCHEME:

$$>N{-}H \xrightarrow[HCO_2H]{HCHO} >N{-}CH_3$$

MECHANISM:

$$-\ddot{N}H + H_2C{=}O \quad H{-}O{-}C(=O){-}H \xrightarrow{-H_2O} \oplus N{=}CH_2 \quad H{-}C(=O){-}O^{\ominus} \longrightarrow >N{-}CH_3 + CO_2$$

NOTES:

1. This reaction is very similar to the **Leuckart** and **Wallach reactions.** In each an intermediate immonium ion is reduced.

2. This reaction is generally used for primary or secondary amines.

3. A modification of this reaction uses $NaBH_4$ or $NaBH_3CN$ as reducing agents.

4. Examples of deaminative fragmentation during the methylation reaction are known (Ref. 4):

CH₂CO₂Me; N; CH₂-OH₂⊕; H-CH; CO₂Me → CO₂Me; N=CH₂; CO₂Me → CO₂Me; HNCH₂OH₂⊕; CO₂Me →

REFERENCES:

1. R.F. Borch and A.I. Hassid, **J. Org. Chem.**, (1972), 37, 1673.

2. M. Tichy, L. Kneizo and S. Vasickova, **Coll. Czech. Chem. Commun.**, (1974), 39, 555.

3. B.L. Sondengam, J. H. Hemo, and G. Charles, **Tetrahedron Letters**, (1973), 261.

4. K. Watanabe and T. Wakabayashi, **J. Org. Chem.**, (1980), 45, 357.

5. W.E. Parham, W.T. Hunter, R. Hanson and T. Lahr, **J. Am. Chem. Soc.**, (1952), 7 5646.

EXAMPLES:

NH_2 —[HCOH, Na BH_3CN / Acetonitrile, 25°, 1 hr]→ NMe_2 ① (92%)

HO, NH_2 —[H_2CO, HCO_2H]→ HO, NMe_2 ② (63%)

HO, N–H, CH_3 —[1.) 35% HCOH, MeOH / 2.) $NaBH_4$]→ N, CH_3, CH_3 ③ (85%) △3

H, COOMe, NH, CH_2COOMe —[HCOOH, CH_2O / △]→ COOMe, COOMe ④ △4 (73%)

CH_2CHCH_3, NH_2 —[CH_2O/HCO_2H]→ CH_2CHCH_3, $N(CH_3)_2$ ⑤ (66%)

REACTION: FAVORSKII REARRANGEMENT

March's **Advanced Organic Chemistry :** 971- 974.

GENERAL SCHEME:

R-C(=O)-C(X)(R')-R'' —($^{\ominus}$OR''')→ R'''-O-C(=O)-C(R)(R')-R''

MECHANISM:

NOTES:

1. The reaction involves the rearrangement of an alpha-halo ketone to produce an ester.

2. Cyclic alpha-halo ketones suffer ring contraction.

REFERENCES:

1. N. Schamp, N. DeKimpe and W. Coppens, **Tetrahedron,** (1975), 31, 2081.

2. A. Abad, M. Arno, J. Pedro and E. Sedane, **Tetrahedron Lett.,** (1981), 1733.

3. T. Sakai, T. Katayama and A. Takeda, **J. Org. Chem.,** (1981), 46, 2924.

4. G. Haufe, **Synthesis,** (1983), 235.

5. R.J. Stedman, L.S. Miller, L.D. Davis and J.R.E. Hoover, **J. Org. Chem.,** (1970), 35, 4169.

EXAMPLES:

ClH_2C / t-Bu-CH-C(=O)-CH_2Cl —($NaOCH_3$ / CH_3OH)→ ClH_2C / t-Bu-CH-C(=O)-OCH_3 (78%) ①

H-CCl / $(CH_2)_9$ C=O / CCl-Br + H-CCl / $(CH_2)_9$ C=O / CBr-Br —(NaOMe, MeOH / C_6H_6, RT, 1 hr)→ $(CH_2)_9$ C(Cl)=C(CO_2CH_3) (80%) ②

O / $(CH_3)_2C$(Br) —($NaCH(CO_2Et)_2$ / Refluxing THF)→ HO, $CH(CO_2Et)_2$, H_3C, CH_3 (51%) ③

O, Cl —(NaOH / EtOH)→ CO_2H (64%) ④ △2

O, Cl, Cl, Cl, Cl —(NaOH)→ Cl, HO_2C–, Cl, Cl (68%) ⑤

REACTION: FEIST- BENIARY FURAN SYNTHESIS

March's Advanced Organic Chemistry : Not Indexed

GENERAL SCHEME:

$RCH_2C(=O)CH_2CO_2Et + ClCH_2C(=O)CH_3 \xrightarrow{\text{Base}}$ 2-RH_2C-3-EtO_2C-4-CH_3-furan

MECHANISM:

$CH_3-C(=O)-\overset{\ominus}{C}H-CO_2Et + ClCH_2-C(=O)-CH_3 \longrightarrow CH_3-C(=O)-C(H)(CO_2Et)-C(CH_3)(OH)-CH_2-Cl$

$\longrightarrow$ dihydrofuran (EtO_2C, CH_3, $-OH$, CH_3) $\xrightarrow{-H_2O}$ furan (EtO_2C, CH_3, CH_3)

NOTES:

REFERENCES:

1. A. Gopalan and P. Magnus, **J. Am. Chem. Soc.**, (1980), 102, 1756.

2. J. Kagen and K.C. Mattes, **J. Org. Chem.**, (1980), 45, 1524.

3. T. Reichstein, H. Zschokke and W. Syz, **Helv.**, (1932), 15, 1112.

4. J.N. Chatterjea and R.R. Ray, **Ber.**, (1959), 92, 998.

5. J.W. Batty, P.D. Howes and C.J.M. Stirling, **J. Chem. Soc.**, Perkin I, (1973), 65.

EXAMPLES:

MeC(O)CH(Cl)COOEt, KOH → (57%) ①

KOH, H_2O, Δ → (NYA) ②

NH_3 → (10%) ③

KOH → (53%) ④

$Me_2\overset{\oplus}{S}CH{=}C{=}CH_2$ $Br^{\ominus}$ + $EtO_2CCH_2C(O)Me$ —NaOEt, EtOH, Δ→ (86%) ⑤

REACTION: FISCHER INDOLE SYNTHESIS

March's Advanced Organic Chemistry : 1032

GENERAL SCHEME:

$C_6H_5NHNH_2 + RCOCH_2R' \longrightarrow RC(=NNHAr)CH_2R' \xrightarrow{ZnCl_2}$ 2-R-3-R'-indole $+ NH_3$

MECHANISM:

NOTES:

1. It is the second step which is technically the named reaction, namely, the formation of an indole from an arylhydrazone of an aldehyde or ketone.

2. Many acid catalysts can be used for this reaction.

3. The mechanism is actually a [3.3]sigmatropic rearrangement. In the acid-catalyzed formation of the enamine-intermediate, (*), the more-substituted enamine is usually formed.

4. **PPA** = Polyphosphoric acid.

REFERENCES:

1. A. Guy and J. Guette, **Synthesis**, (1980), 222.

2. G. Balloliniand P. Todesco, **Chem. Commun.**, (1981), 563.

3. E. Shaw, **J. Am. Chem. Soc.**, (1955), 77, 4319.

4. L.M. Rice, E. Hertz and M.E. Freed, **J. Med. Chem.**, (1964), 7, 313.

5. T. Shono, Y. Matsumura and T. Kanazawa, **Tetrahedron Lett.**, (1983), 1259.

EXAMPLES:

$4\text{-}O_2N\text{-}C_6H_4\text{-}NH\text{-}N{=}C(CO_2C_2H_5)CH_3$ —PPA, 110°, 1hr→ ethyl 5-nitroindole-2-carboxylate, $-CO_2C_2H_5$ (78%) ① ⚠4

$PhCH_2\overset{O}{\overset{\|}{C}}Ph$ + $PhNHNH_2$ —PCl_3→ 2,3-diphenylindole, $-Ph$, Ph (75%) ②

cyclooctanone —$4\text{-}NO_2\text{-}Ph\text{-}NHNH_2$→ O_2N-cyclooctindole, N-H (58%) ③

MeO-C_6H_4-NH-N=C(CH_3)$CH_2CH_2CO_2Me$ —HCl→ MeO-indole, CO_2Me, N-H (85%) ④

$C_6H_5NHNH_2$ + N-benzoyl-2-methoxypyrrolidine (OMe, $O{=}C\text{-}Ph$) —$ZnCl$, Xylene→ indole-3-$CH_2CH_2N(H)C(=O)Ph$, N-H (76%) ⑤

REACTION: FRIEDEL-CRAFTS ACYLATION

March's Advanced Organic Chemistry : 484- 87, 496.

GENERAL SCHEME:

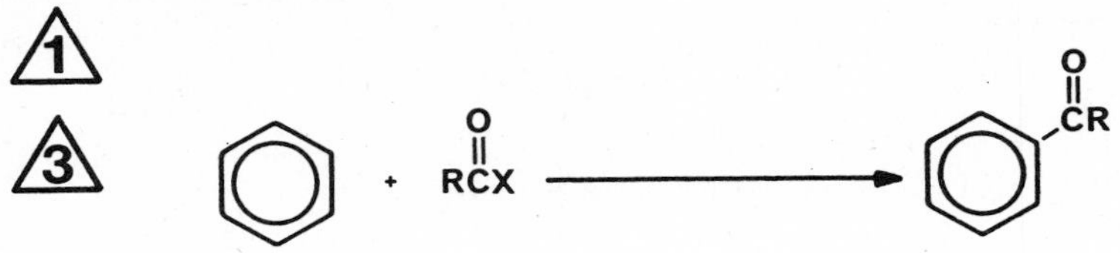

MECHANISM:

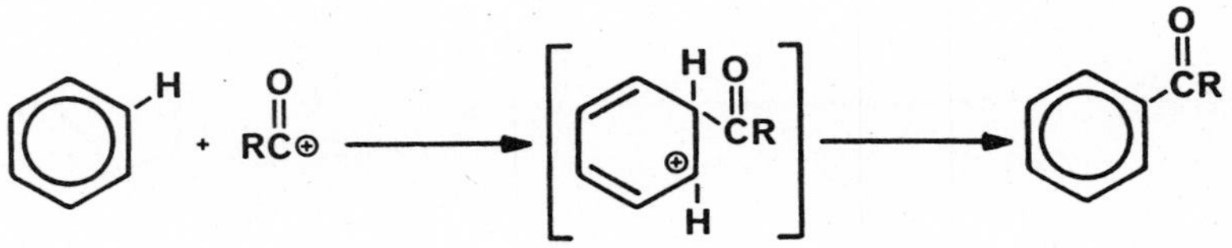

NOTES:

1. Aside from the Friedel-Crafts alkyaltion and acylation, there is an arylation reaction that is sometimes known as the **Scholl Reaction**.

2. The reaction can be carried out on alkenes as well as aromatics; this reaction is sometimes known as the **Darzens-Nenitzescu Ketone Synthesis**.

3. Acids or acid anhydrides can be used in place of the acid halides. Other Lewis acids that have been used include: $SnCl_4$, $ZnCl_2$ and polyphosphoric acid.

4. **TFA** = Trifluroacetic acid

REFERENCES:

1. A.P. Kozikowski and A. Ames, **J. Am. Chem. Soc.**, (1980), 102, 860.

2. H.M.R. Hoffmann and T. Tsushima, **J. Am. Chem. Soc.**, (1977), 99, 6008.

3. M. Jung and R. Brown, **Tetrahedron Lett.**, (1981), 3355.

4. B. Snider and A. Jackson, **J. Org. Chem.**, (1982), 47, 5393.

5. J.H. Babler, **Tetrahedron Lett.**, (1975), 2045.

EXAMPLES:

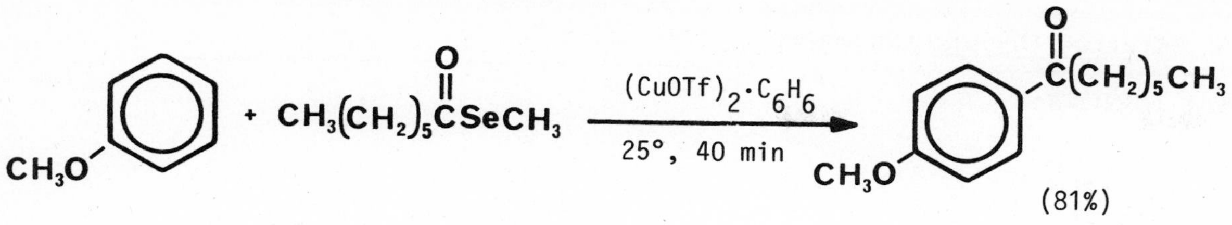

CH_3O-benzene + $CH_3(CH_2)_5C(=O)SeCH_3$ → [$(CuOTf)_2 \cdot C_6H_6$; 25°, 40 min] → CH_3O-C_6H_4-$C(=O)(CH_2)_5CH_3$ (81%) ①

cyclohexene + $CH_3CO^{\oplus}SbF_6^{\ominus}$ → [Tetramethylurea; CH_2Cl_2, -50°] → 3-acetylcyclohexene + 1-acetylcyclohexene (26%) ② △2

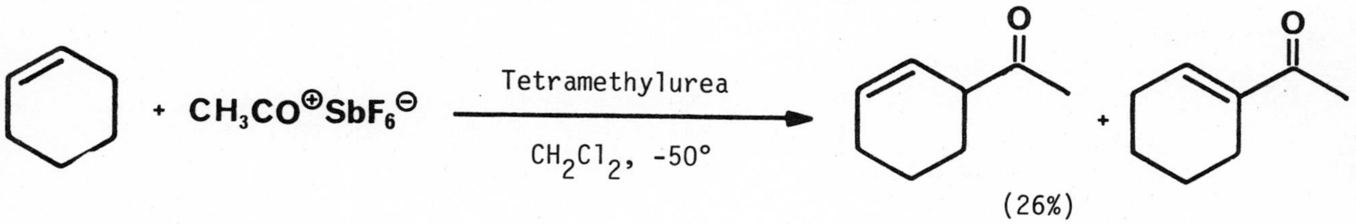

(pyrone) OMe → [Ac_2O, TFA; Δ, 8 hrs] → OMe, O (81%) ③ △4

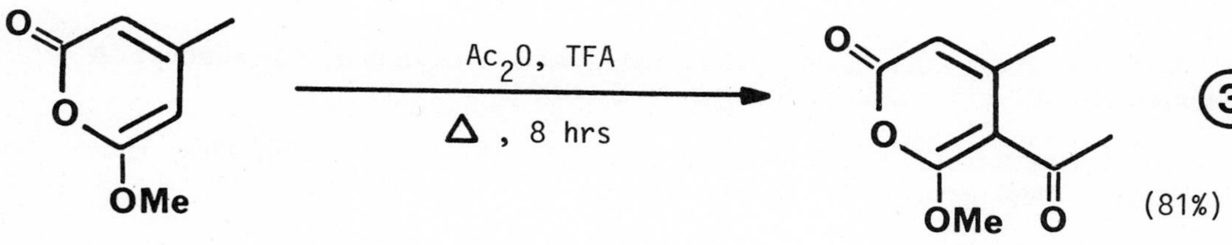

cyclohexene + CH_3COCl → [$EtAlCl_2$; CH_2Cl_2, 25°, 1 hr] → O (73%) ④ △2

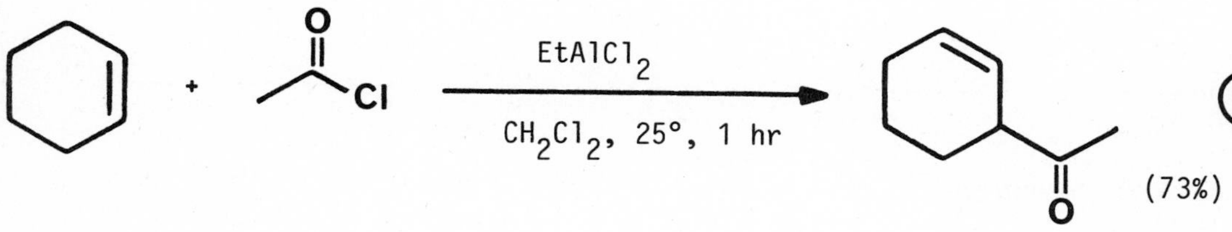

ethylene + CH_3COCl → [$AlCl_3$] → $ClCH_2CH_2COCH_3$ (85%) ⑤ △2

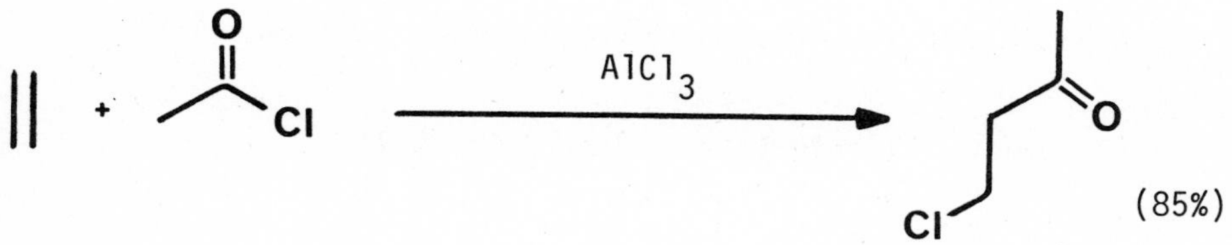

REACTION: FRIEDEL- CRAFTS ALKYLATION

March's **<u>Advanced Organic Chemistry</u> :** 479- 484

GENERAL SCHEME:

△4

+ RX —$AlCl_3$→ R

MECHANISM:

+ $R^{\oplus}$ → [H, R, ⊕, H] → R

NOTES:

1. See: R.M. Roberts and A.A. Khalaf, **Friedel-Crafts Alkylation Chemistry: A Century of Discovery**, Marcel Dekker, Inc. , N.Y. 1984.

2. Since there is formation of a carbonium ion, $\mathbf{R^+}$, it is not surprising that rearrangement of the alkyl group is common.

3. For further information on **carboranes**, see: R.N. Grimes, **Carboranes**, Academic Press, N.Y., 1970.

4. Polyalkylation is a common side-reaction.

REFERENCES:

1. M.E. Jung, A.B. Mossman and M.A. Lyster, **J. Org. Chem.**, (1978), <u>43</u>, 3698.

2. S. Masuda, T. Nakajima and S. Suga, **Bull. Chem. Soc.,Japan**, (1983), <u>56</u>, 1089.

3. J. Plesek, Z. Plzak, J. Stuchlik and S. Hermanek, **Collect. Czech. Chem. Commun.**, (1981), <u>46</u>, 1748.

4. N. Yoneda , T. Fukuhara, Y. Takahashi and A. Suzuki, **Chem. Letters**, (1979), 1003.

5. L.J. Belen'kii and A.P. Yakabov, **Tetrahedron**, (1984), <u>40</u>, 2471.

EXAMPLES:

3,4-(MeO)$_2$C$_6$H$_3$CH$_2$CHO $\xrightarrow[\text{CH}_2\text{Cl}_2\text{, -78°, then 25°}]{\text{ISi(Me)}_3}$ bridged bis(dimethoxybenzo) ether (MeO, MeO, OMe, OMe; O bridge) (85%) ①

benzene + PhCH$_2$CHClCH$_3$ $\xrightarrow[\text{0°, 1 hr}]{\text{AlCl}_3\text{, CS}_2}$ PhCH$_2$–C(H)(CH$_3$)–Ph (60%) ②

HC—CH (B$_{10}$H$_{10}$) + CH$_3$Br $\xrightarrow[\text{25°, 48 hrs}]{\text{AlCl}_3\text{, CS}_2}$ HC—CH (9-CH$_3$B$_{10}$H$_9$) (26%) ③ △3

PhC(O)Me $\xrightarrow[\text{HF / SbF}_5]{\text{EtCl}}$ 3-Et-C$_6$H$_4$C(O)Me (78%) ④

thiophene $\xrightarrow[\text{AlCl}_3]{\text{t-BuCl}}$ t-Bu-thiophenium ⊕ (2-position) + t-Bu-thiophenium ⊕ (3-position) (96%, 97:3) ⑤

REACTION: FRIEDLANDER QUINOLINE SYNTHESIS

March's Advanced Organic Chemistry : 797

GENERAL SCHEME:

MECHANISM:

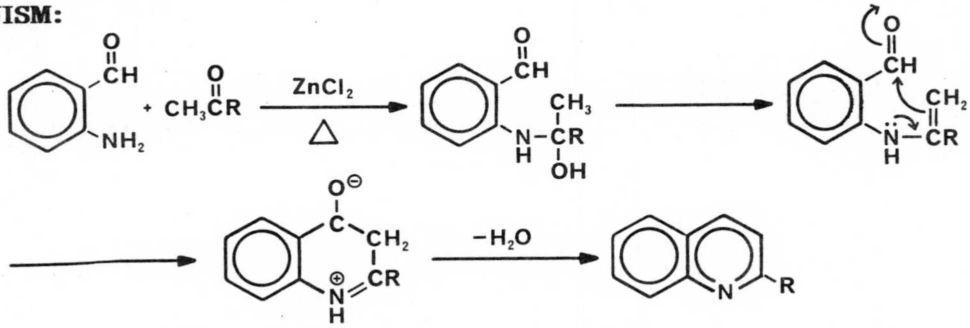

NOTES:

1. The use of this reaction has been limited because of the limited availability of o-aminobenzaldehydes.

2. For a Review, see: C.-C. Cheng and S.-J. Yan, **Org. Reactions**, (1982), 28, 37.

3. The reducing agent used here is the the lithium salt of phthalocyaninecobalt(I). (Pc = **phthalocyanine**).

REFERENCES:

1. H. Eckert, **Angew. Chem. Int. Ed.**, (1981), 20, 208.

2. G. Kempter and S. Hirschberg, **Ber.**, (1965), 98, 419.

3. K.V. Rao and H.-S. Kuo, **J. Heterocylic Chem.**, (1979), 16, 1241.

4. J.H. Markgraf and W.L. Scott, **Chem. Commun.**, (1967), 296.

5. A.D. Settino, G. Primofiore, O. Livi, P.L. Ferrarini and S. Spinelli, **J. Heterocyclic Chem.**, (1979), 16,169.

EXAMPLES:

CHO
NO_2
+ $CH_3\overset{O}{\overset{\|}{C}}Et$
Li[Co(I)Pc]
MeOH
CH_3
N
CH_3
(71%)
① △3

O
NH_2
+
CH_3
N
O
H^+
150°
CH_3
N
$N-CH_3$
(56%)
②

MeO
MeO
CHO
NH_2
Me–C–NH
O
+
O
Me
N
CO_2Me
NO_2
Me
KOH
MeO
MeO
N
NH
C
Me
O
N
CO_2Me
NO_2
Me
(60%)
③

O
+
CHO
NH_2
KOH
N
(55%)
④

O
N
H
N
Br
+
CHO
NH_2
HCl
EtOH
NH
N
N
Br
(59%)
⑤

REACTION: FRIES REARRANGEMENT

==

March's Advanced Organic Chemistry : 499

GENERAL SCHEME:

$$PhOC(O)R \xrightarrow{AlCl_3} p\text{-}RC(O)C_6H_4OH$$

MECHANISM:

1

$$PhOC(O)R \longrightarrow PhO^{\oplus}=C(R)-O-\overset{\ominus}{A}lCl_3 \rightleftharpoons \overset{\oplus}{C}(O)R + PhO-\overset{\ominus}{A}lCl_3 \longrightarrow p\text{-}RC(O)C_6H_4OH$$

NOTES:

1. Although this mechanism is still in question, it is at least a useful way to think of the reaction. The obvious "cross-over" experiments have not lead to a clear decision. The reaction does not work well with m-directing groups on the aromatic ring.

2. **TMEDA** = N,N,N',N'-Tetramethylenediamine

3. The **Photo-Fries** rearrangement is thought to take place by an intramolecular free-radical process.

REFERENCES:

1. J. Gray, Reported in **Reagents**, 8, 14.

2. M. Sibi ad V. Snieckus, **J. Org. Chem.**, (1983), 48, 1935.

3. A. Hallberg, A. Svensson and A.R. Martin, **Tetrahedron Lett.**, (1986), 1959.

4. D.J. Crouse, S.L. Hurlbut and D.M.S. Wheeler, **J. Org. Chem.**, (1981), 46, 374.

5. D. Veierov, Y. Mazur and E. Fiscer, **J. Chem. Soc., Perkin II**, (1980), 1659.

EXAMPLES:

OH, CH_3 — Ac_2O, $Mg(ClO_4)_2$ → $OCOCH_3$, CH_3 — $AlCl_3$ → OH, $COCH_3$, CH_3 (86-92%) ① △1

$OCNEt_2$ (O) — s-BuLi, TMEDA, THF, -78°, 1 hr → [$OCNEt_2$ (O), Li] — RT, 10-12 hrs → OH, $CNEt_2$ (O) (75%) ② △2

S, N, CF_3, O, CF_3 — LDA → S, N, H, CF_3, O, CF_3 (44%) ③

OMe, O, O — hν, EtOAc → OMe, OH, O (71%) ④ △3

O, CH_2, O — hν → O, O (25%) ⑤

REACTION: FRITSCH-WIECHELL REARRANGEMENT

==

March's Advanced Organic Chemistry : 978- 980

GENERAL SCHEME:

$$Ph_2C{=}CHX \xrightarrow{:B} PhC{\equiv}CPh$$

MECHANISM:

$$Ph_2C{=}CHX \xrightarrow{:B} Ph_2C{=}\overset{\ominus}{C}{-}X \longrightarrow Ph_2C{=}C{:} \longrightarrow PhC{\equiv}CPh$$

NOTES:

1. Typical bases include sodium amide, alkoxide ion, and organolithium reagents.

2. The reaction is inhibited by electron-withdrawing groups on the aromatic rings.

3. There does not have to be an anti orientation of the migrating and leaving group; however, this view is not without argument. G. Kobrich, **Angew Chem. Internat. Ed.**, (1972), 11, 473.

REFERENCES:

1. B. Sket, M. Zupan and A. Pollak, **Tetrahedron Lett.**, (1976), 784.

2. P.J. Stang, D.P. Fox, C.J. Collins, and C.R. Watson, Jr., **J. Org. Chem.**, (1978), 43, 364.

3. A.A. Bothner-By, **J. Am. Chem. Soc.**, (1955), 77, 3293.

4. G. Kobrich ad D. Merkel, **Angew. Chem. internat. Ed.**, (1970), 9, 243.

5. A. Merz and G. Thumm, **Ann.**, (1978), 1526.

EXAMPLES:

① $Ph_2C{=}CHCl \xrightarrow[Et_2O]{h\nu} PhC{\equiv}CPh\ (25\%) + Ph_2C{=}CH_2\ (23\%) + Ph_2C{=}CH{-}CH{=}CPh_2\ (10\%)$

② $Me(Ph)\overset{*}{C}{=}CH(OSO_2CF_3) + Ph(Me)\overset{*}{C}{=}CH(OSO_2CF_3) \xrightarrow{BuLi} [Me(Ph)\overset{*}{C}{=}C:] \longrightarrow Me\overset{*}{C}{\equiv}CPh$ ③ (triangle)

$^*C^{14}$

③ $(p\text{-}BrC_6H_4)(Ph)C{=}CHBr \xrightarrow{t\text{-}BuO^-} Br{-}C_6H_4{-}C{\equiv}C{-}C_6H_5$ (Quant.)

④ $(c\text{-}C_3H_5)_2C{=}CHCl \xrightarrow{BuLi} c\text{-}C_3H_5{-}C{\equiv}C{-}c\text{-}C_3H_5$ (80%)

⑤ $(p\text{-}MeOC_6H_4)(Ph)C{=}CBr_2 \xrightarrow{Electrolysis} C_6H_5{-}C{\equiv}C{-}C_6H_4{-}OMe$ (80%)

REACTION: GABRIEL SYNTHESIS

==

March's **Advanced Organic Chemistry :** 377, 591

GENERAL SCHEME:

$$\text{Phthalimide (NH)} + RX \longrightarrow RNH_2 + \text{C}_6\text{H}_4(CO_2H)_2$$

MECHANISM:

$$\text{Phthalimide (NH)} \xrightarrow{:B} \text{Phthalimide anion } (\ominus N) + RX \longrightarrow \text{N-alkylphthalimide (NR)} + X^{\ominus}$$

$$\xrightarrow[H_2O]{H^{\oplus}} \text{C}_6\text{H}_4(CO_2H)_2 + RNH_3^{\oplus}$$

NOTES:

1. This is a simple way of preparing primary amines.

2. The **Ing- Manske Modification** of this reaction involves the use of hydrazine for the removal of the primary amine from the imide.

$$\text{N-alkylphthalimide (NR)} \xrightarrow{H_2NNH_2} \text{phthalhydrazide (NH–NH)} + RNH_2$$

REFERENCES:

1. M. Sato and S. Ebine, **Synthesis**, (1981), 472.

2. J.C. Sheehan and W.A. Bolhofer, **J. Am. Chem. Soc.**, (1950), 72, 2786.

3. D. Landini, F. Montanari ad F. Rolla, **Synthesis**, (1978), 223.

4. A. Zwierzak and S.P. Pilichowska, **Synthesis**, (1982), 922.

5. B. Dietrich, M.W. Hosseini, J.M. Lehn and R.B. Sessions, **J. Am. Chem. Soc.**, (1981), 103, 1282.

EXAMPLES:

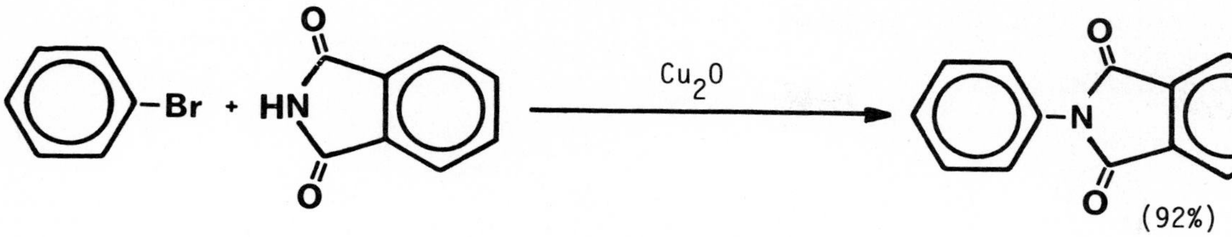

① Bromobenzene + phthalimide, Cu_2O → N-phenylphthalimide (92%)

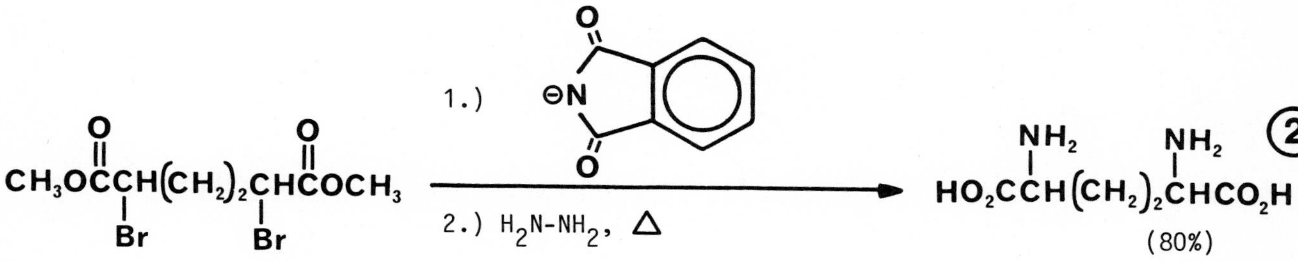

② $CH_3OCCH(CH_2)_2CHCOCH_3$ (C=O on each ester carbon; Br on each CH), 1.) ⊖N (phthalimide anion); 2.) $H_2N\text{-}NH_2$, Δ → $HO_2CCH(CH_2)_2CHCO_2H$ (NH$_2$ on each CH) (80%)

③ bis(phthalimidoethoxy)cyclohexane, $H_2N\text{-}NH_2$, Δ → 1,2-bis(2-aminoethoxy)cyclohexane (NH_2, NH_2) (60%)

④ EtBr + $Na^{\oplus}$ $^{\ominus}N(P(O)(OEt)_2)(CO_2t\text{-}Bu)$ $\xrightarrow[C_6H_6,\ 80°]{(n\text{-}Bu)_4N^+Br^-}$ $Et\text{-}N(P(O)(OEt)_2)\text{-}CO_2t\text{-}Bu$ $\xrightarrow[RT]{HCl\ /\ C_6H_6}$ $EtNH_2$ (NYA)

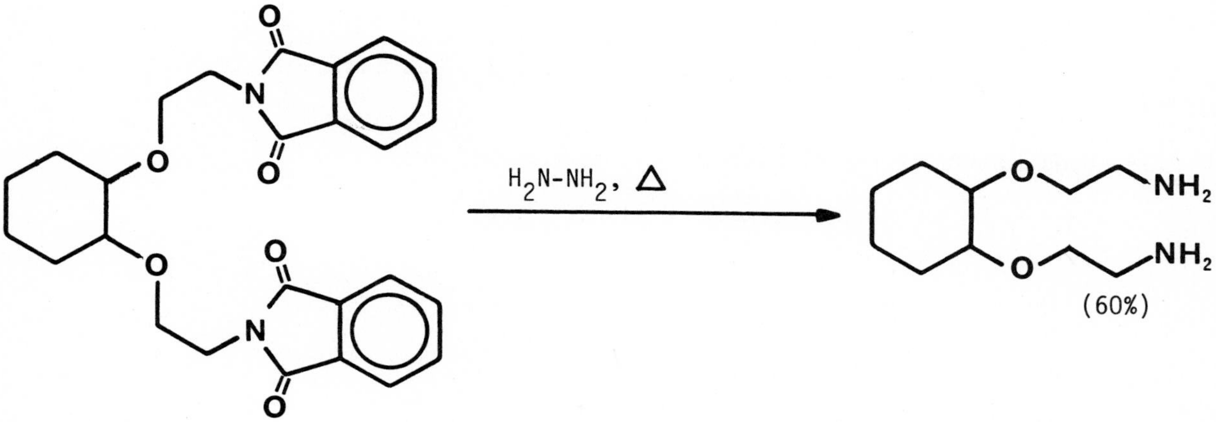

⑤ $(TsOCH_2CH_2N(Ts)CH_2CH_2\text{-})_2O$, 1.) ⊖N (phthalimide anion); 2.) $H_2N\text{-}NH_2$ → $(H_2NCH_2CH_2N(Ts)CH_2CH_2\text{-})_2O$ (82%)

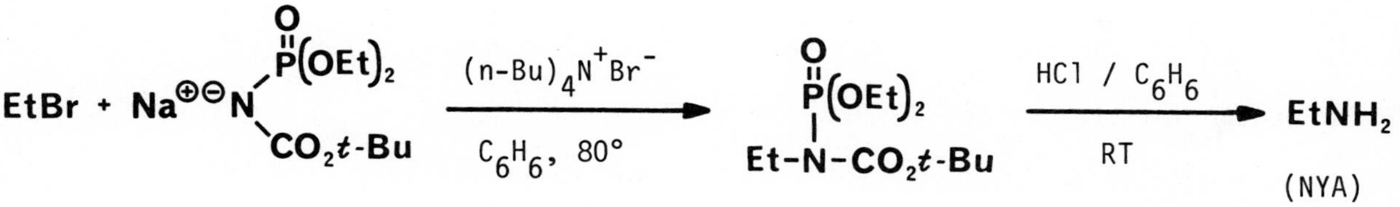

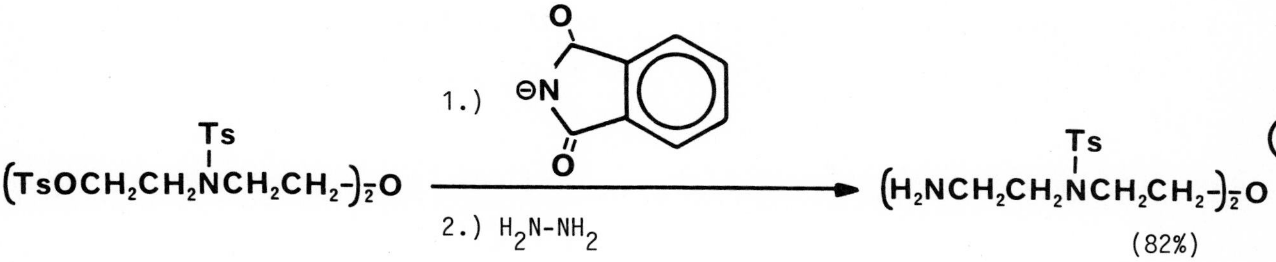

REACTION: GATTERMAN FORMYLATION

===

March's Advanced Organic Chemistry: 488, 497

GENERAL SCHEME:

$$C_6H_6 + Zn(CN)_2 \xrightarrow[H_2O]{HCl} C_6H_5CHO$$

MECHANISM:

$$C_6H_6 + Zn(CN)_2 + H^{\oplus} \longrightarrow C_6H_5CH{=}\overset{\oplus}{N}H_2 \xrightarrow{H_2O} C_6H_5CHO$$

NOTES:

1. Sometimes known as the **Gatterman Aldehyde Synthesis**.

2. Formally known as the **Gatterman- Koch Reaction**.

REFERENCES:

1. A. Rahm, R. Guilhemat and M. Pereyre, **Synth. Commun.**, (1982), 12, 485.

2. R. Adams and I. Levine, **J. Am. Chem. Soc.**, (1923), 45, 2373.

3. L. Toniold and M. Graziani, **J. Organometallic Chem.**, (1980), 194, 221.

4. F.M. Aslam, P.H. Gore and M. Jehangir, **J. Chem. Soc.**,Perkin 1, (1972), 892.

5. A.H. Gorwin and G.G. Kleinspehn, **J. Am. Chem. Soc.**, (1953), 75, 2089.

EXAMPLES:

Mesitylene + $Me_2C(CN)OH$ $\xrightarrow[\text{ClCH}_2\text{CH}_2\text{Cl, 83°}]{\text{AlCl}_3}$ 2,4,6-trimethylbenzaldehyde (48%) ①

Resorcinol $\xrightarrow[\text{2.) H}_2\text{O}]{\text{1.) Zn(CN)}_2\text{, HCl}}$ 2,4-dihydroxybenzaldehyde (95%) ②

Benzene $\xrightarrow[\text{AlCl}_3 + \text{Cu}_2\text{Cl}_2]{\text{CO + HCl}}$ benzaldehyde (78%) ③ △2

2,3-Dimethylnaphthalene $\xrightarrow[\text{2.) HCl, 50°}]{\text{1.) HCl, Zn(CN)}_2\text{, AlCl}_3}$ 2,3-dimethyl-1-naphthaldehyde (89%) ④

Ethyl 4-methylpyrrole-3-carboxylate (Me, CO_2Et, NH) $\xrightarrow{\text{Zn(CN)}_2\text{, HCl}}$ ethyl 5-formyl-4-methylpyrrole-3-carboxylate (Me, CO_2Et, OHC, NH) (73%) ⑤ △1

REACTION: GLASER COUPLING REACTION
CADIOT- CHODKIEWICZ COUPLING REACTION
EGLINGTON REACTION

March's <u>Advanced Organic Chemistry</u>: 640

GENERAL SCHEME:

$$R-C\equiv C-H \xrightarrow{Cu,\ O_2} R-C\equiv C-C\equiv C-R$$

MECHANISM:

$$R-C\equiv C-R \xrightarrow[O_2]{Cu^{\oplus}} R-C\equiv C\cdot \longrightarrow R-C\equiv C-C\equiv C-R$$

NOTES:

1. When the reaction involves a terminal alkyne it is often called the **Cadiot-Chodkiewicz Reaction.** A reaction of a terminal halogen-substituted alkyne is often called an **Eglington Reaction.**

2. For a Review, see: G. Eglington and W. McCrae, <u>Adv. in Organic Chemistry</u>, (1963), <u>4</u>, 252.

REFERENCES:

1. H.A. Stansbury, Jr., and W.R. Proops, **J. Org. Chem.**, (1962), <u>27</u>, 320.

2. W. Chodkiewicz, **Ann.**,(1957), <u>2</u>, 819.

3. T. Ando and M. Nakagawa, **Bull. Chem. Soc. Japan**, (1967), <u>40</u>, 363.

4. K. Yamamoto and F. Sondheimer, **Angew. Chem. Int. Ed.**,(1973), <u>12</u>, 68.

EXAMPLES:

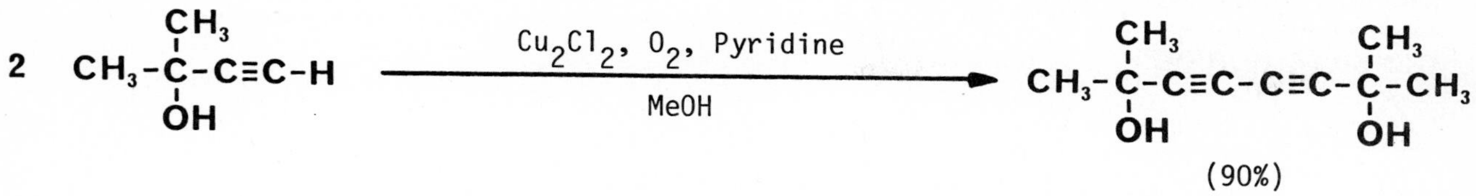
2
Cu_2Cl_2, O_2, Pyridine
MeOH
(90%)

(1)

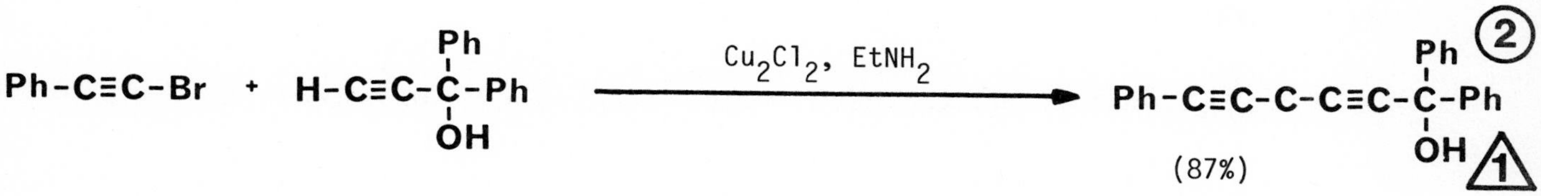
Cu_2Cl_2, $EtNH_2$
(87%)

(2)

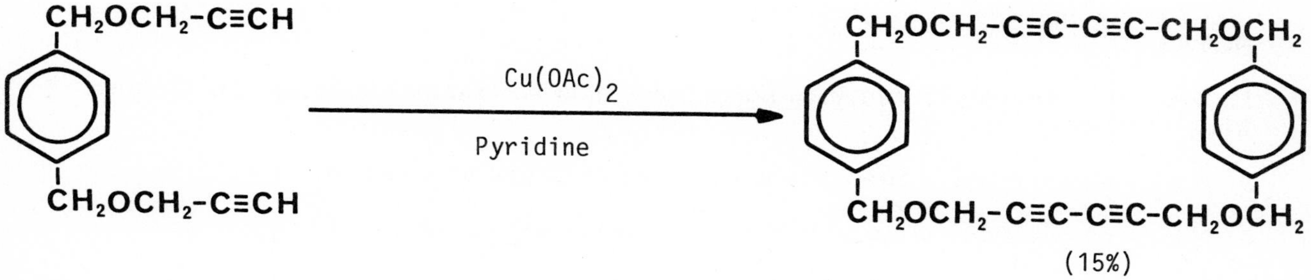
$Cu(OAc)_2$
Pyridine
(15%)

(3)

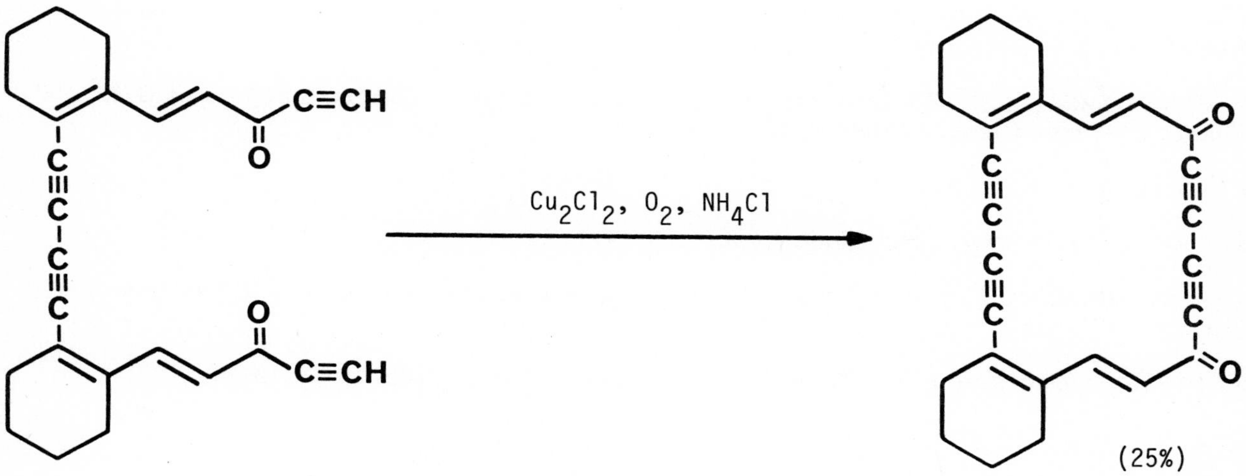
Cu_2Cl_2, O_2, NH_4Cl
(25%)

(4)

REACTION: GOMBERG- BACHMANN REACTION

==

March's Advanced Organic Chemistry : 640- 641

GENERAL SCHEME:

$$ArH + Ar'N_2^{\oplus} \xrightarrow{OH^{\ominus}} Ar-Ar'$$

MECHANISM:

$$2\ Ar'N_2^{\oplus} + OH^{\ominus} \longrightarrow Ar'N{=}N{-}O{-}N{=}NAr' \longrightarrow Ar'\cdot + N_2 + \cdot ON{=}NAr'$$

$$\xrightarrow{Ar} [\text{cyclohexadienyl radical: Ar, H, R, R, R, R, R}] \xrightarrow{\cdot ON=NAr'} Ar-Ar' + Ar'N_2OH$$

NOTES:

1. When the reaction is run intramolecularly it is sometimes known as the **Pschorr Ring Closure.**

2. An example of the related **Vorlander- Meyer Coupling Reaction.**

3. Yields of reactions often suffer due to side reactions of the diazonium salts.

REFERENCES:

1. J.A. Beadle, S.H. Korzeniowski, D.E. Rosenberg, B.J. Garcia-Slanga and G.W. Gokel, **J. Org. Chem.**, (1984), 49, 1594.

2. Ibid

3. Reported in **Org. Reactions**, (1957), 9, 421.

4. E.R. Atkinson, C.R. Morgan, H.H. Warren and T.J. Manning, **J. Am. Chem. Soc.**, (1945), 67, 1513.

5. P.A.S. Smith and B.B. Brown, **J. Am. Chem. Soc.**, (1951), 73, 2435.

EXAMPLES:

$BF_4^{\ominus}$ $N_2^{\oplus}$ PhOCH$_3$ —KOAc, 18-Crown-6, PhOMe, 25°→ OCH$_3$ (23%) ①

$N_2^{\oplus}$ —KOAc, Freon 113, 18-Crown-6→ (73%) ② △1

$N_2^{\oplus}Cl^{\ominus}$ —H_2O, Cu→ (25%) ③ △2

Cl, Cl, CO_2H, $^{\oplus}N_2$ + CO_2H, $^{\oplus}N_2$ —$NaNO_2$ / NaOH, H_2O→ Cl, CO_2H, CO_2H, Cl (15-20%) ④

N_3 —180°→ N H (76%) ⑤

REACTION: GRIGNARD REACTION

March's **Advanced Organic Chemistry** : 816- 822

GENERAL SCHEME:

$$\mathrm{RC(=O)R(H)} + \mathrm{R'MgX} \longrightarrow \mathrm{R{-}C(OMgX)(R'){-}R(H)} \xrightarrow{\text{Hyd}} \mathrm{R{-}C(OH)(R'){-}R(H)}$$

MECHANISM:

$$\mathrm{RC(=O)R(H)} + \mathrm{R'^{\ominus}MgX^{\oplus}} \longrightarrow \mathrm{R{-}C(OMgX)(R'){-}R(H)} + \mathrm{H_2O} \longrightarrow \mathrm{R{-}C(OH)(R'){-}R(H)} + \mathrm{XMgOH}$$

NOTES:

1. Any addition of a **Grignard reagent** (RMgX) to a carbonyl-containing compound is classified as a Grignard Reaction.

2. There are many named variations of the Grignard reaction:

a. **Benary Reaction:**

$$\mathrm{RMgX} + \mathrm{{>}N{-}CH{=}CH{-}C(=O){-}R'} \longrightarrow \left[\mathrm{{>}N{-}CHCHC(OMgX)R'}\right] \xrightarrow{H^{\oplus}} \mathrm{RCH{=}CHC(=O)R'}$$

b. **Bodroux-Chichibabin Aldehyde Synthesis:**

$$\mathrm{RMgX} + \mathrm{H{-}C(OR')(OR''')\text{-}OR''} \longrightarrow \mathrm{R{-}CH(OR')(OR''')} \xrightarrow{H^{\oplus}} \mathrm{RC(=O)H}$$

c. **Bouveault Aldehyde Synthesis:**

$$\mathrm{R_2N{-}CHO} + \mathrm{R'MgX} \longrightarrow \left[\mathrm{R_2N{-}C(R')(H){-}OMgX}\right] \xrightarrow{H^{\oplus}} \mathrm{R'CHO}$$

d. **Boord Olefin Synthesis:**

$$\mathrm{RCH_2CHO} \xrightarrow[\text{HCl}]{\text{R'OH}} \mathrm{RCH_2CH(Cl)(OR')} \longrightarrow \mathrm{RCH{=}CH{-}OR'} \xrightarrow{Br_2}$$

$$\mathrm{RCH(Br)CH(Br)OR'} \xrightarrow{R''MgX} \mathrm{RCH(Br){-}CH(R'')(OR')} \xrightarrow{\text{Zn dust}} \mathrm{RCH{=}CHR'}$$

REFERENCES:

1. D. Holt, **Tetrahedron Lett.**, (1981), 2243.

2. C. Descoins, C. Henrick and J.B. Siddall, **Tetrahedron Lett.**, (1972), 3777.

3. J. F. Normant, A. Commercon and J. Villieras, **Tetrahedron Lett.**, (1975), 1465. Note that in this reaction the Grignard reagent, in the presence of Cu ion, is really transformed into an organocuprate. These reagents undergo smooth displacement of halogens in organohalogen compounds and also undergo conjugate addition to enones.

EXAMPLES:

+ CH_2=CHMgBr → [OMgBr, CH=CH_2, Cl → O, CH=CH_2]

+ CH_2=CHMgBr → OH, CH_2, CH_2 ①

(69%, 85:15 cis/trans)

O, HC, OMe, OMe —H-C≡C-MgCl→ OH, HC≡C, OMe, OMe ②

(90%)

MgBr —1.) CuBr; 2.) H-C≡C-H→ $CuMgBr_2$

③

+ BrC≡CCH_2OTMS —H^+→ C≡C, OH

(65%)

REACTION: GROB FRAGMENTATION

March's Advanced Organic Chemistry : 926- 928

GENERAL SCHEME:

R

R'H

Base

+ =R'

MECHANISM:

R

:B

R'—H

+ =R'

NOTES:

1. The reaction usually involves chemistry separated by three carbon atoms.

2. The initial fragmentation results in an aldehyde; however, under the reducing conditions, this is converted into an alcohol.

REFERENCES:

1. M. Kato, H. Kurihara and A. Yoshikoshi, **J. Chem. Soc., Perkin I,** (1979), 2740.

2. B.M. Trost and W.J. Frazee, **J. Am. Chem. Soc.**, (1977), 99, 6124.

3. H.A. Patel and S. Dev., **Tetrahedron,** (1981), 37, 1577.

4. M. Kodama, T. Takahashi, T. Kurihara and S. Ito, **Tetrahedron Lett.**, (1980), 2811.

5. M. Shimizu, R. Ando and I. Kuwajima, **J. Org. Chem.**, (1981), 46, 5246.

EXAMPLES:

MsO, CH_3, H–O, H, CH_3 — $LiAlH_4$, DME, Δ — CH_3, H_2C, HO, H, CH_3 (90%) ① △2

MeO⊖, O, H, Me, MeO_2C, CO_2Me — NaOMe — CO_2Me, MeO_2C, CO_2Me, CH_3 (84%) ②

Me, OTs, O, H — t-BuO$^-$K$^+$, THF — Me, O (77%) ③

O–H, OMs, OCOPh — t-BuO$^-$ — CH_2, Me, O=, O, C–Ph, O (75%) ④

H, O, n-Bu, O, Se–Ph, O — MeO$^-$ — O, n-Bu, OMe (86%) ⑤

REACTION: HANTZSCH PYRIDINE SYNTHESIS

March's Advanced Organic Chemistry: Not indexed.

GENERAL SCHEME:

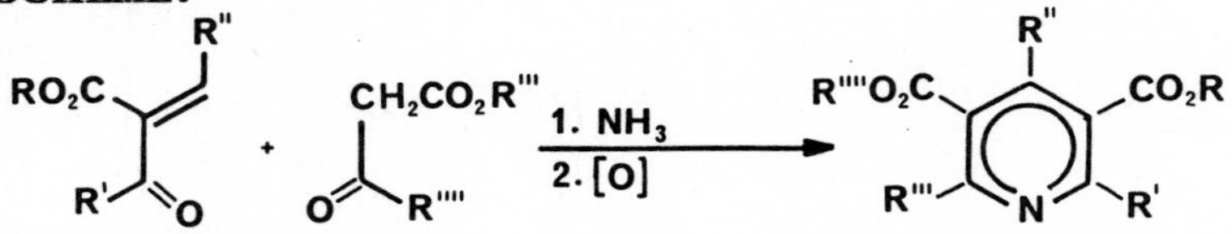

RO$_2$C, R'', R', O + CH$_2$CO$_2$R''', O, R'''' — 1. NH$_3$ 2. [O] → R''''O$_2$C, R'', CO$_2$R, R''', N, R'

MECHANISM:

NH$_2$, O, R''', R'''' + O, O, R', OR, R'' → O, R'', O, R''''O, OR, R''', N, H, O, R' →

O, R'', O, R''''O, OR, R', R''', N, H, OH — [O] → O, R'', O, R''''O, OR, R''', N, R'

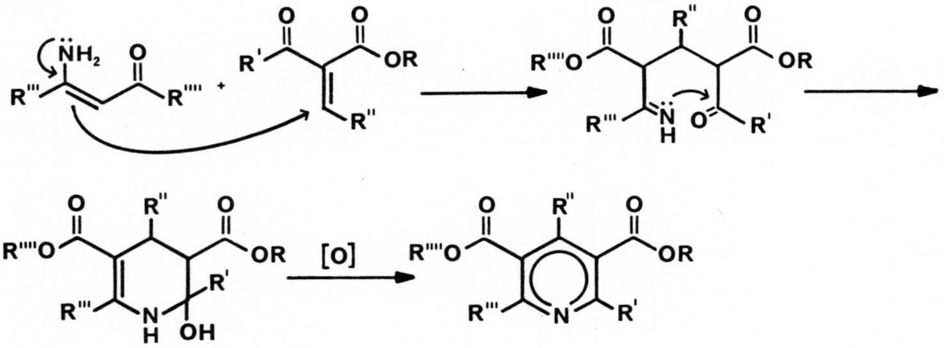

NOTES:

REFERENCES:

1. A. Singer and S.M. McElvain, **Org. Synth.**, Coll. Vol. II., (1943), 214.

2. E.H. Hunters and E.N. Shaw, **J. Org. Chem.**, (1948), 13, 674.

3. Y. Watanabe, K. Shiota, T. Hoshiko and S. Ozaki, **Synthesis**, (1983), 761.

4. B. Loev and K.M. Snader, **J. Org. Chem.**, (1965), 30, 1914.

EXAMPLES:

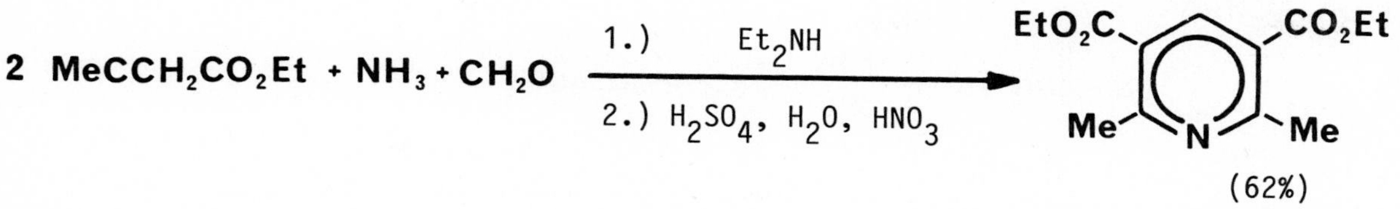

(1)

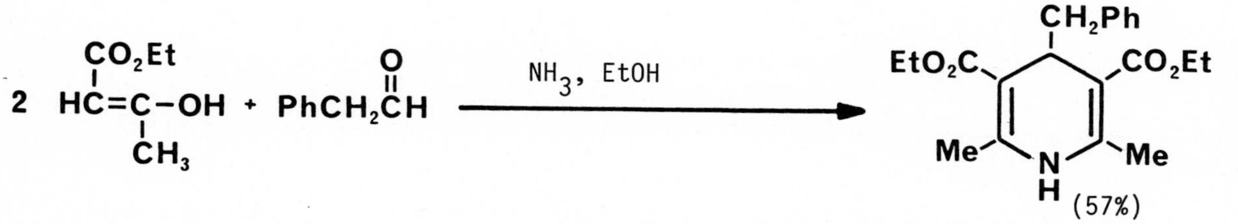

(2)

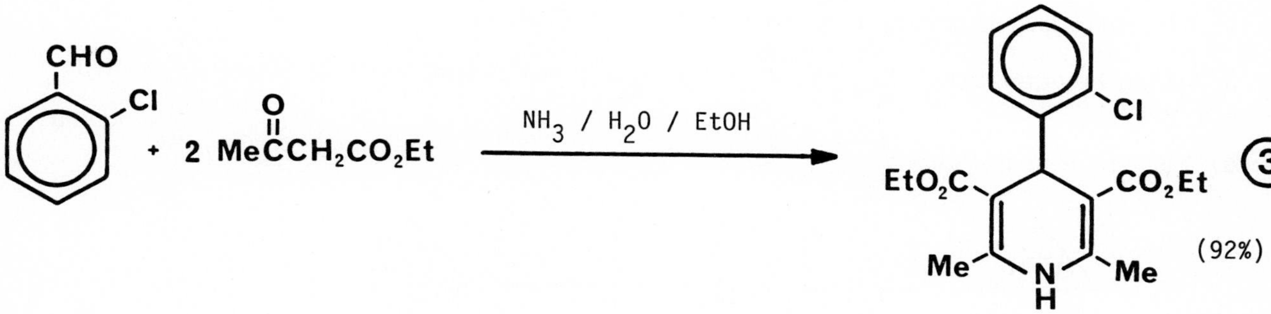

(3)

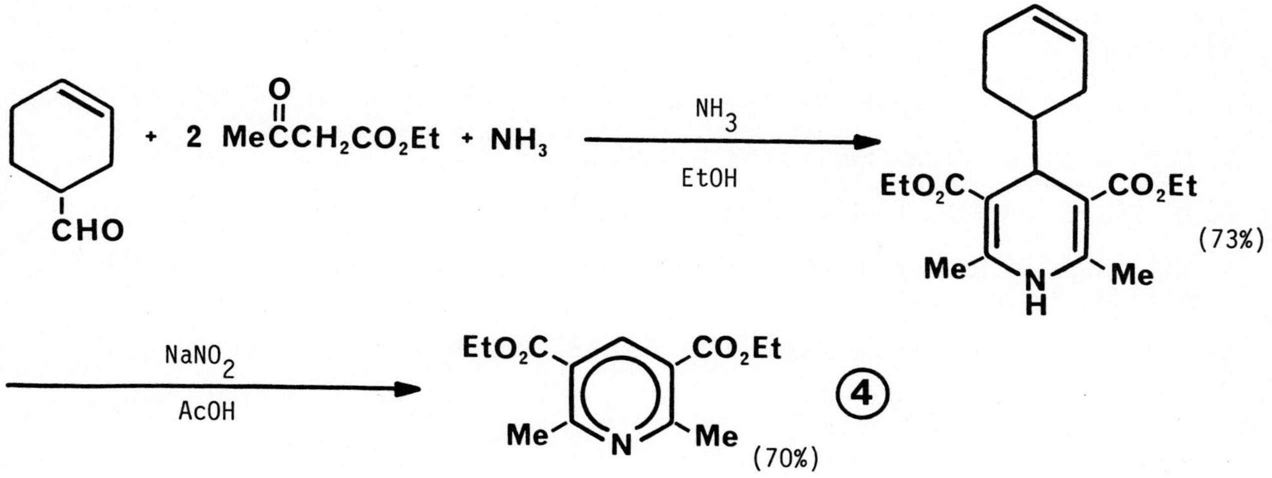

(4)

REACTION: HELL- VOLHARD- ZELINSKI REACTION

March's Advanced Organic Chemistry : 531

GENERAL SCHEME:

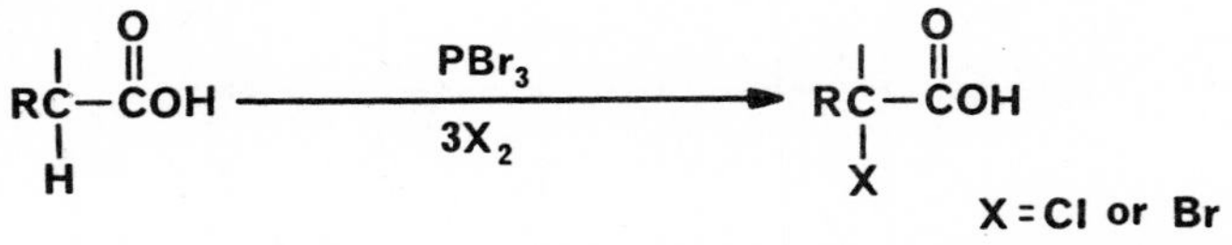

MECHANISM:

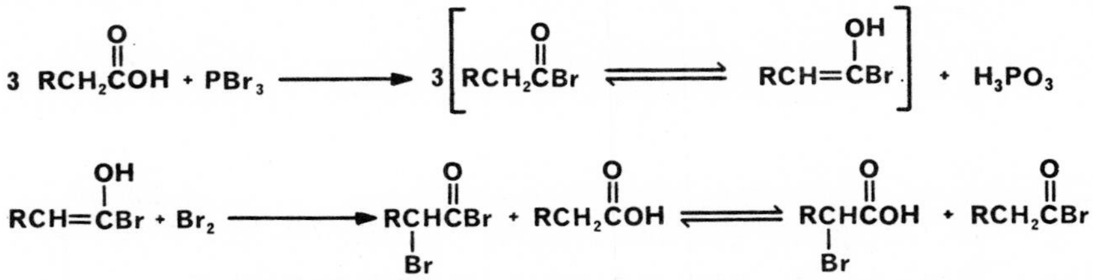

NOTES:

1. Only catalytic amounts of PBr_3 are required for this reaction, since the acid bromide reacts with the starting acid to form the alpha-substituted acid.

2. The reaction does not work with halogen = F, I.

3. Alpha halogenation (X = Br, Cl) can take place if **NBS** or **NCS** with HBr or HCl are used.

4. Via the intermediates:

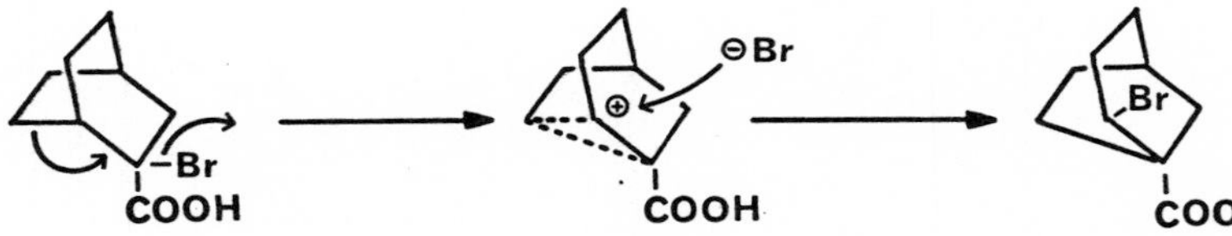

REFERENCES:

1. L. Carpino and L. McAdams, **Org. Synthesis**, (1970), Coll. Vol. 2, 50, 31.

2. C.F. Ward, **J. Chem. Soc.**, (1922), 1164.

3. H. Kwart and F.V. Scalzi, **J. Am. Chem. Soc.**, (1964), 86, 5496.

4. J.C. Little, A.R. Sexton, Y.-L. C. Tong, ad T.E. Zurawic, **J. Am. Chem. Soc.**, (1969), 91, 7098.

5. A.W. Chow, D.R. Jakas and J.R.E. Hoover, **Tetrahedron Lett.**, (1966), 5427.

EXAMPLES:

$C_6H_5CH_2COOH + Br_2$ —— PCl_3 (Catalyst), Benzene, Reflux ——> $C_6H_5CH(Br)COOH$ (62%) ①

$CH_3CH_2CH_2COOH$ —— Br_2, PCl_3 ——> $CH_3CH_2CH(Br)COOH$ (82%) ②

Cyclohexane-COOH —— Br_2, $SOCl_2$ ——> 1-Br-cyclohexane-1-COOH (95%) ③

Cyclohexane-COOH —— Cl_2, PCl_3 ——> 1-Cl-cyclohexane-1-COOH (70-75%) ④

Bicyclo-COOH —— Br_2, PCl_3 ——> Br, COOH ⑤ △4

REACTION: HOFMANN DEGRADATION

==========

March's Advanced Organic Chemistry : 927- 28

GENERAL SCHEME:

$$HC-C-NR_3^{\oplus}OH^{\ominus} \longrightarrow >C=C< + NR_3$$

MECHANISM:

$$-C-C-NR_3^{\oplus} \quad (H,\ OH^{\ominus}) \longrightarrow >C=C< + NR_3 + H_2O$$

NOTES:

1. **Hofmann elimination** products are generally the least-substituted alkenes.
2. **Emde degradation** is often used where the **Hofmann degradation** does not work:

$\oplus NMe_2$, $Cl^{\ominus}$ $\xrightarrow{Na/Hg}$ NMe

REFERENCES:

1. G. Delodts, G. Dressaire and Y. Langlois, **Synthesis**, (1979), 510.
2. K. Hayakawa, I. Fuji and K. Kanematsu, **J. Org. Chem.**, (1983), 48, 166.
3. A.R. Katritsky and A.M.E.- El Mowafy, **Chem. Commun.**, (1981), 96.
4. J.-P. Maffrand and M. Lucas, **Heterocycles**, (1980), 14, 325.
5. R.K. Hill, J.A. Joule and L.J. Loeffler, **J. Am. Chem. Soc.**, (1962), 84, 4951.

EXAMPLES:

KOH

Reflux under Argon, 4 hrs

(65%) ① ⚠1

MeOH

RT, 5 Min

$+ MeCO_2C{\equiv}CCO_2Me$

(86%) ②

150°

n-Bu (97%) ③

$CF_3SO_3^{\ominus}$

30% NaOH in MeOH

△

$MeOSO_3^{\ominus}$

(85%) ④

Dowex, OH^-

$I^{\ominus}$

(40%) ⑤

REACTION: HOFMANN- LOFFLER- FREYTAG REACTION

March's Advanced Organic Chemistry : 1043

GENERAL SCHEME:

X(R)N–CH₂CH₂CH CH₂–R' —(1. $H^{\oplus}$, Δ; 2. $OH^{\ominus}$)→ N-R-2-R'-pyrrolidine

MECHANISM:

$X\text{-}N(R)\text{-}(CH_2)_4\text{-}R' \xrightarrow{H^{\oplus}} X\text{-}\overset{\oplus}{N}H(R)\text{-}(CH_2)_4\text{-}R' \longrightarrow R\text{-}\overset{\oplus}{N}H\text{-}(CH_2)_4\text{-}R' + X\cdot$

$\longrightarrow \{ X\text{-}\overset{\oplus}{N}H(R)\text{-}CH_2CH_2CH_2CH_2\text{-}R' + R\text{-}\overset{\oplus}{N}H_2\text{-}CH_2CH_2CH_2\dot{C}H\text{-}R' \} \longrightarrow R\text{-}\overset{\oplus}{N}H\text{-}(CH_2)_4\text{-}R' \longrightarrow$ etc. $+ R\text{-}\overset{\oplus}{N}H_2\text{-}CH_2CH_2CH_2CH(R')X$

$\xrightarrow{OH^{\ominus}} R\text{-}\ddot{N}H\text{-}CH_2CH_2CH_2CR'(H)\text{-}X \longrightarrow$ N-R-2-R'-pyrrolidine

NOTES:

1. **NCS** = N-Chlorosuccinimide

REFERENCES:

1. M. Kimura and Y. Ban, **Synthesis**, (1976), 201.
2. R. Dupeyre and A. Rassat, **Tetrahedron Lett.**, (1973), 2699.
3. E.J. Corey and W.R. Hertler, **J. Am. Chem. Soc.**, (1960), 82, 1657.
4. R. Furstoss, P. Teissier and B. Waegell. **Tetrahedron Lett.**, (1970), 1263.
5. S.L. Titouani, J.-P. Lavergne and Ph. Viallefont, **Tetrahedron**, (1980), 36, 2961.

EXAMPLES:

NCS, $h\nu$

Ether

(50%) ① △1

H_2SO_4

65°, 30 min

(25%) ②

H_2SO_4

△

(43%, optically inactive, k_H/k_D =3.54) ③

F_3C-COOH

$h\nu$

(NYA) ④

H_2SO_4

$h\nu$

(80%) ⑤

REACTION: HOFMANN REARRANGEMENT

March's **Advanced Organic Chemistry :** 945, 949, 983- 84.

GENERAL SCHEME:

$$\mathrm{R\overset{O}{\overset{\|}{C}}NH_2 + NaOBr \longrightarrow R{-}N{=}C{=}O \xrightarrow[-CO_2]{Hyd.} RNH_2}$$

MECHANISM:

$$\mathrm{R\overset{O}{\overset{\|}{C}}\ddot{N}H_2 + \overset{+1}{Br}\,\overset{-2}{O} \longrightarrow R\overset{O}{\overset{\|}{C}}{-}\underset{Br}{NH} \xrightarrow{OH^{\ominus}} (R){-}\overset{O}{\overset{\|}{C}}{-}\ddot{N}^{\ominus}{-}Br \longrightarrow O{=}C{=}N{-}R}$$

$$\mathrm{\xrightarrow{H_2O} R{-}NH{-}COOH \longrightarrow \left[R{-}\overset{H}{\overset{|}{N}}{-}\underset{\underset{O{-}H}{|}}{C}{=}O\right] \longrightarrow RNH_2 + CO_2}$$

NOTES:

1. The reagents for the reaction are often bromine and sodium or potassium hydroxide.

REFERENCES:

1. P. Radlick and L. Brown, **Synthesis**, (1974), 290.

2. G.C. Finger, L.D. Starr, A. Roe, and W.J. Link, **J. Org. Chem.**, (1962), 27, 3965.

3. S.S. Simon, Jr., **J. Org. Chem.**, (1973), 38, 414.

4. M. Waki, Y. Kitajma and N. Izumiya, **Synthesis**, (1981), 266.

5. A.O. Sy and J.W. Raksis, **Tetrahedron Lett.**, (1980), 2223.

EXAMPLES:

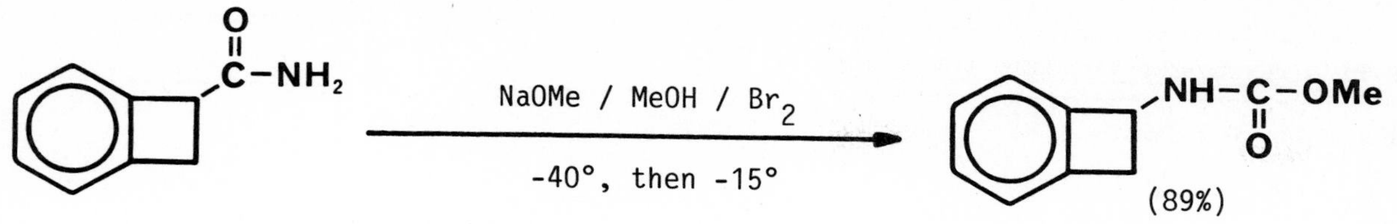

Reaction ①: NaOMe / MeOH / Br_2, −40°, then −15° (89%)

Reaction ②: Br_2, KOH (87%)

Reaction ③: $Pb(OAc)_4$, Pyridine (79%)

Reaction ④: $PhCH_2OC(O)NHCH(CH_2C(O)NH_2)CO_2H$ → $Ph{-}I(OC(O)CF_3)_2$, Pyridine → $PhCH_2OC(O)NHCH(CH_2NH_2)CO_2H$ (40-60%)

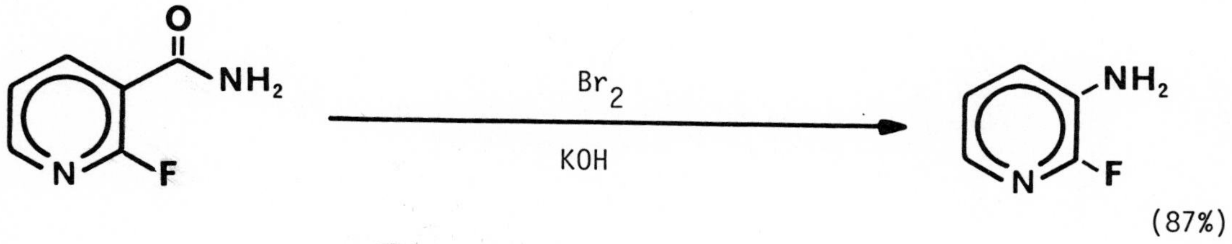

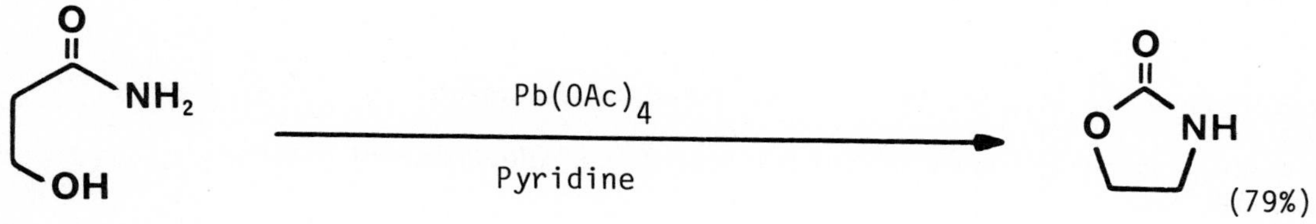

Reaction ⑤: Br_2 / NaOH, $Bu_4N^+HSO_4^-$, CH_2Cl_2 → N=C=O (70-80%)

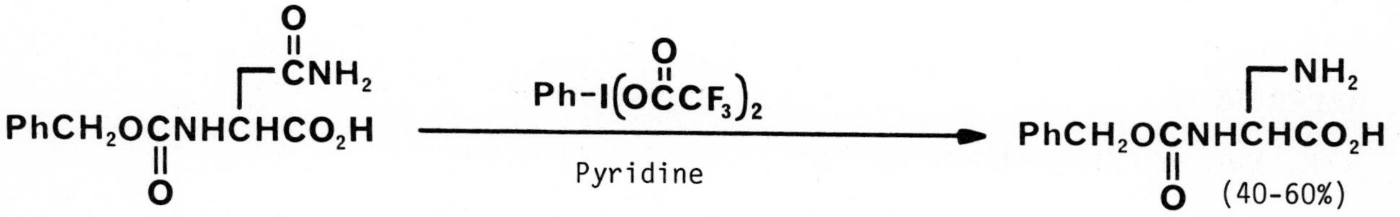

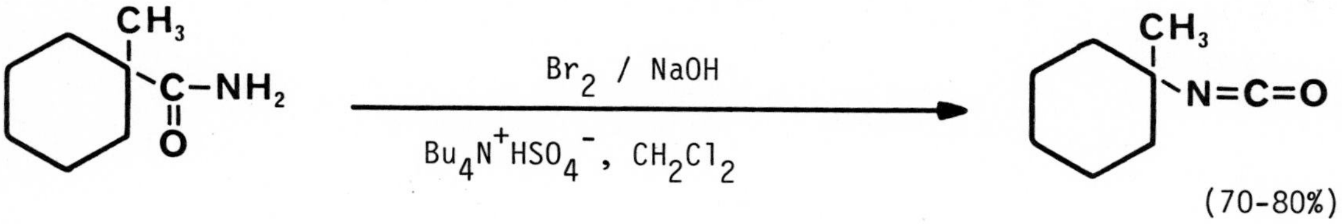

REACTION: HORNER-EMMONS (WITTIG- HORNER, WADSWORTH-EMMONS)

March's Advanced Organic Chemistry : 848

GENERAL SCHEME:

1

2

$$(RO)_2\overset{O}{\overset{\|}{P}}CH\begin{matrix}R\\R'\end{matrix} \xrightarrow{:B} (RO)_2\overset{O}{\overset{\|}{P}}C^{\ominus}\begin{matrix}R\\R'\end{matrix} \xrightarrow{>C=O} >C=C\begin{matrix}R\\R'\end{matrix} + (RO)_2PO_2^{\ominus}$$

MECHANISM:

NOTES:

1. The precursor ylide is prepared by the **Arbusov Reaction.**

2. This procedure is particularly useful for transfer of electron-withdrawing groups with the alkene.

REFERENCES:

1. K. Mori and H. Ueda, **Tetrahedron Lett.**, (1981), 461.

2. S. Hannessian and P. Lavalle, **Can. Jour. Chem.**, (1977), 55, 562.

3. W.R. Roush, **J. Am. Chem. Soc.**, (1980), 102, 1390.

4. Y.K. Lee and A.G. Schultz. **J. Org. Chem.**, (1979), 44, 719.

5. C. Schmidt, N.H. Chishti and T. Breining, **Synthesis,** (1982), 391.

EXAMPLES:

$(EtO)_2P(O)CH(COOEt)Me$ → EtO_2C ... OH (53%) ①

$(MeO)_2P(O)CH_2COOMe$ → OBz, $=CHCO_2Me$, MeO, O, Ph, –OSi*t*-Bu, Ph (92%) ②

$(EtO)_2P(O)CH_2CHCHCOOMe$ / $^{+}Li^{-}N(TMS)_2$ → MeO_2C (84%) ③

OEt, $(EtO)_2P(O)C^{-}HCN$ → OEt, =CHCN (NYA) ④

$(EtO)_2P(O)C^{-}HCOOMe$ → CO_2Me (95%) ⑤

REACTION: HUNSDIECKER REACTION

March's <u>Advanced Organic Chemistry</u> : 613, 654- 55

GENERAL SCHEME:

1 2 3

$$R-\overset{O}{\overset{\|}{C}}-OAg \xrightarrow{Br_2} RBr + CO_2 + AgBr$$

MECHANISM:

$$R\overset{O}{\overset{\|}{C}}O^{\ominus}Ag^{\oplus} + Br_2 \longrightarrow R\overset{O}{\overset{\|}{C}}OBr \longrightarrow R-\overset{O}{\overset{\|}{C}}-O\cdot + Br\cdot \longrightarrow$$

$$R\cdot + CO_2 + Br \xrightarrow{R\overset{O}{\overset{\|}{C}}OBr} R\overset{O}{\overset{\|}{C}}O\cdot + RBr \longrightarrow \text{etc.}$$

NOTES:

1. This reaction can be summarized as a decarboxylative bromination.

2. The **Cristol-Firth Modification** of this reaction uses a mixture of the acid and mercuric oxide.

$$R\overset{O}{\overset{\|}{C}}OH + HgO + Br_2 \xrightarrow{H^{\oplus}/CCl_4} RBr$$

3. When iodine is used with a two-fold excess of the acid salt, the reaction is known as the **Simonini Reaction**, and the product is an ester.

$$2\,R\overset{O}{\overset{\|}{C}}O^{\ominus}Ag^{\oplus} \xrightarrow{I_2} R\overset{O}{\overset{\|}{C}}OR$$

4. A thallium modification.

5. Without a 100 W source of light this reaction did not go.

REFERENCES:

1. J. Meek and D. Osuga, **Org. Synth.,V,** (1973), 126.

2. R. Cambie, R. Hayward, J. Jurlina, P. Rutledge and P. Woodgate, **J. Chem. Soc., Perkin I,** (1981), 2608.

3. A. Meyers and M. Fleming, **J. Org. Chem.**, (1979), <u>44</u>, 3405.

4. a. J.-P. Barnier, G. Rousseau and J.-M. Conia, **Synthesis**, (1983), 915.
   b. For a similar experiment, see: L.A. Paquette, M. Hoppe, L.J. Johnston and K.U. Ingold, **Tetrahedron Lett.**, (1986), 411.

5. A. Hammond and C. Descoins, **Bull Soc. Chim. Fr.**, (1978), 299.

EXAMPLES:

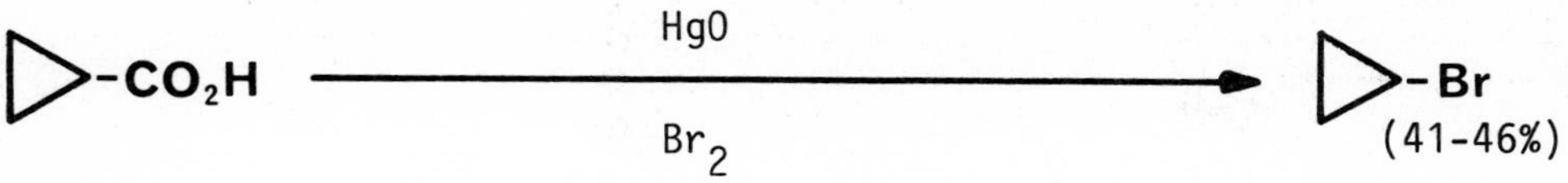

$$\triangleright\text{-}CO_2H \xrightarrow[Br_2]{HgO} \triangleright\text{-}Br \quad (41\text{-}46\%)$$ ① △2

$$2\ CH_3(CH_2)_nCO_2H \xrightarrow{Tl_2CO_3} 2\ CH_3(CH_2)_nCO_2Tl \xrightarrow{Br_2} 2\ CH_3(CH_2)_nBr$$

$(n = 10\text{-}16)$ (80-85%) ② △4

p-Cl-C$_6$H$_4$-CO$_2$H → (1.) HgO / Light; 2.) Br_2) → p-Cl-C$_6$H$_4$-Br (80%) ③ △5

Cyclopropane-1-CO$_2$H-1-CO$_2$Et → (Br_2 / HgO / CCl_4, hν) → Cyclopropane-1-Br-1-CO$_2$Et → (1.) KOH; 2.) Br_2 / HgO / hν) → 1,1-dibromocyclopropane (60%) ④

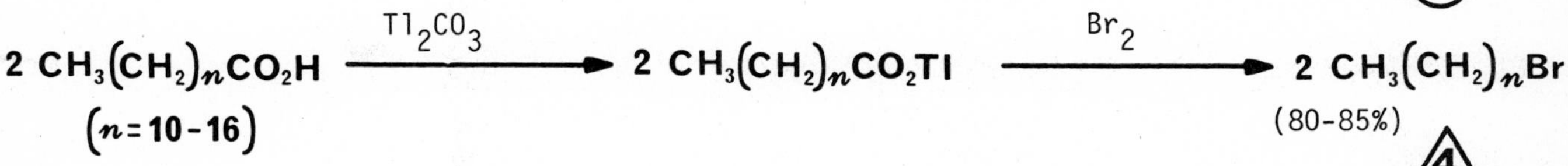

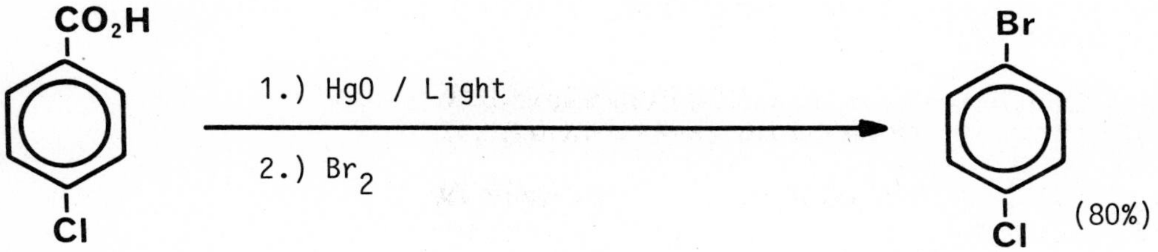

$$HO_2C(CH_2)_6CO_2Et \xrightarrow[Br_2\ /\ CCl_4\ /\ h\nu]{HgO} BrCH_2(CH_2)_5CO_2Et$$

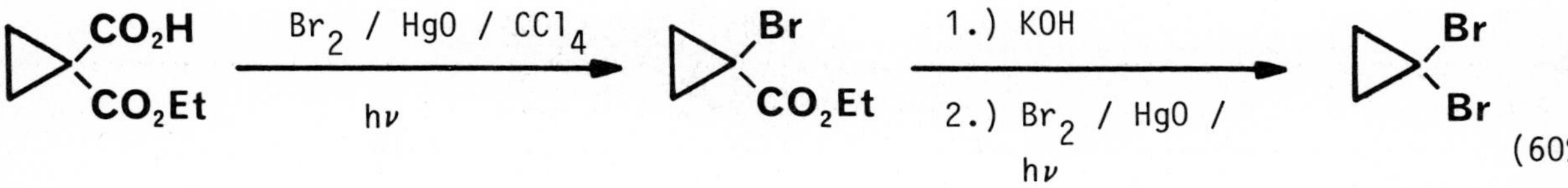

(50%) ⑤

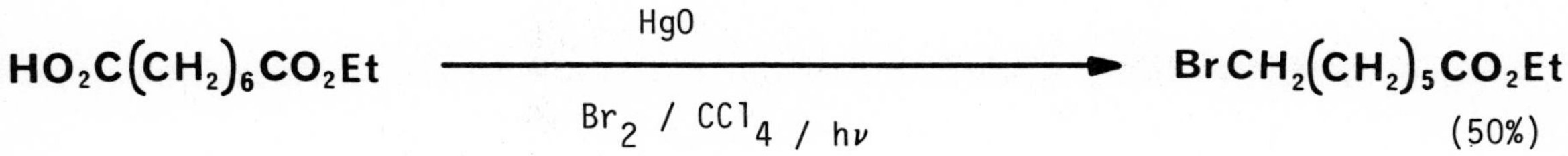

REACTION: JONES OXIDATION

March's **Advanced Organic Chemistry :** 1057

GENERAL SCHEME:

$$R-\overset{OH}{\underset{H}{C}}-R(H) \xrightarrow[H_2SO_4]{CrO_3} R-\overset{O}{\overset{\|}{C}}-R(H)$$

MECHANISM:

$$R-\overset{H}{\underset{R(H)}{C}}-O-H \xrightarrow{\text{"}H_2CrO_4\text{"}} R-\overset{H}{\underset{R(H)}{C}}-O-\overset{O}{\underset{O}{Cr}}-OH \xrightarrow{:B} R-\overset{O}{\overset{\|}{C}}-R(H)$$

NOTES:

1. The **Jones Reagent** is generally prepared from CrO_3 and H_2SO_4. Other similar oxidizing reagents include:

Fieser's Reagent: CrO_3/ Acetic acid
Sarrett Reagent: CrO_3/ Pyridine
Cornforth Reagent: CrO_3/ Pyridine/ Water
Thiele Reagent: CrO_3/ Ac_2O/ H_2SO_4

2. The alcohol is often stirred in acetone while the oxidizing agent is added dropwise. The progress of the oxidation is easily monitored visually by formation of the green-colored lower layer that forms as the oxidation takes place. When the color of the oxidizing solution persists, the oxidation is considered complete.

3. **Via:**

$$RCHO \longrightarrow RCH(OH)_2 \longrightarrow R\overset{OH}{C}H-O\overset{O}{\underset{O}{Cr}}-OH \longrightarrow$$

REFERENCES:

1. P.A. Grieco, M. Nishizawa, T. Oguri, S.D. Burke and N. Marinovich, **J. Am. Chem. Soc.**, (1977), 99, 5573.

2. E.J. Eisenbraun, **Org. Synth.**,(1965), 45, 28.

3. J. Meinwald, J. Crandall and W.E. Hymans, **Org. Synth.**, (1965), 45, 77.

4. K. Bowden, I.M. Heilbron, E.R.H. Jones and B.C.L. Weedon, **J. Chem. Soc.**, (1946), 39.

5. C. Djerrasi, R.R. Engle and A. Bowers, **J. Org. Chem.**, (1956), 21, 1547.

EXAMPLES:

OMe, OAc, O, O, H, OAc, CHO → CrO_3 / H_2SO_4, H_2O → OMe, OAc, O, O, H, OAc, COOH (NYA) ① △3

OH → H_2CrO_4, Acetone → =O (92-96%) ②

OH → CrO_3 / H_2SO_4, H_2O → =O (79-88%) ③

OH, CHC≡CH → CrO_3, H_2SO_4, H_2O, Acetone → O, CC≡CH (75%) ④

O, HO → CrO_3 / N_2, H_2SO_4 → O (90%) ⑤

REACTION: KNOEVENAGEL REACTION

March's Advanced Organic Chemistry : 835- 841

GENERAL SCHEME:

$$RC(=O)R' + G{-}CH_2{-}G' \xrightarrow{\text{Base}} RR'C{=}CGG'$$

MECHANISM:

$$R(H){-}\overset{R}{C}{=}O + {}^{\ominus}CHG_2 \longrightarrow (H)R{-}\underset{O^{\ominus}}{\overset{R}{C}}{-}CHG_2 \xrightarrow{H^{\oplus}} (H)R{-}\underset{HO}{\overset{R}{C}}{-}CHG_2 \xrightarrow{^{\ominus}B}$$

$$(H)R{-}\underset{HO}{\overset{R}{C}}{-}\overset{\ominus}{C}G_2 \longrightarrow (H)RRC{=}CG_2 + OH^{\ominus}$$

NOTES:

1. The reaction is usually limited to aldehydes and ketones that do not have an alpha hydrogen.

2. The base is often a secondary amine.

3. Typical electron-withdrawing groups include:

G = CHO, $-\underset{O}{\underset{\|}{C}}-R$, COOR, CN, NO_2, $-\underset{O}{\underset{\|}{S}}-R$, SO_2R or other EWG's

REFERENCES:

1. F. Texier-Boullet and A. Foucand, **Tetrahedron Lett.**, (1982), 4927.

2. H. Iio, M. Isobe, T. Kawaiand T. Goto, **J. Am. Chem. Soc.**, (1979), 101, 6076.

3. C.G. Butler, R.K. Callow and N.C. Johnson, **Proc. Royal Soc.**, (1961), B, 155, 417.

4. Reported in the Review by G.G. Yakobson and N.E. Akhmetova, **Synthesis**, (1983), 173.

5. D.A.R. Happer and B.E. Steenson, **Synthesis**, (1980), 806.

EXAMPLES:

CH_3COCH_3 + $NC{-}CH_2{-}CO_2CH_3$ —(Al_2O_3)→ $(CH_3)_2C{=}C(CN)CO_2CH_3$ (53%) ①

—($CH_2(COOt\text{-}Bu)_2$, $TiCl_4$)→ (70%) ②

—($CH_2(COOH)_2$)→ (NYA) ③

—($CN{-}CH_2{-}COOEt$, KF)→ (36%) ④

—($SO_2Me{-}CH_2{-}COOEt$, piperidine, HOAc, Δ)→ (50%) —(LiI, DMF)→ (73%) ⑤

REACTION: KNORR PYRROLE SYNTHESIS

==

March' Advanced Organic Chemistry: Not Indexed

GENERAL SCHEME:

R C=O, EtO_2C, NH_2 + CO_2Et, O, R → R, CO_2Et, EtO_2C, N H, R

MECHANISM:

R, C=O, CH, EtO_2C, $\ddot{N}H_2$; CO_2Et, CH_2, C=O, R —$-H_2O$→ R, O, CO_2Et, EtO_2C, $\ddot{N}$ H, R (an enamine) →

HO H, R, CO_2Et, EtO_2C, H, $\overset{\oplus}{N}$ H, R —$-H_2O$→ R, CO_2Et, EtO_2C, $\ddot{N}$ H, R

NOTES:

REFERENCES:

1. J.-Y. Valnot, **Synthesis**, (1978), 590.

2. Y. Tamura, S. Kato and M. Ikeda, **Chem. Ind.**, (1971), 767.

3. Contemporary Heterocyclic Chemistry, G.R. Newkome and W.W. Paudler, Wiley, New York, 1982, 22.

4. H. Plieninger, P. Hess and J. Ruppert, **Ber.**, (1968), 101, 240.

5. H. Fischer, **Org. Synth.** Coll. Vol. 2, (1943), 202.

EXAMPLES:

$CH_3CH(NH\text{-}i\text{-}Pr)CH{=}N\text{-}i\text{-}Pr$ + $CH_3COCH_2CO_2Et$ —(Molecular sieve / Dry benzene)→ ethyl 2,5-dimethylpyrrole-3-carboxylate (H_3C, CO_2Et, CH_3) (75%) ①

$MeCOCH_2CO_2Et$ —(NH_2OSO_3H / aq. K_2CO_3)→ [$MeCOCH(NH_2)CO_2Et$] —($MeCOCH_2CO_2Et$)→ pyrrole (Me, CO_2Et, EtO_2C, Me, N–H) (34%) ②

Me(C=O)CH(Me)NH_2 + $MeCOCH_2COMe$ → pyrrole (Me, CMe(=O), Me, Me, N–H) (40-60%) ③

$MeCOCH(CH_2CO_2Et)COCH_2CH_2CO_2Et$ + $(EtO_2C)_2C{=}NOH$ —(Zn / AcOH)→ pyrrole ($EtO_2CH_2CH_2C$, CH_2CO_2Et, EtO_2C, Me, N–H) (41%) ④

$MeCOC(N{=}O)HCO_2Et$ —(Zn / $MeCO_2H$)→ [$MeCOCH(NH_2)CO_2Et$] + $MeC(OH){=}CHCO_2Et$ → pyrrole (Me, CO_2Et, EtO_2C, Me, N–H) (apx. 57%) ⑤

REACTION: KOCH-HAAF REACTION

March's Advanced Organic Chemistry : 432, 720

GENERAL SCHEME:

R-OH or C=C → (H_2SO_4, HCO_2H) → [$R^{\oplus}$ → $R'^{\oplus}$ Rearrangement to most stable carbonium ion.] → R'COOH

MECHANISM:

R-OH or C=C → ($H^{\oplus}$) → carbocation + :C≡O: ($HCO_2H + H_2SO_4$) → acylium ion + H_2O → carboxylic acid

NOTES:

1. The acid-catalyzed hydrocarboxylation of an alkene is known as the **Koch Reaction**. When the source of both the CO and the H_2O is formic acid, the process is called the **Koch-Haaf reaction**.

2. Reagent ratio: Alcohol: formic acid: H_2SO_4 = 1:1:24

REFERENCES:

1. Y. Takahashi, N. Yoneda and H. Nagai, **Chem. Letters**, (1982), 1187.

2. H. Langhals, I. Mergelsberg and C. Ruchardt, **Tetrahedron Lett.**, (1981), 2365.

3. J.A. Peters, J. Rog and H. van Bekkum, **Recueil**, (1974), 93, 248.

4. F.J. McQuillin and D.G. Parker, **J.Chem. Soc., Perkin I**, (1974), 809.

5. Y. Takahashi, N. Yoneda and H. Nagai, **Chem. Lett.**, (1982), 1187.

EXAMPLES:

$CH_3-C(OH)(CH_3)-CH_2CH_2CH_2OH$ —[HCOOH, H_2SO_4]→ (product, 94%) ①

(3-methylcyclohexanol) —[HCOOH, H_2SO_4; Slow stirring, 25°]→ CO_2H, CH_3 (66%) ② ⚠2

CH_3, OH —[D_2SO_4; HCOOH]→ H_3C, CO_2H; 37*, 75*, 4*, 4* ③

* percent of deuterium exchange

(NYA)

—[H_2SO_4; HCOOH]→ CO_2H (64%) ④

OH, $CH_2CH_2CH_2OH$ —[HCOOH, CH_2O; Δ]→ O, O (88%) ⑤

REACTION: KOENIGS-KNORR SYNTHESIS

==

March's Advanced Organic Chemistry : Not Indexed

GENERAL SCHEME:

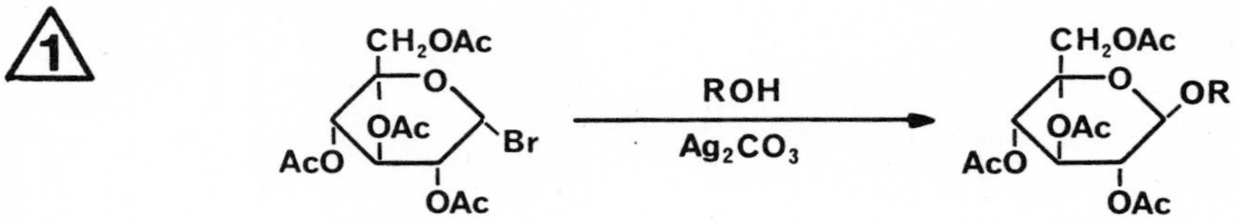

1

Tetra-O-acetyl glycosyl bromide (ring bearing CH_2OAc, AcO, OAc, OAc, Br) $\xrightarrow[Ag_2CO_3]{ROH}$ tetra-O-acetyl glycoside (ring bearing CH_2OAc, AcO, OAc, OAc, OR)

MECHANISM:

Glycosyl bromide (ring bearing CH_2OAc, AcO, OAc, OAc, Br) + H–Ö–R ⟶ glycoside (ring bearing CH_2OAc, AcO, OAc, OAc, OR)

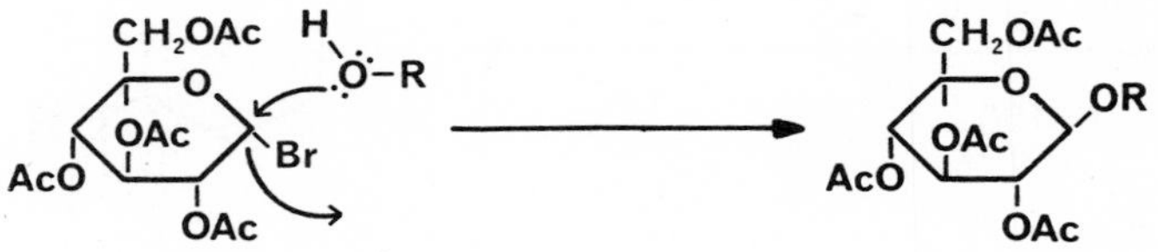

NOTES:

1. The reaction of the chloride with a phenolate is known as the **Michael method**.

REFERENCES:

1. R.B. Conrow and S. Bernstein, **J. Org. Chem.**, (1971), 36, 863.

2. A. Knochel, G. Rudolph and J. Thiem, **Tetrahedron Lett.**, (1974), 55.

3. M. Hagedorn, R.R. Sauers and A.E. Eichholz, **J. Org. Chem.**, (1978), 43, 2070.

4. F. Imperato, **J. Org. Chem.**, (1976), 41, 3478.

EXAMPLES:

CdCO$_3$

Toluene

(54%)

(1)

AgNO$_3$

Crown ether

R=Me, 81%

R=iPr, 65%

R=t-Bu, 57%

R= C_6H_{11}, 43%

(2)

n=10, 46%

n=6, 17%

n=3, 52%

(3)

$Hg(OAc)_2$

C_6H_6

(4)

(Yield described as "low")

REACTION: KOLBE ELECTROLYSIS REACTION

March's Advanced Organic Chemistry: ***

GENERAL SCHEME:

$$R{-}CO_2^{\ominus} \xrightarrow{\text{Electrolysis}} R{-}R$$

MECHANISM:

$$RC(=O){-}\ddot{O}:^{\ominus} \xrightarrow{-1e^{\ominus}} R{-}C(=O){-}\ddot{O}\cdot \longrightarrow R\cdot \; (-CO_2) \xrightarrow{R\cdot} R{-}R$$

NOTES:

REFERENCES:

1. D.A. White, **Org. Synth.**, (1981), 60.
2. J. Knolle and H.J. Schaefer, **Angew Chem. Int. Ed.**, (1975), 14, 758.
3. E.J. Corey and R.R. Saures, **J. Am. Chem. Soc.**, (1959), 81, 1739.
4. G. Stork, A. Meisels and J.E. Davies, **J. Am. Chem. Soc.**, (1963), 85, 3419.
5. A.F. Vellturo and G.W. Griffin, **J. Am. Chem. Soc.**,(1965),87, 3021.

EXAMPLES:

2 $MeO_2C(CH_2)_4CO_2H$ —($-e^-$; MeOH, MeO^-)→ $MeO_2(CH_2)_8CO_2Me$ (71%) ①

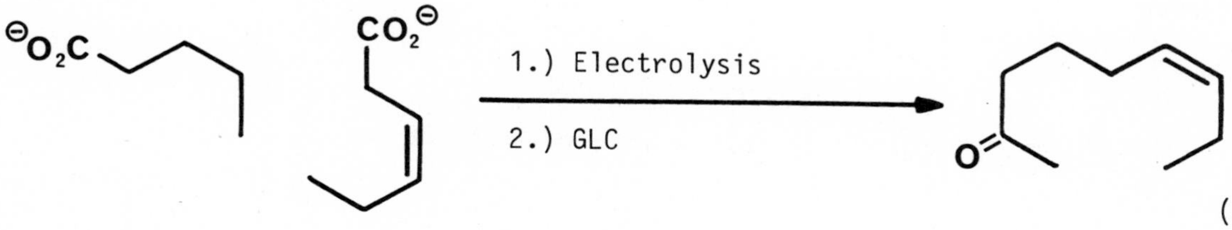

1.) Electrolysis
2.) GLC
(33%) ②

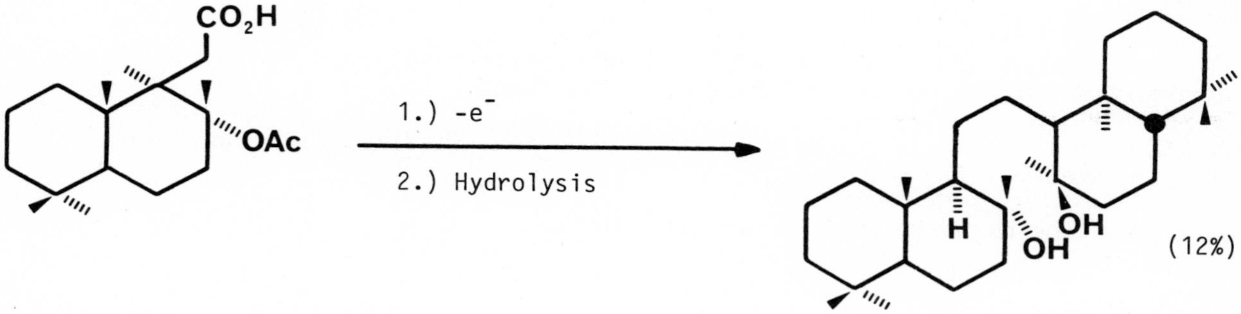

1.) $-e^-$
2.) Hydrolysis
(12%) ③

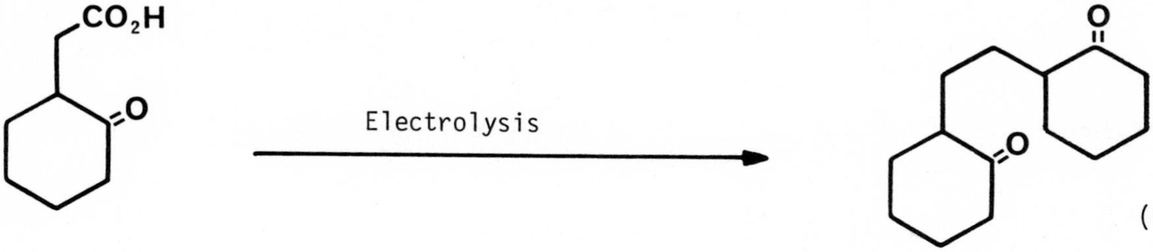

Electrolysis
(NYA) ④

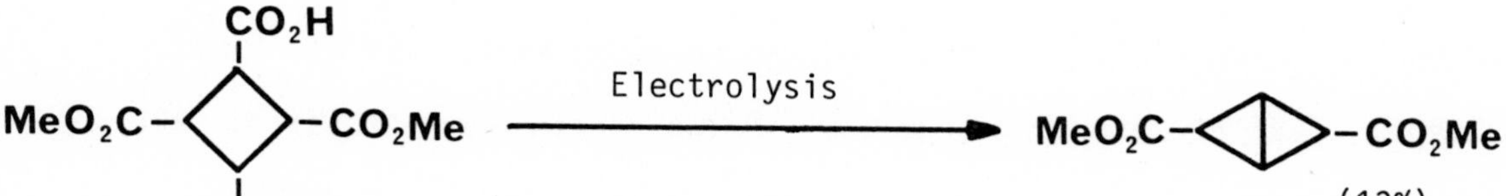

Electrolysis
(13%) ⑤

REACTION: LOSSEN REARRANGEMENT

March's Advanced Organic Chemistry : 945, 985

GENERAL SCHEME:

$$\mathrm{RC(=O)NHOC(=O)R} \xrightarrow{:B} \mathrm{RN{=}C{=}O} \xrightarrow{H_2O} \mathrm{RNH_2}$$

MECHANISM:

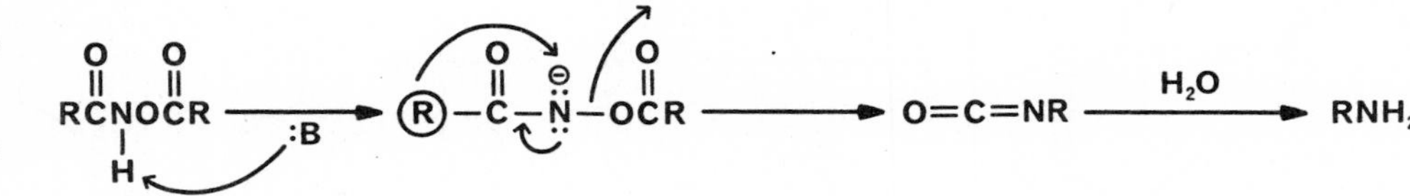

NOTES:

1. The reaction involves the formation of isocycanates from hydroxamic acid derivatives. It can be initiated by base or by heating.

2. The mechanism is similar to that for the **Curtius** and the **Hofmann** rearrangements.

3. Loss of the carboxylate anion leaves a **nitrene**.

$$\mathrm{R{-}C(=O){-}\ddot{N}:} \longrightarrow \mathrm{O{=}C{=}\ddot{N}R}$$

4. Nitromethane and PPA is an old in situ preparation of hydroxylamine.

$$\mathrm{CH_3NO_2} \xrightarrow[\Delta]{PPA} \mathrm{NH_2OH} + \mathrm{CO}$$

REFERENCES:

1. S. Bittner, S. Grinberg and I. Kartoon, **Tetrahedron Lett.**, (1974), 1965.

2. H.R. Snyder, C.T. Elston and D.B. Kellom, **J. Am. Chem. Soc.**, (1953), 75, 2014.

3. G.B. Bachman and J.E. Goldmacher, **J. Org. Chem.**, (1964), 29, 2576.

4. H. Ulrich, B. Tucer and R. Richter, **J. Org. Chem.**, (1978), 43, 1544.

EXAMPLES:

$PhC(=O)NHOH + EtO_2CN{=}NCO_2Et + Ph_3P + EtOH \xrightarrow{THF} PhN{=}C{=}O + EtO_2CNHNHCO_2Et$

$\xrightarrow{PhC(=O)NHOH}$ PhCNHOCNH-Ph (72%) ①

2-Naphthoic acid (CO_2H) $\xrightarrow[2.)\ OH^-]{1.)\ H_2NOH,\ PPA,\ \Delta}$ 2-naphthylamine (NH_2) (82%) ②

4-Cl-$C_6H_4CO_2H$ $\xrightarrow[2.)\ OH^-]{1.)\ MeNO_2,\ PPA,\ \Delta}$ 4-Cl-$C_6H_4NH_2$ (80%) ③ ⚠4

Cyclic amidine N-OSO_2Me $\xrightarrow[H_2O,\ RT]{NaOH}$ cyclic urea (NH, NH, C=O) (95%) ④

REACTION: MADELUNG INDOLE SYNTHESIS

March's Advanced Organic Chemistry : Not Indexed

GENERAL SCHEME:

Me

NH-Ac

XS Base

-Me

N H

MECHANISM:

2

CH_3 :B $CH_2^{\ominus}$ CH_3

N-C-CH_3 H O N-C-CH_3 H O N H $O^{\ominus}$

-CH_3 N H

NOTES:

1. The reaction will take place preferentially at a methyl group than a more substituted group.

2. This reaction is very similar to the **Reissert Indole Synthesis**:

CH_3 NO_2 + EtO-C(=O)-C(=O)-OEt Base Zn/HOAc CO_2Et O NH_2

-CO_2Et N H

REFERENCES:

1. Reported in **The Chemistry of Indoles**, R.J. Sundberg, Academic Press, (1970), New York, page 190. This is an excellent review of indole chemistry.

2. J.W. Schulenberg, **J. Am. Chem. Soc.**, (1968), 90, 7008.

3. W.J. Houlihan, V.A. Parrino and Y. Uike, **J. Org. Chem.**, (1981), 46, 4511.

4. P. Rosenmund and W.H. Haase, **Ber.**, (1966), 99, 2504.

5. C.F.H. Allen and J. Van Allan, **Org. Synth.**, (1955), Coll. Vol. 3, 597.

EXAMPLES:

Me, NHAc, CH_2Me —XS NH_2^-→ (–Me, N H, Et) + (Me, –Me, N H, Me) ① ⚠1

(NYA, 35:1)

CH_2CO_2Me, N–C(=O)–Ph, MeO_2C —XS MeO^-→ CO_2Me, –Ph, N, MeO_2C ②

(13%)

Me, N, N(H)–C(=O)–Ph —2 eq BuLi→ –Ph, N, N H ③

(22%)

NO_2, O, O —$SnCl_2$ / HCl→ N H, O ④

(72%)

CH_3, NH–C(=O)–CH_3 —1.) $NaNH_2$, 240-60° 2.) aq. EtOH→ –CH_3, N H ⑤

(80-83%)

March's Advanced Organic Chemistry: 412, 596

GENERAL REACTION:

$$EtOCOCH_2COOEt \xrightarrow[3.\ Hyd.,\ \Delta]{1.\ :B,\ RX;\ 2.\ :B,\ R'X} HC(R)(R')COOH$$

MECHANISM:

2

$$EtOCOCH_2COOEt \xrightarrow{:B} EtOCOCH{=}C(O^{\ominus})OEt \longleftrightarrow EtOCOCH^{\ominus}COOEt \xrightarrow{RX} EtOCOCH(R)COOEt$$

$$\xrightarrow{:B} EtOCOC^{\ominus}(R)COOEt \xrightarrow{R'X} EtOCOC(R)(R')COOEt \xrightarrow{Hyd.} HOCOC(R)(R')COOH \xrightarrow{\Delta} HC(R)(R')COOH$$

NOTES:

1. This reaction is similar to the **acetoacetic ester synthesis** in that both require a methylene group flanked by two electron-withdrawing groups.

2. Decarboxylation of a malonic acid derivative has a mechanism very similar to that for decarboxylation of a β-ketoacid.

$$>C(COOH)(COOH) \xrightarrow{\Delta} >C{=}C(OH)_2 \longrightarrow -\overset{H}{\underset{|}{C}}-COOH$$

3. This example is a combination of the **malonic ester synthesis** and the **Gabriel Synthesis**.

REFERENCES:

1. R.G. Riley and R.M. Silverstein, **Tetrahedron**, (1974), 30, 1171.

2. J.W. Cook and C.A. Lawrence, **J. Chem. Soc.**, (1935), 1637.

3. J.E. McMurry and J.M. Musser, **J. Org. Chem.**, (1975), 40, 2556.

4. M.S. Dunn and B.W. Smart, **Org. Synth.**, (1950), 30, 7.

EXAMPLES:

$$CH{=}CHCH_2Cl + MeCH(CO_2Et)_2 \xrightarrow{NaOEt} CH{=}CHCH_2C(Me)(CO_2Et)-CO_2Et \quad (63\%)$$ ①

$$\text{1-cyclohexenyl-}CH_2Br \xrightarrow[\text{2.) } EtO^- \text{ 3.) Hydrolysis, } \Delta]{\text{1.) } CH_2(COOEt)_2} \text{1-cyclohexenyl-}CH_2CH_2CO_2H \quad (65\%)$$ ②

$$MeCH_2CH_2CH_2Br + LiCH(CO_2Li)CO_2Et \xrightarrow[-78°]{THF / N_2} Me(CH_2)_4CO_2Et \quad (80\%)$$ ③

$$\text{Phthalimido-}N{-}\overset{\ominus}{C}(CO_2Et)_2 + ClCH_2CO_2Et \xrightarrow[150\text{-}60°]{} \text{Phthalimido-}N{-}C(CO_2Et)_2CH_2CO_2Et$$

$$\xrightarrow[HCl / CH_3CO_2H]{5\ H_2O} H_2OCCH_2CH(NH_2)CO_2H \ (43\%) + C_6H_4(CO_2H)_2$$ ④

REACTION: MANNICH REACTION

March's Advanced Organic Chemistry : 496, 800- 802, 834, 855

GENERAL SCHEME:

$$HCHO + HN(CH_3)_2 + RCH_2COR' \xrightarrow{H^{\oplus}} (H_3C)_2NCH_2-CH(R)-CO-R'$$

MECHANISM:

$$RCH_2COR' \underset{}{\overset{H^{\oplus}}{\rightleftharpoons}} RCH{=}C(OH)R'$$

$$HCHO + HN(CH_3)_2 \rightleftharpoons HOCH_2-N(CH_3)_2 \overset{H^{\oplus}}{\rightleftharpoons} CH_2{=}N^{\oplus}(CH_3)_2 \longrightarrow R'CO-CH(R)-CH_2-N(CH_3)_2$$

NOTES:

1. This reaction, like other amine- carbonyl condensations, can be applied to biomimetic synthetic strategies.

2. The reaction can be carried out with other aldehydes and ketones, however, is most popular and common with formaldehyde.

3. Quaternization of the Mannich base, followed by elimination, provides a method for placing an **exocyclic** methylene next to a carbonyl group.

$$-\overset{|}{\underset{|}{C}}(H)-CH_2-\overset{|}{\underset{|}{N}}- \xrightarrow{^{\ominus}HO} \;>C{=}CH_2$$

4. Via:

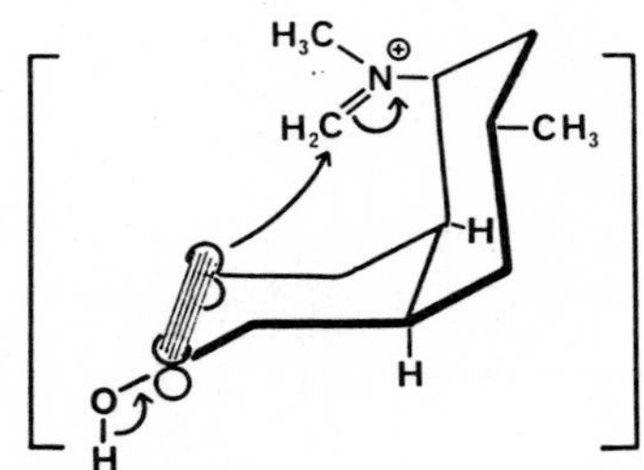

REFERENCES:

1. J. Schreiber, C.-G. Wermuth and A. Meyer, **Bull. Soc. Chim. Fr.**, (1973), 625.

2. A. Kozikowski and H. Ishida, **Heterocycles**, (1980), 14, 55.

3. W.L. Scott and D.A. Evans, **J. Am. Chem. Soc.**, (1972), 94, 4779.

4. S. Danishefsky, T. Kitahara, P.F. Shuda and S.J. Etheridge, **J. Am. Chem. Soc.**, (1977), 99, 6066.

5. P.A. Wender and J.C. Lechleiter, **J. Am. Chem. Soc.**, (1980), 102, 6340.

EXAMPLES:

$+ HC(=O)-COOH + 2$ morpholine

1.) 50°

2.) CH_3COOH

(55%) ①

$H_2CN^+(Me)_2Cl^-$

CH_2NMe_2

(95-100%) ②

HCOH

△

(NYA) ③ ⚠4

1.) $H_2CN^+(Me)_2I^-$, LDA

2.) MeI

3.) $NaHCO_3$

(31%) ④ ⚠3

1.) $Me_2{}^+NCH_2I^-$, LDA

2.) MeI

3.) K_2CO_3

(Quant.) ⑤

REACTION: McFAYDEN- STEVENS REDUCTION

March's Advanced Organic Chemistry : 398

GENERAL SCHEME:

$$R-\overset{O}{\overset{\|}{C}}-NH-NH-SO_2-Ar \xrightarrow{\text{Base}} R-\overset{O}{\overset{\|}{C}}-H$$

MECHANISM:

$$R-\overset{O}{\overset{\|}{C}}-\underset{}{\overset{H}{\overset{|}{N}}}-\underset{H}{\underset{|}{N}}-SO_2-Ar \xrightarrow{:B} R-\overset{O}{\overset{\|}{C}}-N{=}NH \xrightarrow{-N_2} R-\overset{O}{\overset{\|}{C}}-H$$

NOTES:

1. The sulfonylhydrazide derivatives are prepared from acids; thus, the reaction can be considered as the reduction of an acid to an aldehyde.

REFERENCES:

1. H. Babao, W. Herbert and A. Stiles, **Tetrahedron Lett.**, (1966), 2927.

2. S. Rozen, I. Shahak and E. Bergmann, **Tetrahedron Lett.**, (1973), 2327.

3. M.S. Newman and E.G. Calflisch, Jr., **J. Am. Chem. Soc.**, (1958), 80, 862.

4. C.C. Dudman, P. Grice and C.B. Reese, **Tetrahedron Lett.**, (1980), 4645.

EXAMPLES:

$$Me_2CHCCl(=O) + Me\text{-}C_6H_4\text{-}SO_2NHNH_2 \xrightarrow[\text{Ether}]{\text{Pyridine}} Me_2CHC(=O)NHNHSO_2\text{-}C_6H_4\text{-}Me$$

$$\xrightarrow[HOCH_2CH_2OH]{HOCH_2CH_2O^-Na^+} \xrightarrow[H_2O]{H^+} Me_2CHC(=O)\text{-}H \quad (34\%) \quad ①$$

1.) $SOCl_2$
2.) $TsNHNH_2$
3.) Na_2CO_3, $HOCH_2CH_2OH$, 160°

(50%) ②

$$\text{3-pyridyl-}C(=O)\text{-}NHNH_2 \xrightarrow[\text{Hydrazine, 160-65°}]{Na_2CO_3} \text{3-pyridyl-}CHO \quad (64\%) \quad ③$$

$$Me_2CH\text{-}C_6H_2(CHMe_2)_2\text{-}SO_2NHNH\text{-}C(=O)Ph \xrightarrow[\text{Hydrazine hydrate, }\Delta]{K_2CO_3\ /\ MeOH} HC(=O)Ph \quad (84\%) \quad ④$$

REACTION: McMURRY OLEFINATION REACTION

==

March's **Advanced Organic Chemistry :** 1112

GENERAL SCHEME:

$$>C=O \xrightarrow{TiCl_3/K} >C=C<$$

MECHANISM:

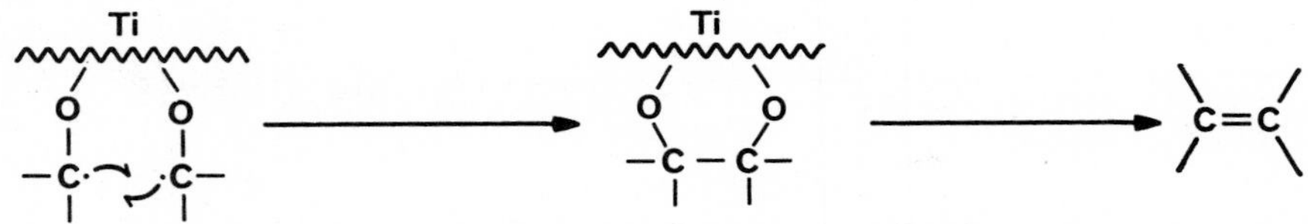

NOTES:

1. R. Dams, M. Malinowski, I. Westdorp and H.Y. Giese, **J. Org. Chem.**, (1982), 47, 248, suggest that the reduction of $TiCl_3$ with K, Li, or Mg give a Ti[0] metal while $LiAlH_4$ gives a Ti[I].

2. A mixed coupling usually uses one carbonyl compound in excess. See, however, B.P. Mundy, D.R. Bruss. Y. Kim, R.D. Larsen and R.J. Warnet, **Tetrahedron Lett.**, (1985), 3927, and references cited for an examination of the Ti-mediated pinacol coupling reaction.

3. This method of ring closer seems to be superior to the **acyloin** or **Thorpe** methods for all ring sizes.

REFERENCES:

1. J.E. McMurry and M.P. Fleming, **J. Org. Chem.**, (1976), 41, 896.

2. J.E. McMurry and L.R. Krepski, **J. Org. Chem.**, (1976), 41, 3925.

3. A.L. Baumstark, E.J.H. Bechara and M.J. Semigram, **Tetrahedron Letters**, (1976), 3265.

4. J.E. McMurry and K.L. Kees, **J. Org. Chem.**, (1977), 42, 2655.

5. J.E. McMurry, **Accts. Chem. Res.**, (1974), 7, 281.

EXAMPLES:

Ti(0) (85%) ①

Ti(0) (54%) ② △2

$TiCl_3$-$LiAlH_4$ (40-61%) ③

n-Bu-C(CH$_2$)$_8$C-Bu-n (O, O) $TiCl_3$ [Zn-Cu] Bu-n, Bu-n (76%) ④

CHO $TiCl_3$ / K ( )$_2$ (85%) ⑤

REACTION: MEERWEIN ARYLATION

March's Advanced Organic Chemistry : 641

GENERAL SCHEME:

EWG(R)C=C(H)R' —($CuCl_2$ / $ArN_2^{\oplus}$)→ EWG(R)C=C(Ar)R'

MECHANISM:

$ArN_2^{\oplus}X^{\ominus}$ + CuX_2 → Ar· + N_2 + CuX + X_2

Ar· + EWG(R)C=C(H)R' → ·C(EWG)(R)–C(H)(R')–Ar —($CuCl_2$)→ EWG(R)C=C(Ar)R' + CuCl

NOTES:

1. Via the intermediate:

REFERENCES:

1. H. Fritz, G. Rihs, P. Sutter and C.D. Weis, **J. Heterocyclic Chemistry**, (1981), 18, 1571.

2. Experimental reported in **Org. Reactions**, (1976), 24, 239. This is a Review written by C.S. Rondestvedt, Jr.

3. See Ref. 2, page 250.

4. S. Oae, K. Shinhama and Y.H. Kim, **Bull. Chem. Soc. Jpn.**, (1980), 53, 1065.

5. Ibid.

EXAMPLES:

+ CH_2=CHCN → (CH_2CHCN, $CuCl_2$) (65%) ①

+ CH_2=CHCHO → ($CuCl_2$, Acetone, pH=4) $CH_2CHClCHO$ (53%) ②

+ → ($CuCl_2$) (25%) ③ △1

N_2Cl + CH_2=CHCN → ($CuCl_2$, t-$BuSNO_2$) CN, Cl (31%) ④

NO_2, N_2Cl + CO_2Et → ($CuCl_2$, t-$BuSNO_2$) NO_2, CO_2Et, Cl (35%) ⑤

REACTION: MEERWEIN- PONNDORF- VERLEY REDUCTION

March's Advanced Organic Chemistry : 811, 813

GENERAL SCHEME:

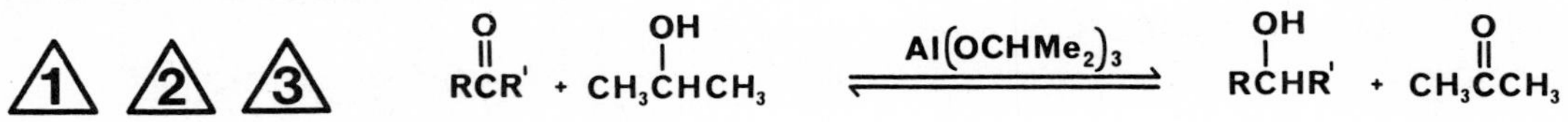

$$RC(=O)R' + CH_3CH(OH)CH_3 \xrightleftharpoons{Al(OCHMe_2)_3} RCH(OH)R' + CH_3C(=O)CH_3$$

MECHANISM:

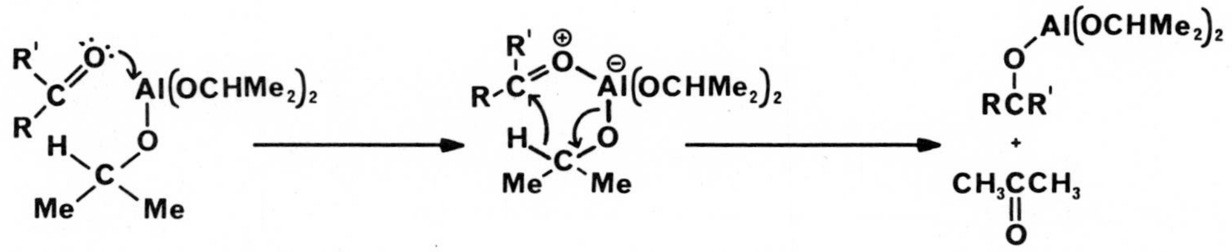

NOTES:

1. This reaction is reversible; the reverse process is known as the **Oppenauer Oxidation.** The driving force for the reaction is the removal of volatile products as they form.

2. Alkene bonds will not reduce under these conditions.

3. For a recent modification:

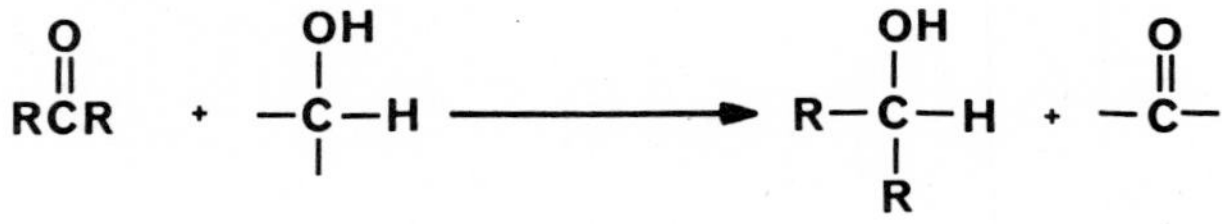

See: G.P. Boldrini, D. Savoia, E. Taliavini, C. Trombini and A. Umani-Ronchi, **J. Org. Chem.**, (1985), 50, 3082.

4. An example of the **Oppenauer Oxidation.**

REFERENCES:

1. V. Hach, **J. Org. Chem.**, (1973), 38, 293.

2. C.S. Marvel and G.L. Schertz, **J. Am. Chem. Soc.**, (1943), 65, 2055.

3. W.E. Doering and R.W. Young, **J. Am. Chem. Soc.**, (1950), 72, 631.

4. M. Botta, F. DeAngelis, A. Gamacorta, L. Labbiento and R. Nicoletti, **J. Org. Chem.**, (1985), 50, 1916.

5. P.D. Bartlett and W.P. Giddings, **J. Am. Chem. Soc.**, (1960), 82, 1240.

EXAMPLES:

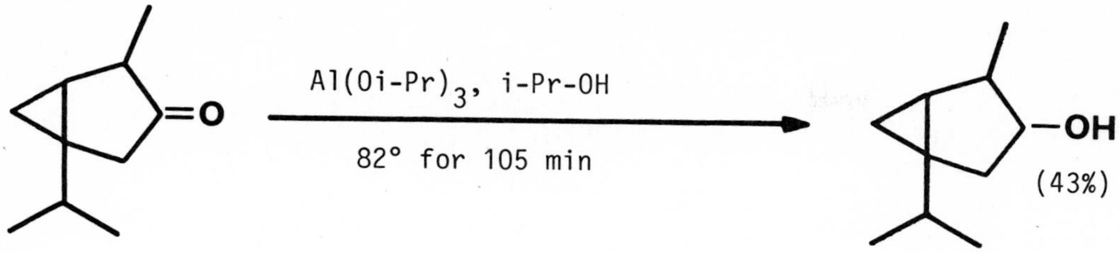

$Al(Oi\text{-}Pr)_3$, i-Pr-OH

82° for 105 min

-OH (43%) ①

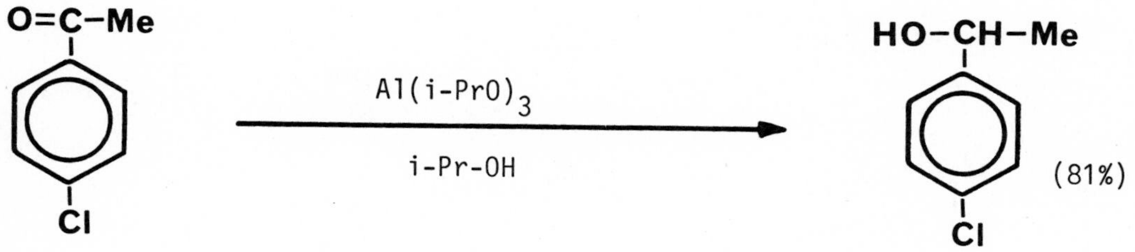

O=C-Me ... Cl

$Al(i\text{-}PrO)_3$

i-Pr-OH

HO-CH-Me ... Cl (81%) ②

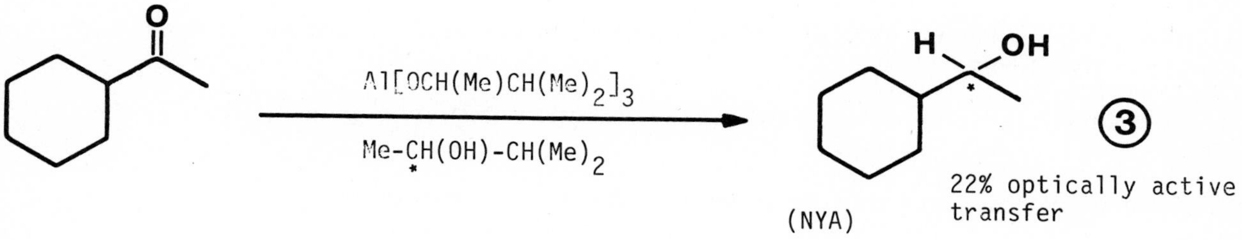

$Al[OCH(Me)CH(Me)_2]_3$

Me-C*H(OH)-CH(Me)$_2$

H OH

(NYA) 22% optically active transfer ③

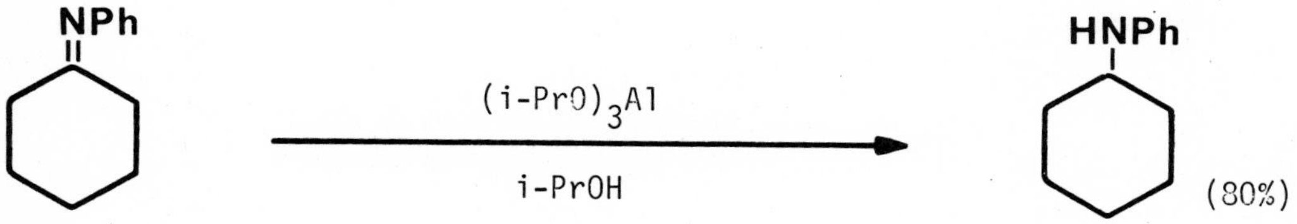

NPh

$(i\text{-}PrO)_3Al$

i-PrOH

HNPh (80%) ④

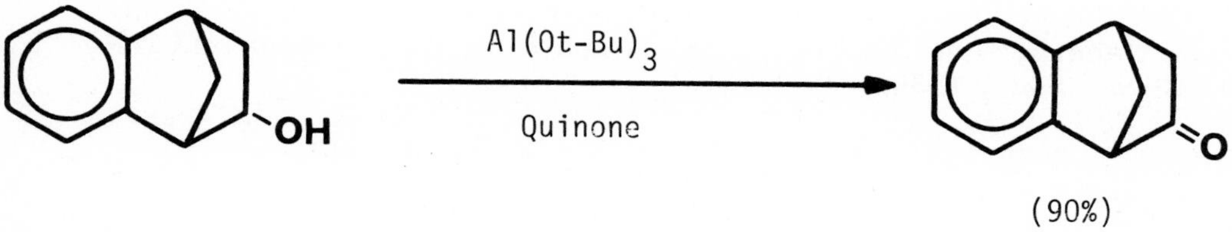

OH

$Al(Ot\text{-}Bu)_3$

Quinone

O (90%) ⑤ 4

REACTION: MEYERS ALDEHYDE SYNTHESIS

March's Advanced Organic Chemistry : 424- 26

GENERAL SCHEME:

3

R = H, Ph, or CO_2Et

dihydro-1,3-oxazine–CH_2R → (1. :B; 2. R'X; 3. $NaBH_4$; 4. Hyd.) → $H(O=)C–CH(R)R'$

MECHANISM:

oxazine–CH_2 → (BuLi) → oxazine–$\overset{\ominus}{C}HR$ → (R'X) → oxazine–CHR(R') → ($NaBH_4$) → tetrahydro-oxazine (N–H)–CHR(R') → (H_2O) → $H(O=)C–CH(R)R'$

NOTES:

1. Dihydro-1,3-oxazine derivatives are commercially available, but are also easy to prepare: **Synthesis**, (1972), 333.

2. If **R-** contains a halogen a cyclization reaction may take place [see: **J. Am. Chem. Soc.**, (1969), 91, 765.

3. Ketones may be prepared by a modification of this reaction:

oxazine–R → (MeI) → N-Me oxazinium ($\oplus$)–R → (R'MgX) → N-Me oxazine–C(R)(R') → ($H^{\oplus}$) → RCR' (C=O)

REFERENCES:

1. R.V. Stevens, J.M. Fitzpatrick, M. Kaplan and R.L. Zimmerman, **Chem. Commun.**, (1971), 857.

2. A.I. Meyers, A. Nabeya, H.W. Adickes, I.R. Yolitzer, G.R. Malone, K.C. Kovelesky, R.L. Nolen and R.C. Portnoy, **J. Org. Chem.**, (1973), 38, 36.

3. A.I. Meyers, A. Nabeya, H.W. Adickes, J.M. Fitzpatrick, G.R. Malone and I.R. Politzer, **J. Am. Chem. Soc.**, (1969), 91, 764.

4. A.I. Meyers and N. Nazarenko, **J. Org. Chem.** , (1973), 38, 175.

EXAMPLES:

1.) BH_4^-
2.) H_3O^+

CHO
CO_2Et
(64%) ①

1.) BuLi
2.) $Br(CH_2)_3Br$

1.) $NaBH_4$
2.) H_3O^+

—CHO
(20%) ②

1.) BuLi
2.) PhCHCHCHO

1.) $NaBH_4$
2.) H_3O^+

③

1.) BuLi
2.)
3.) MeI

4.)
5.) H_3O^+

CHO

(30% overall) ④

REACTION: MICHAEL REACTION

==

March's Advanced Organic Chemistry : 541, 665, 671, 674, 680, 682, 713- 719, 724, 729, 736, 773, 778, 845, 884.

GENERAL SCHEME:

1

$$EWG{-}CH_2R + R'(R'')C{=}C(EWG')R''' \xrightarrow{:B} EWG{-}CH(R){-}C(R')(R''){-}CH(EWG')(R''')$$

MECHANISM:

$$EWG{-}CH(H){-}R \xrightarrow{:B} EWG{-}\overset{\ominus}{C}HR \; + \; R'(R'')C{=}C(EWG')R''' \longrightarrow H{-}C(EWG)(R){-}C(R')(R''){-}C(EWG')(R'''){-}H$$

NOTES:

1. The reaction generally involves alpha-beta unsaturated carbonyl systems; however, other electron-withdrawing groups, such as nitriles and nitro compounds also work.

2. **VTB** = Vinyltriphenylphosphonium bromide

3. This is an example of the recently investigated **Michael- Michael Ring Closures (MIMIRC Reaction)**[3].

REFERENCES:

1. M. Binns, R. Haynes, T. Houston and W. Jackson, **Tetrahedron Lett.**, (1980), 573.

2. S. Pelletier, A. Venkov, J. Finer-Moore and N. Mody, **Tetrahedron Lett.**, (1980), 809.

3. G.H. Posner, S.-B. Lu and E. Asirvatham, **Tetrahedron Lett.**, (1986), 659.

4. G.H. Posner and E. Asirvatham, **Tetrahedron Lett.**, (1986), 663.

5. Y. Houbrechb, P. Laszlo and P. Pennetreau, **Tetrahedron Lett.**, (1986), 705.

EXAMPLES:

+ Ph-S-$CH_2CH=CH_2$ —LDA, THF, HMPT→ CHCH=CH_2 (SPh) + SPh, $CH_2CH=CH_2$ (1)

(89%, 95:5)

CH_2 —$HN(Et)_2$ / Al_2O_3→ CH_2NEt_2 (2)

(Quant.)

OLi —1.) $H_2CCHC(O)Me$ / 2.) Et_3B / 3.) $H_2CCHP^+(Ph)_3Br^-$ (VTB)→ (3) △2 △3

(21%)

—1.) $LiSnBu_3$ / 2.) $H_2CCHC(O)CH_2Me$ / 3.) $MeOC(O)CHCH_2$→ MeO, HO, $SnBu_3$ (4)

(64%)

+ —THF / t-BuO^-K^+→ (5)

(80%)

REACTION: NAZAROV CYCLIZATION

March's Advanced Organic Chemistry : Not indexed.

GENERAL SCHEME:

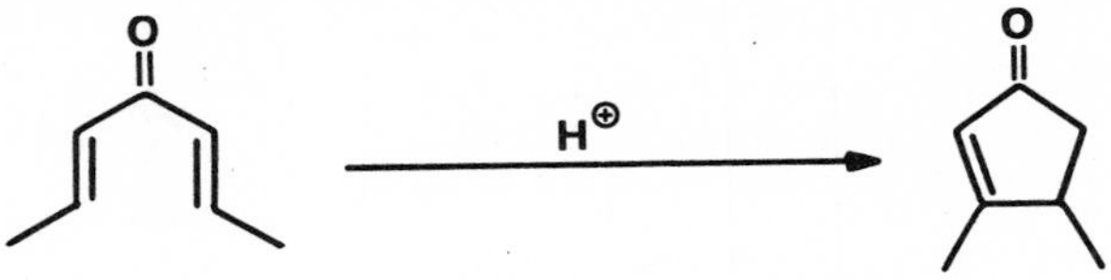

MECHANISM:

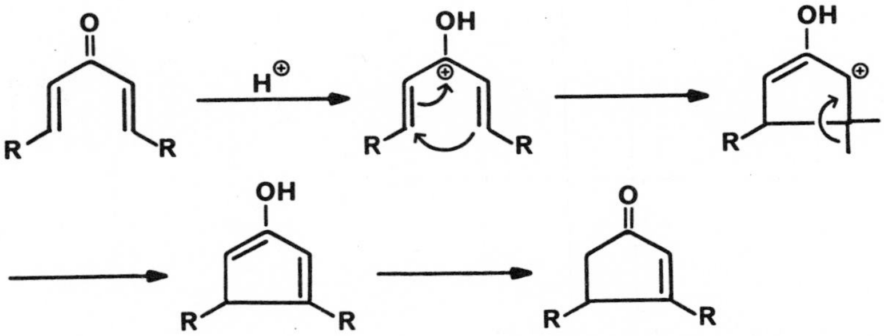

NOTES:

1. For a review, see: C. Santelli-Rouvier and M. Santelli, **Synthesis**, (1983), 429.

2. A **"Silicon-directed" Nazarov Cyclization**: S.E. Denmark, K.L. Habermas, G.A. Hite and T.K. Jones, **Tetrahedron**, (1986), 42, 2821.

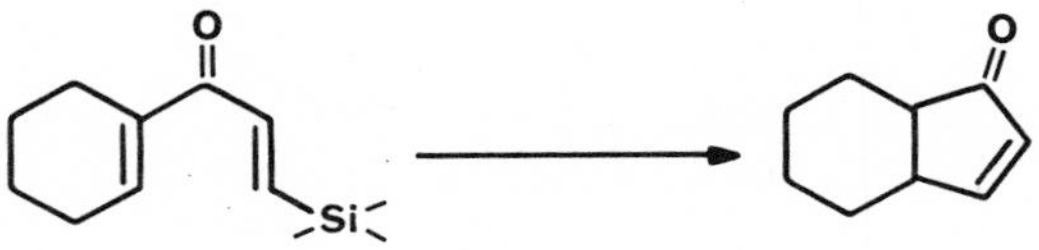

REAGENT	YIELD
$AlCl_3$	30%
$SnCl_4$	30%
$TiCl_4$	0%
$BF_3 \cdot Et_2O$	23%
$FeCl_3$	84%

REFERENCES:

1. I.N. Nazarov and I.N. Pinkina, **Chem. Abstr.**,(1948), 42, 7731. Reported in Review shown in **NOTE** 1.

2. P.E. Eaton, C. Giordano, G. Schloemer, and U. Vogel, **J. Org. Chem.**, (1976), 41, 2238.

3. S. Dev, **J. Indian Chem. Soc.**, (1957), 34, 169.

4. R.M. Jacobson, G.P. Lahm and J.W. Clader, **J. Org. Chem.**, (1980), 45, 395.

5. M.A. Tius, D.P. Astrab, A.H. Fauq, J.-B. Ousset and S. Trehan, **J. Am. Chem. Soc.**, (1986), 108, 3438.

EXAMPLES:

$HCOOH$ / H_3PO_4

90°, 7 hrs

(70%) ①

PPA

100°

(62%) ②

HO +

PPA

60°

(60%) ③

AcO

TsOH / PhMe

(66%) ④

SiMe₃

$FeCl_3$

CH_2Cl_2

(84%, 78:22) ⑤ △2

REACTION: NEBER REACTION

March's **Advanced Organic Chemistry:** 981

GENERAL SCHEME:

$$\mathrm{RCH_2(R')C{=}NOTs} \xrightarrow[H_2O]{:B} \mathrm{RCH(NH_2)C(=O)R'}$$

MECHANISM:

$$\mathrm{RCH_2(R')C{=}NOTs} \longrightarrow \mathrm{R\overset{\ominus}{C}H(R')C{=}\ddot{N}{-}OTs} \xrightarrow{H_2O:} \mathrm{R'{-}C{=}N\ (2H\text{-azirine},\ CH{-}R)}$$

$$\longrightarrow \mathrm{R'{-}C(OH){-}NH\ (aziridine,\ CH{-}R)} \xrightarrow{:B} \mathrm{RCH(NH_2)C(=O)R'}$$

NOTES:

1. The reaction is generally carried out with **R** = aryl, but examples are known with **R** = alkyl or H.

REFERENCES:

1. Y. Tamura, H. Fujiwara, K. Sumoto, M. Ikeda and Y. Kita, **Synthesis**, (1973), 215.

2. T.A. Geissmann and A. Armen, **J. Am. Chem. Soc.**, (1955), 77, 1623.

3. S. Ueda, S. Naruto, T. Yoshida, T. Sawayama and H. Uno, **Chem. Commun.**, (1985), 218.

4. R.F. Parcell and J.P. Sanchces, **J. Org. Chem.**, (1981), 46, 5229.

5. J.L. LaMattina and R.T. Suleske, **Synthesis**, (1980), 329.

EXAMPLES:

$PhCH_2(Ph)C{=}NOSO_2$–(2,4,6-trimethylphenyl) —NaOEt / EtOH→ $PhCH(NH_3^{\oplus}Cl^{\ominus})C(=O)Ph$ (34%) ①

2,4-$(O_2N)_2C_6H_3$–$CH_2C(=N{-}OTs)CH_3$ —Pyridine→ 2,4-$(O_2N)_2C_6H_3$–$HC{-}C(CH_3){=}N$ (NYA) ②

Ph–CH₂–CH(CO_2Me)–(1,2-benzisoxazol-3-yl) —NaH, DMF→ 2H-azirine with $PhCH_2$, CO_2Me and 2-HOC₆H₄ substituents (90%) ③

2-phenylcyclohexanone $N{-}\overset{\oplus}{N}Me_3$ $^{\ominus}I$ hydrazonium salt —NaOEt / EtOH; HCl→ 2-amino-2-phenylcyclohexanone (74%) ④

Br–C_6H_4–$C(=N{-}OTs){-}Me$ —K^+OAc^- / EtOH→ Br–C_6H_4–$C(OEt)_2{-}CH_2NH_2$ (92%) ⑤

REACTION: NEF REACTION

==

March's <u>Advanced Organic Chemistry</u> : 786

GENERAL SCHEME:

$$\text{RR'CH-NO}_2 \xrightarrow[\text{2.) Acid}]{\text{1.) Base}} \text{RR'C=O}$$

MECHANISM:

$$\text{RR'CH-}\overset{\oplus}{\text{N}}\text{(=O)O}^{\ominus} \xrightarrow{:B} \text{RR'C=}\overset{\oplus}{\text{N}}\text{(O}^{\ominus}\text{)O}^{\ominus} \xrightarrow{H^{\oplus}} \text{RR'C=}\overset{\oplus}{\text{N}}\text{(OH)O}^{\ominus} \xrightarrow{H^{\oplus}} \text{RR'C=}\overset{\oplus}{\text{N}}\text{(OH)}_2 \xrightarrow[-H^{\oplus}]{H_2O}$$

$$\text{HO-C(R)(R')-N(OH)}_2 \xrightarrow{-H_2O} \text{HO-C(R)(R')-N=O} \longrightarrow \text{RR'C=O}$$

NOTES:

1. The reaction works for primary and secondary aliphatic and alicyclic nitro compounds.

2. Via:

$[$2-methyl-2-(2-nitropropyl)cyclohexane-1,3-dione 1-ethylene ketal, $NO_2$$]$

3. See the review by J. McMurry (**Acc. Chem. Res.**,(1974), <u>7</u>, 281) for discussions of the novel uses of titanium reagents in organic synthesis.

REFERENCES:

1. M. Gabobardes and H. Pinnick, **Tetrahedron Lett.**, (1981), 5235.

2. E. Keinan and Y. Mazur, **J. Am. Chem. Soc.**, (1977), <u>99</u>, 3861.

3. J. Nokami, T. Sonoda and S. Wakabayashi, **Synthesis**, (1983), 763.

4. T. Yanami, A. Ballatore, M. Miyashita, M. Kato and A. Yoshikoshi, **Synthesis**, (1980), 407. See also: Y. Nakashita, T. Watanabe, E. Benkert, A. Lorenzi-Riatsch and M. Hesse, **Helv.**, (1984), <u>67</u>, 1207.

5. J.E. McMurry and J. Melton, **J. Org. Chem.**, (1973), <u>38</u>, 4367.

EXAMPLES:

NO_2, CH_3 (cyclohexane)

1.) LDA

2.) $MoO_5 \cdot Py \cdot HMPT$

O, CH_3 (85%) (1)

NO_2

SiO_2, NaOH

O (99%) (2)

$n\text{-}C_7H_{15}\text{-}\overset{NO_2}{CH}CH_2CH_2\overset{O}{C}CH_3$

Electrolysis

MeOH, $Na^+HCO_2^-$

$n\text{-}C_7H_{15}\text{-}\overset{O}{C}CH_2CH_2\overset{O}{C}CH_3$ (3)

(87-90%)

O, O + NO_2

1.) Bu_3P

2.) $HOCH_2CH_2OH$, TsOH

50% NaOH

$Bu_4N^+Br^-$

O, O, O, O (90%) (4) (2)

NO_2 +

NO_2

$TiCl_3$

O (35%) (5) (3)

REACTION: PAAL- KNORR PYRROLE SYNTHESIS

March's Advanced Organic Chemistry : Not indexed.

GENERAL SCHEME:

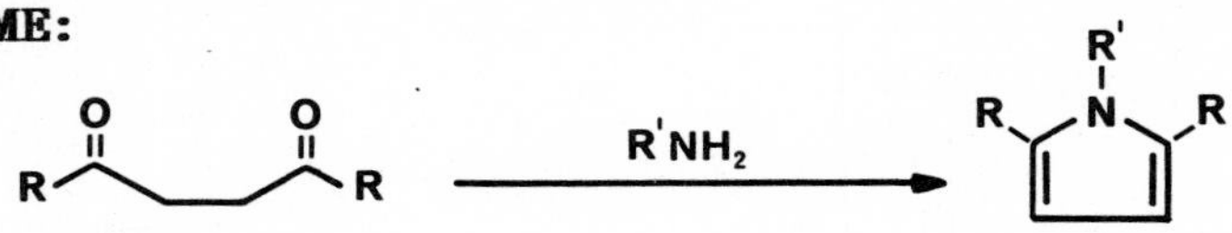

MECHANISM:

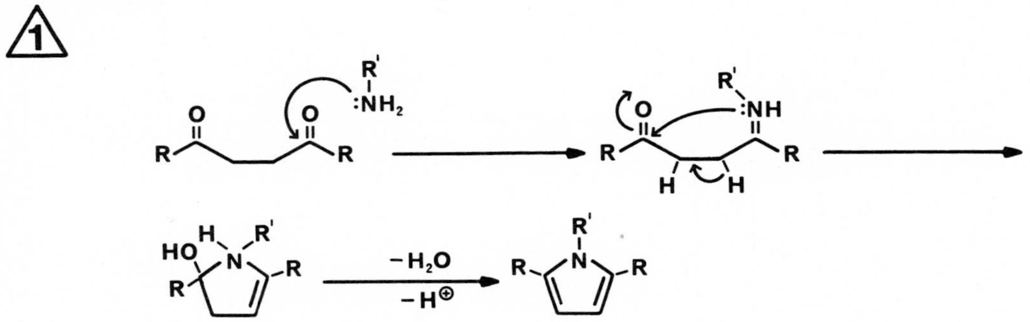

NOTES:

1. The reaction is limited by the access to appropriate diketones.

REFERENCES:

1. H.H. Wasserman, E. Gosselink, D.D. Kieth, J. Nadelson and R.J. Sykes, **Tetrahedron**, (1976), 32, 1863.

2. Ng. Ph. Buu-Hoi and Ng. D. Xuong, **J. Org. Chem.**, (1955), 20, 639.

3. Ng. Ph. Buu-Hoi, Ng. D. Xuong, and J.M. Gazave, **J. Org. Chem.**, (1955), 20, 850.

4. D.N. Young and C.F. Allen, , **Org. Synth.**, (1936), 16, 25.

EXAMPLES:

$(NH_4)_2CO_3$

DMF, Δ

H

O

O

N
H

(40%)

①

O O

$CH_3CCH_2CH_2CCH_3$

$H_2N(CH_2)_3NEt_2$

Δ

CH_3 N CH_3

(71%) $CH_2CH_2CH_2NEt_2$

②

O O

$MeCCH_2CH_2CMe$ + NH_2 NH_2

Δ

Me Me

N N

Me Me

(NYA)

②

NH_2

O O

$CH_3CCH_2CH_2CCH_3$ + OH

Δ

CH_3 N CH_3

(Quant.)

OH

③

O O

$CH_3CCH_2CH_2CCH_3$

$(NH_4)_2CO_3$

100°

CH_3 N CH_3
H

(81%)

④

REACTION: PERKIN REACTION

March's Advanced Organic Chemistry: 842- 43

GENERAL SCHEME:

2

$$ArCHO + RCH_2COOCOCH_2R \xrightarrow{RCH_2CO_2^{\ominus}} (H)(Ar)C{=}C(R)(COOH)$$

MECHANISM:

$$RCH_2COOCOCH_2R \xrightarrow{:B} RCH_2COOCO\overset{\ominus}{C}HR + ArCHO \longrightarrow RCH_2COOC(=O)CH(R)CH(O^{\ominus})Ar$$

$$\longrightarrow \text{[cyclic intermediate: } RCH_2C(O^{\ominus})\text{–O–CH(Ar)–CHR–C(=O)–O–]} \longrightarrow RCH_2COOCH(Ar)CH(R)CO_2^{\ominus} + RCH_2COOCOCH_2R$$

$$\longrightarrow RCH_2COOCH(Ar)CH(R)COOCOCH_2R + RCH_2CO_2^{\ominus} \longrightarrow RCH_2CO{-}\overset{\ominus}{C}(Ar)(H){-}C(R)COOCOCH_2R$$

$$\longrightarrow (Ar)(H)C{=}C(R)(COOCOCH_2R) + RCH_2CO_2^{\ominus} \longrightarrow (Ar)(H)C{=}C(R)(COOH)$$

NOTES:

1. The base used is often the salt of the corresponding anhydride.

2. The use of α-acylamido acids leads to the formation of azlactones.

REFERENCES:

1. J.R. Johnson, **Org. Synth.**, Coll. Vol. 3, (1955), 426.

2. R.W. Maxwell and R. Adams, **J. Am. Chem. Soc.**,(1930), 52, 2967.

3. G.E. Vandenberg, J.B. Harrison, H.E. Carter and B.J. Magerlein, **Org. Synth.**, (1967), 47, 101.

4. Reported in **Org. Reactions**, (1942), 1, 257.

5. J. Merchant and A. Gupta, **Chem & Ind.**, (1978), 628.

EXAMPLES:

(1) 2-furaldehyde (CHO) → [Ac_2O / K^+OAc^-] → 3-(2-furyl)acrylic acid (COOH) (65%)

(2) 3-nitrobenzaldehyde (CHO, NO_2) → [EtC(O)OC(O)Et / $Na^+CO_2Et^-$, Δ] → 3-nitrocinnamic acid (COOH, NO_2) (100%)

(3) 3,4-dimethoxybenzaldehyde (MeO, MeO, CHO) + CH_2CO_2H–NHC(=O)Ph → [Ac_2O / Na^+OAc^- / 100°] → MeO(MeO)C_6H_3–CH=(oxazolone: O, O, N=, Ph) (70%)

(4) 4-nitrobenzaldehyde (CHO, NO_2) → [Ac_2O / NaOAc] → $CH=CHCO_2H$ (NO_2) (90%)

(5) salicylaldehyde (OH, CHO) + R–C_6H_4–$OCH_2CO_2^{\ominus}Na^{\oplus}$ → [Ac_2O] → 3-aryloxycoumarin (O, =O, O–C_6H_4–R) (Apx. 40%)

R = –Me, –OMe, –NO_2

REACTION: PETERSON OLEFINATION REACTION

March's Advanced Organic Chemistry : 839

GENERAL SCHEME:

$$R-\overset{O}{\overset{\|}{C}}-R' \xrightarrow[\text{2. AcOH or } H_3O^{\oplus} \text{ then NaH}]{\text{1. } Me_3Si\frown MgCl} R-\overset{CH_2}{\overset{\|}{C}}-R'$$

MECHANISM:

$$R-\overset{O}{\overset{\|}{C}}-R' \xrightarrow{Me_3Si\frown MgCl} MgCl-O-C(R)(R')-CH_2SiMe_3 \xrightarrow[\Delta]{AcOH} H-O-C(R)(R')-CH_2SiMe_3 \longrightarrow R-\overset{CH_2}{\overset{\|}{C}}-R'$$

NOTES:

1. Advantages of this reaction over the **Wittig** reaction include: (1) By-products are more easily removed, and (2) the reaction suffers less from steric effects.

2. For example, there was no reaction under Wittig conditions.

3. Different products can be formed depending on whether the hydroxysilane is treated with KH or acid. [See: P.F. Hurdlick and D. Peterson, **J. Am. Chem. Soc.**, (1975), 97, 1464 and P.F. Hurdlick, D. Peterson and R.J. Rona, **J. Org. Chem.**, (1975), 40, 2263.

Me₃Si(H)(n-Pr)C–C(n-Pr)(H)OH: KH/THF (syn) → trans n-Pr–CH=CH–n-Pr; $BF_3 \cdot EtOH$ (anti) → cis n-Pr–CH=CH–n-Pr

REFERENCES:

1. R.K. Boeckman, Jr. and S.M. Silver, **Tetrahedron Lett.**, (1973), 3497.

2. D.J. Peterson, **J. Org. Chem.**, (1968), 33, 780.

3. D.J. Ager, **Tetrahedron Lett.**, (1981), 2923.

4. D.J.S. Tsai and D.S. Matteson, **Tetrahedron Lett.**, (1981), 2751.

5. D.J. Peterson, **J. Org. Chem.**, (1968), 33, 780.

EXAMPLES:

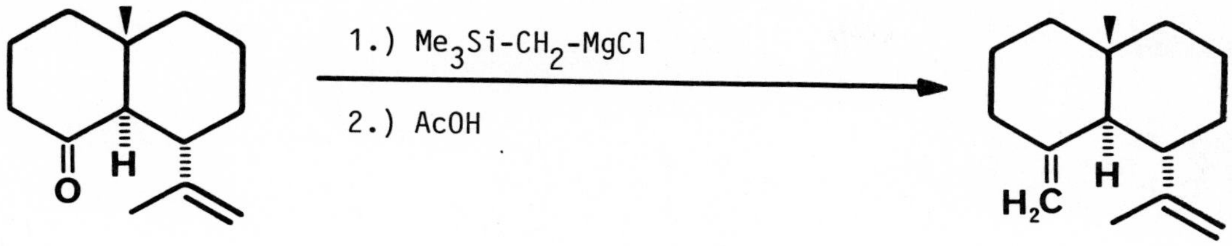

① △1 △2

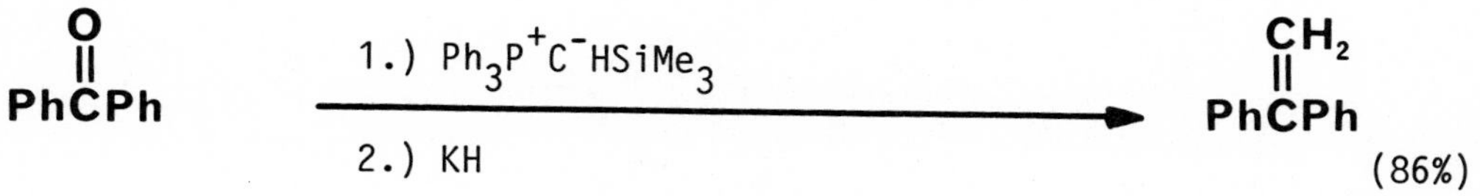

②

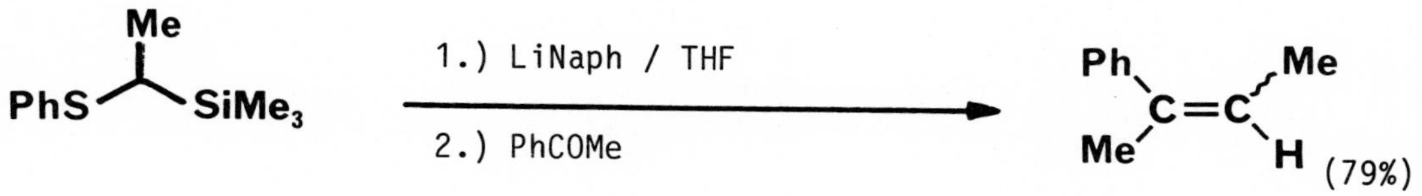

③

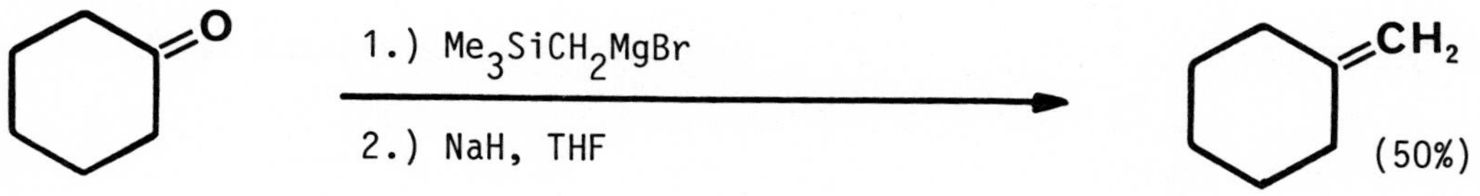

④

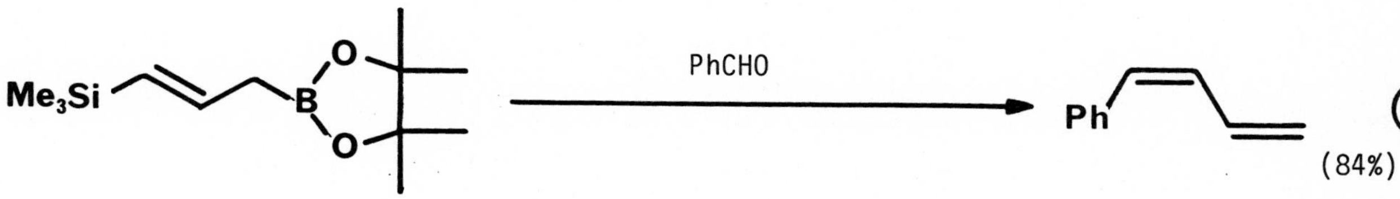

⑤

REACTION: PFITZNER- MOFFATT OXIDATION

March's Advanced Organic Chemistry : Not Indexed

GENERAL SCHEME:

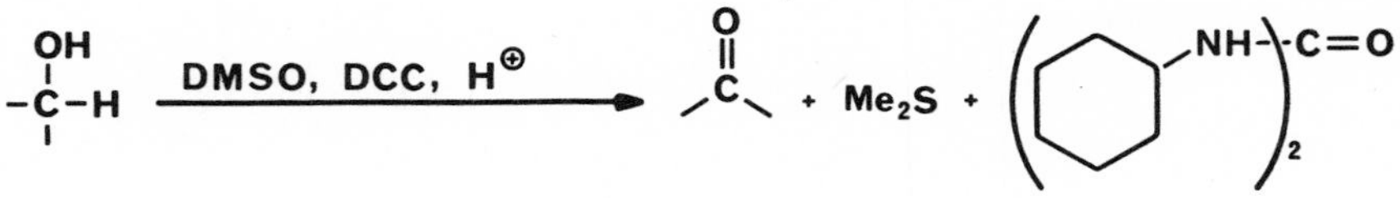

1

OH–C–H —(DMSO, DCC, $H^{\oplus}$)→ C=O + Me_2S + (cyclohexyl–NH)$_2$C=O

MECHANISM:

CH_3–S(=O)–CH_3 + cyclohexyl–N=C=N–cyclohexyl, $H^{\oplus}$ → CH_3–$\overset{\oplus}{S}$–CH_3 –O–C(=N–cyclohexyl)–NH–cyclohexyl, $H^{\oplus}$; H–Ö–C(H) →

$(CH_3)_2\overset{\oplus}{S}$–O–C(H)– , :B → O=C

+

cyclohexyl–NH–C(=O)–NH–cyclohexyl

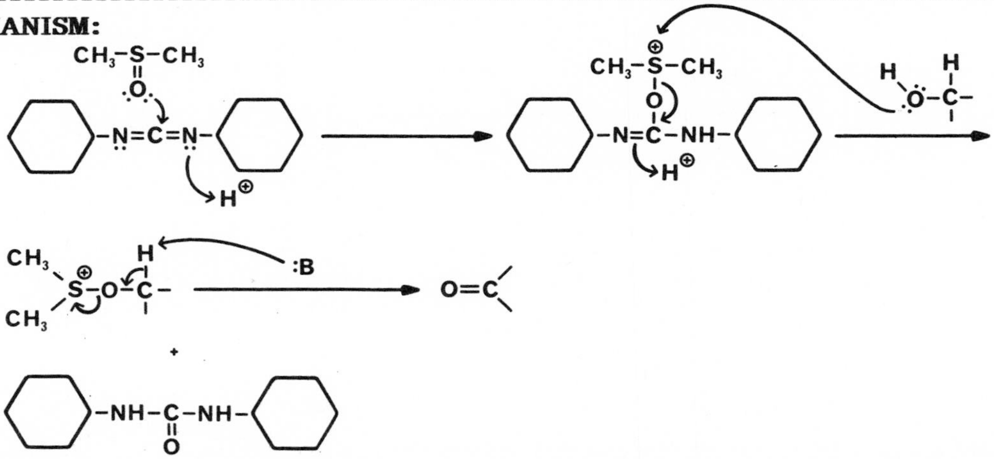

NOTES:

1. See: N.M. Weinshenker and C.-M. Shen, **Tetrahedron Lett.**, (1972), 3281 for the preparation of the insoluble, polymeric carbodiimide.

REFERENCES:

1. H.-J. Liu, H.-K. Hung, G.L. Mhene and M.L.D. Weinberg, **Can. J. Chem.**, (1978), 56, 1368.

2. K.E. Pfitzner and J.G. Moffatt, **J. Am. Chem. Soc.**. (1963), 85, 3027.

3. I. Dyong, R. Hermann, and G. von Kiedrowski, **Synthesis**, (1979), 526.

4. N.M. Weinshenker and C.-M. Shen, **Tetrahedron Lett.**, (1972), 3285.

5. N.M. Weinshenker and F.D. Green, **J. Am. Chem. Soc.**, (1968), 90, 506.

EXAMPLES:

DMSO, DCC

H_3PO_4

(80%)

①

DCC, DMSO

H_3PO_4

(61% as hydrazone derivative)

②

DMSO / DCC

(60%)

③

DMSO

Cross-linked carbodiimide/
polystyrene matrix

(91%)

④

Diisopropoly-
carbodiimide

DMSO

(NYA)

⑤

REACTION: PICTET- SPENGLER ISOQUINOLINE SYNTHESIS

==

March's **Advanced Organic Chemistry**: Not indexed

GENERAL REACTION:

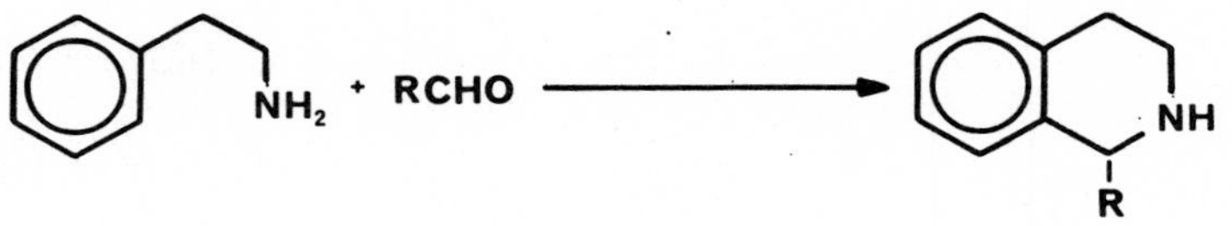

MECHANISM:

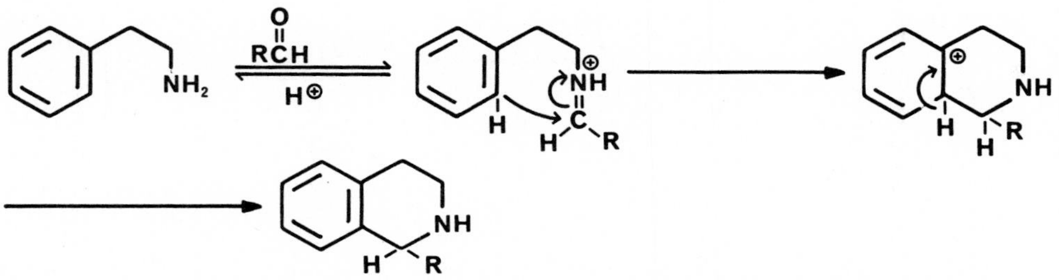

NOTES:

1. The α,β-epoxyacid is a precursor to benzaldehyde.

2. Along with alkylation, the reaction conditions afforded methylation (See: **Eschweiler- Clarke**).

3. This reaction was carried out without acid catalysis.

REFERENCES:

1. Reported in a Review: D. Valentine, Jr. and J.W. Scott, **Synthesis**, (1978), 329.

2. O.M. Friedman, K.N. Parameswaram and S. Burstein, **J. Med. Chem.**, (1963), 6, 227.

3. D.G. Harvey, E.J. Miller and W. Robson, **J. Chem. Soc.**, (1941), 153.

4. D. Soerens, J. Sandrin, F. Ungemach, P. Mokry, G.S. Wu, E. Yamanaka, L. Hutchins, M. DiPierro and J.M. Cook, **J. Org. Chem.**, (1979), 44, 535.

EXAMPLES:

$CO_2^{\ominus}Na^{\oplus}$

HO, HO, CO_2Me, H, $H_2N{\cdot}HCl$ + O, OMe, OMe → HO, HO, CO_2Me, NH, MeO, MeO (76%) (1) [triangle 1]

HO, MeO, MeO, OMe, NH_2 —HCOH / HCOOH→ HO, MeO, MeO, OMe, N, Me (28-30%) (2) [triangle 2]

COOH, NH_2, N H —pH=6.5 / CH_2O→ COOH, NH, N H (80%) (3)

COOMe, N, H, CH_2Ph, N, Me + Me, O, O, H —C_6H_6 / Δ→ COOMe, N, CH_2Ph, N, Me, O, Me (94%) (4) [triangle 3]

REACTION: PINACOL REARRANGEMENT

March's **Advanced Organic Chemistry :** 945, 949-50, 963-64

GENERAL SCHEME:

$$R{-}C(OH)(R'){-}C(OH)(R''){-}R''' \longrightarrow R{-}C(=O){-}C(R''')(R'){-}R''$$

MECHANISM:

$$R{-}C(OH)(R'){-}C(OH)(R''){-}R''' \longrightarrow R{-}C(OH)(R'){-}C(OH_2^{\oplus})(R''){-}R''' \longrightarrow R{-}C(OH)(R'){-}\overset{\oplus}{C}(R''){-}R'''$$

$$\longrightarrow R{-}\underset{\oplus}{C}(OH){-}C(R'')(R'){-}R''' \longrightarrow R{-}C(=O){-}C(R'')(R'){-}R'''$$

NOTES:

1. **R-** is usually, alkyl, aryl or H. If different **R-** groups are present, a number of different products may result. The reaction conditions may greatly influence the product composition.

2. There are a number of reactions classified as "**Pinacol- type**" rearrangements. These are generally 1,2-diol derivatives that are rearranged under basic conditions.

REFERENCES:

1. E. Keinan and Y. Mazur, **J. Org. Chem.**, (1978), 43, 1020.

2. J.E. Horan and R.W. Schiessler, **Org. Synth.**, (1961), 41, 53.

3. B.P. Mundy, Y. Kim, and R.J. Warnet, **Heterocycles**, (1983), 20, 1727.

4. B.P. Mundy, R. Srinivasa, R.D. Otzenberger and A.R. DeBernardis, **Tetrahedron Letters**, (1979), 2673.

5. G. Buchi, W. Hofheinz and J.V. Paukstelis, **J. Am. Chem. Soc.**, (1966), 88, 4113.

EXAMPLES:

$Me_2C(OH)-C(OH)Me_2$ $\xrightarrow{FeCl_3\text{-}SiO_2}$ $Me_3CC(=O)Me$ (70%) ①

$\xrightarrow{H_2SO_4}$ (69-81%) ②

$\xrightarrow{H_2SO_4}$ (91%) ③

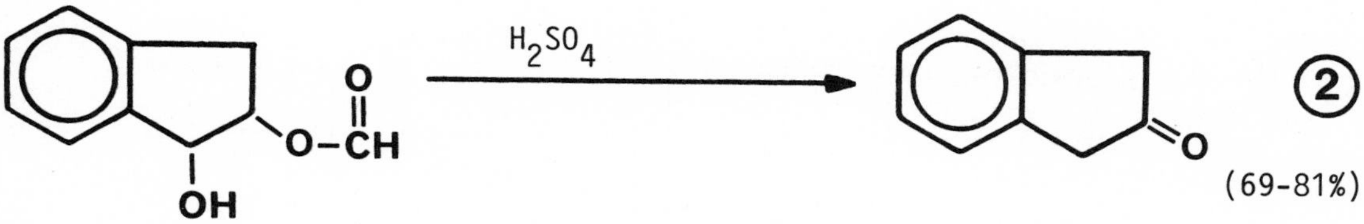

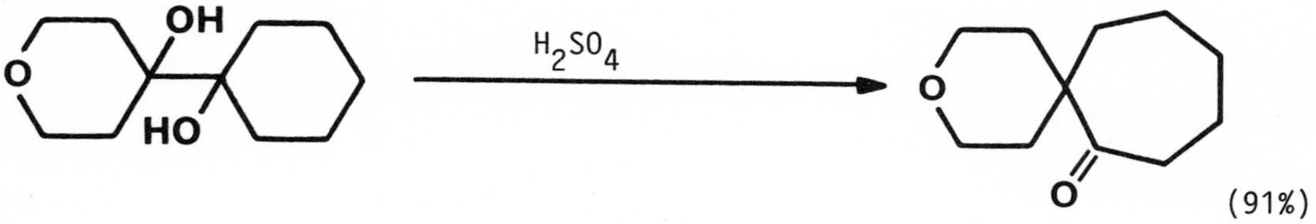

$\xrightarrow{H_2SO_4}$ (38%) ④

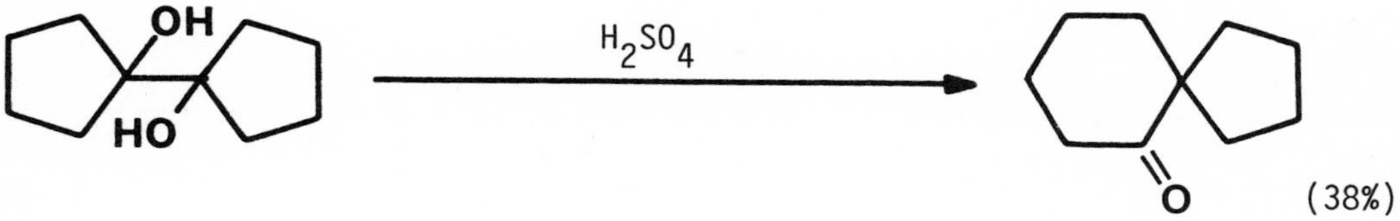

HO OTs $\xrightarrow{t\text{-}BuO^-}$ (85%) ⑤ △2

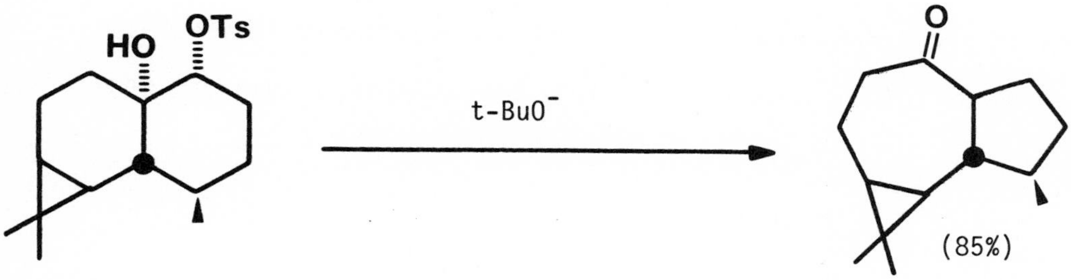

REACTION: POMERANZ- FRITSCH REACTION

===

March's Advanced Organic Chemistry: Not indexed

GENERAL SCHEME:

MECHANISM:

NOTES:

1. An isoquinoline synthesis.

2. **PPA** = polyphosphoric acid.

3. This reaction shows that the intermediate imine-acetal can be taken on to isoquinoline.

4. The **Schlittler- Muller modification** gives 1-substituted isoquinolines.

REFERENCES:

1. E. Schlittler and J. Muller, **Helv.**,(1948), 31, 914.

2. T. Kametani, S. Shibuya and I. Noguchi, **J. Pharm. Soc. Jpn**, (1965), 85, 667.

3. See: W.J. Gensler, **Org. Reactions**, (1951), 6, 191 for a review of this reaction.

4. M.J. Bevis, E.J. Forbes, N.N. Naik and B.C. Uff, **Tetrahedron**, (1971), 27, 1253.

EXAMPLES:

+ $HC(=O)-CH(OEt)_2$ → (NH$_2$, Me, OEt, OEt, N, Me)

H_2SO_4 → N, Me (15%) ① △4

$CH_2CH(OEt)_2$, N, MeO, O, MeO, OMe — PPA → MeO, N, O, MeO, OMe ② (NYA) △2

CHO + $H_2NCH_2CH(OEt)_2$ — H^+ → N ③ (NYA)

$CH(OEt)_2$, NH, MeO, OMe — BF_3 / CH_3CO_2H, $(CF_3CO)_2O$ → MeO, OMe, N (82%) ④

REACTION: PREVOST REACTION

March's Advanced Organic Chemistry : 733

GENERAL SCHEME:

$$>C=C< \xrightarrow{I_2,\ PhCOOAg} -\underset{|}{\overset{OH}{C}}-\underset{OH}{\overset{|}{C}}-$$

MECHANISM:

$>C=C< \ + \ I_2 \ + \ PhCOO \longrightarrow$ [iodonium-opened cyclic intermediate: $-C-C-I$, O, C=O, Ph] $\longrightarrow$

[cyclic benzoxonium ion] $+\ ^{\ominus}OCPh$ (O) $\longrightarrow$ PhCO(O)–C–C–OCPh(O) $\xrightarrow{\text{Hyd.}}$ HO–C–C–OH

NOTES:

1. The ratio of iodine to silver benzoate is 1 : 2.

2. A thallium-based version of the **Woodward modification.**

3. If the molar ratio of iodine to silver salt is 1 : 1 in acetic acid containing water, the product diol is primarily <u>cis</u>. This is known as the **Woodward modification** of the Prevost reaction. The water traps the intermediate to give the <u>cis</u> diol.

[cyclic acetoxonium ion, Me, + H_2O] $\longrightarrow$ [orthoester hydrate, Me, O–H] $\longrightarrow$ AcO–C–C–OH

REFERENCES:

1. R.G. Harvey, P.P. Fu, C. Cortez and J. Papak, **Tetrahedron Lett.**, (1977), 3533.

2. R. Cambie and P. Rutledge, **Org. Synthesis**, (1979), 59, 169.

3. R.B. Woodward and F.V. Brutcher, Jr., **J. Am. Chem. Soc.**, (1958), 80, 209.

4. P.S. Ellington, D.G. Hey and G.D. Meakins, **J. Chem. Soc., C**, (1966), 1327.

EXAMPLES:

1.) $PhCOOAg, I_2$

2.) MeO^-, C_6H_6

HO OH (94%) ①

1.) $TlOAc, I_2, HOAc, H_2O$

2.) $NaOH, H_2O, EtOH$

OH OH ② △2 (70-75%)

Me Me O

I_2, AcO^-Ag^+ (2 eq)

$HOAc, H_2O$

Me Me OH OH O (71%) ③

Me C_8H_{17} Me

Prevost conditions

with $PhCO_2Hg$

$PhCO_2$ $PhCO_2$ Me Me C_8H_{17} (67%)

④

Woodward conditions

followed by reduction

HO HO Me Me C_8H_{17} (81%)

REACTION: PRINS REACTION

===

March's Advanced Organic Chemistry : 856- 58.

GENERAL SCHEME:

$$HCHO + RCH{=}CH_2 \xrightarrow[H_2O]{H^{\oplus}\ \text{catalyst}} RCH(OH)CH_2CH_2OH \ \text{or} \ RCH{=}CHCH_2OH \ \text{or} \ \text{4-R-1,3-dioxane}$$

MECHANISM:

$$HCHO \longrightarrow H_2C^{\oplus}OH \xrightarrow{RCH=CH_2} R\overset{\oplus}{C}H-CH_2CH_2OH$$

$$\xrightarrow{-H^{\oplus}} RCH{=}CHCH_2OH$$

$$\xrightarrow{H_2O} RCH(OH)CH_2CH_2OH \longrightarrow \text{4-R-1,3-dioxane}$$

NOTES:

1. The product composition is very dependent on the specific alkene and the reaction conditions.

2. The mechanism for the dioxane formation is not well understood. See: **J. Org. Chem.**,(1968), 33, 4155 and references cited.

3. The hexakis(acetato)trihydrato-μ_3-oxotrisrhodium acetate can be made by the ozonolysis of rhodium diacetate in acetic acid. See: **J. Chem. Soc.**,Dalton, (1973), 2665.

4. This method eliminates the need for an aqueous system.

REFERENCES:

1. J. Thivolle- Cazat and I. Tkatchenko, **Chem. Commun.**, (1982), 1128.

2. I. Tomoskozi, L. Gruber, G. Kovacs, I. Szekely and V. Simonidesz, **Tetrahedron Letters**, (1976), 4639.

3. a. M. Delmas and A. Gaset, **Synthesis**, (1980), 871.
   b. R. El Gharbi, M. Delmas and A. Gaset, **Synthesis**, (1981), 361.

4. B.M. Mitzner, S. Lemberg, V. Mancini and P. Barth, **J. Org. Chem.**, (1966), 31, 2022.

5. B.B. Snider, D.J. Rodini, T.C. Kirk and R. Cordova, **J. Am. Chem. Soc.**, (1982), 104, 555.

EXAMPLES:

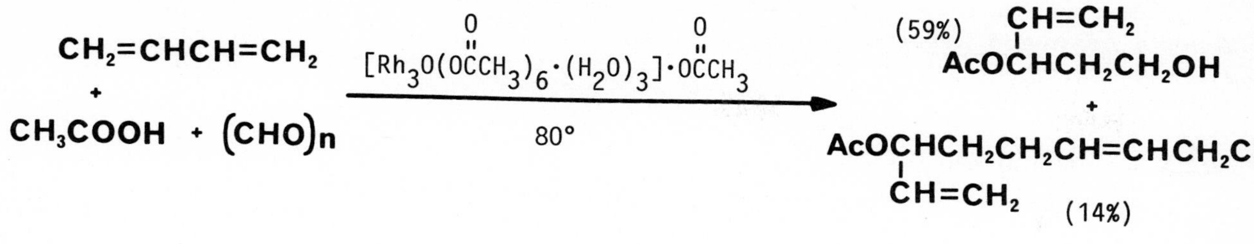
$CH_2=CHCH=CH_2$
+
CH_3COOH + $(CHO)_n$
$[Rh_3O(OCCH_3)_6 \cdot (H_2O)_3] \cdot OCCH_3$
80°
(59%)
CH=CH₂
$AcOCHCH_2CH_2OH$
+
$AcOCHCH_2CH_2CH=CHCH_2CH_2OH$
$CH=CH_2$
(14%)

① Δ1 Δ3

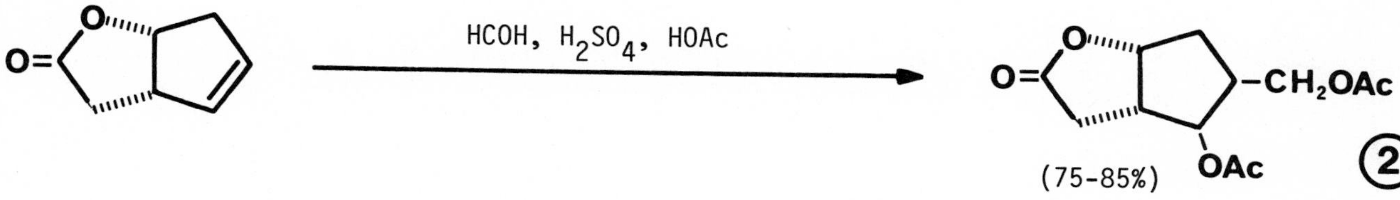
$HCOH, H_2SO_4, HOAc$
$-CH_2OAc$
OAc
(75-85%)

②

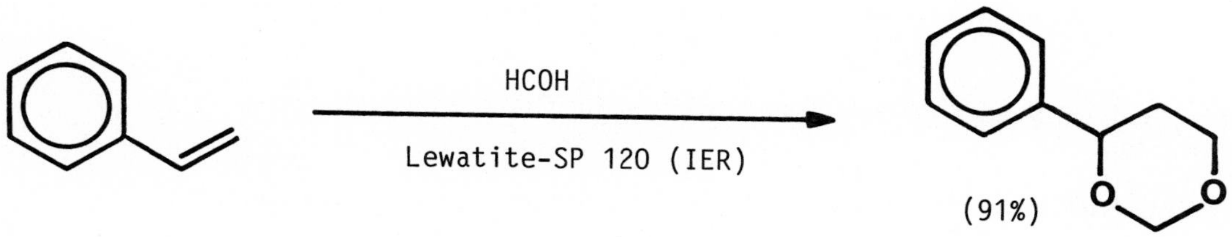
HCOH
Lewatite-SP 120 (IER)
(91%)

③ Δ4

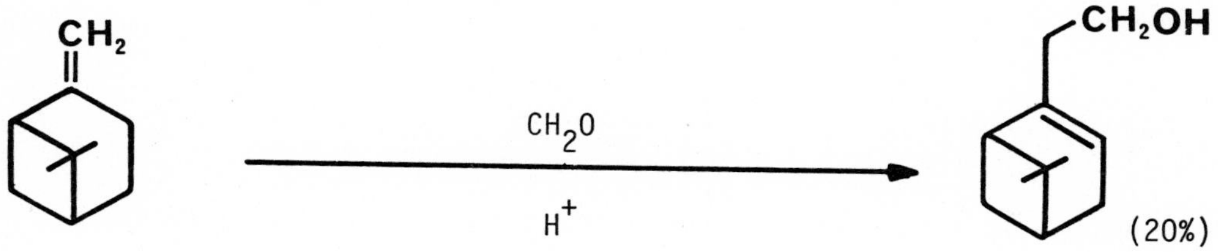
CH_2
CH_2O
H^+
CH_2OH
(20%)

④

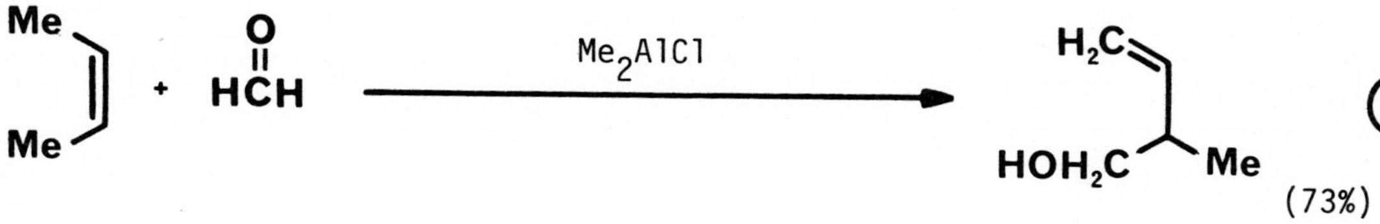
Me
Me
HCH
Me_2AlCl
H_2C
HOH_2C
Me
(73%)

⑤

REACTION: PUMMERER REARRANGEMENT

===

March's <u>Advanced Organic Chemistry</u> : Not Indexed

GENERAL SCHEME:

$$RC(=O)CH_2S(=O)R' \xrightarrow[H_2O]{H^{\oplus}} RC(=O)CH(OH)SR' \dashrightarrow RC(=O)CH(=O)$$

MECHANISM:

2

$$RC(=O)CH_2S(=O)R' \xrightarrow{H^{\oplus}} RC(=O)CH(H)S^{\oplus}(OH)R' \longrightarrow RC(=O)CH{=}S(OH_2^{\oplus})R' + H_2O$$

$$\longrightarrow RC(=O)CH(OH)SR'$$

NOTES:

1. There is a similarity to the **Nef Reaction** in that there is an internal oxidation -reduction process.

2. The keto-sulfoxides are often available by the general use of the nucleophile:

$$R-S(=O)-CH_2^{\ominus}$$

REFERENCES:

1. K. Konno, K. Hashimoto, H. Shirahama and T. Matsumoto, **Tetrahedron Lett.**, (1986), 3865.

2. M. Watanabe, S. Nakamori, H. Hasegawa, K. Shirai and T. Kumamoto, **Bull. Chem. Soc. Japan**, (1981), <u>54</u>, 817.

3. K. Ogura, J. Watanabe and H. Iida, **Tetrahedron Lett.**, (1981), 4499.

4. H.-D. Becker, G.J. Mikol and G.A. Russell, **J. Am. Chem. Soc.**, (1963), <u>85</u>, 3410.

5. Reported in **Reagents**, Vol. 6, 369.

EXAMPLES:

①
OMEM; 1.) TFAA; 2.) Pyridine; SPh; O; O; HC; O; O; MeOH; NH_3; N
(52%)

②
O; *n*-Bu O; $PhSCH_2CH_2CHC-OH$; Ac_2O, TsOH; PhS; O; *n*-Bu; O
(68%)

③
SOMe; $PhCH_2CH$; CO_2Me; H_2O_2, HOAc; $PhCH_2COOH$
(75%)

④
O; O; S; Me; H^+; O SMe; CH; OH
(82%)

⑤
H; O; CH_2SMe; Me; Me; O; I_2; MeOH; H; O; $CH(OMe)_2$; Me; Me; O; +; Me; O; OMe; Me; O
(1:1, NYA)

REACTION: RAMBERG- BACKLUND REACTION

March's Advanced Organic Chemistry : 921- 23

GENERAL SCHEME:

$$RCH_2SO_2-\underset{Cl}{\overset{R}{C}}-H \xrightarrow{\text{Base}} \underset{R}{\overset{H}{}}\!\!>C=C<\!\!\underset{R}{\overset{H}{}}$$

MECHANISM:

$$R-\underset{H}{\overset{H}{C}}-SO_2-\underset{Cl}{\overset{R}{C}}-H \quad :B \longrightarrow R-\overset{\ominus}{C}H-SO_2-\underset{Cl}{\overset{R}{C}}-H \longrightarrow RHC\underset{SO_2}{-}CHR$$

$$\xrightarrow{-SO_2} RHC{=}CHR$$

NOTES:

1. Halogen reactivity: **I > Br >> Cl**

2. For a Review of this reaction, see: L.A. Paquette, **Org. Reactions**, (1977), 25, 1.

REFERENCES:

1. L.A. Paquette and M.P. Trova, **Tetrahedron Lett.**, (1986), 1895.

2. H. Matsuyama, Y. Miyazawa, Y. Takei and M. Kobayashi, **Chem. Letters**, (1984), 833.

3. H. Matsuyama, Y. Miyazawa and M. Kobayashi, **Chem. Letters**, (1986), 433.

4. G.D. Hartman and R.D. Hartman, **Synthesis**, (1982), 504.

5. L.A. Paquette and R.W. Houser, **J. Am. Chem. Soc.**, (1969), 91, 3870.

EXAMPLES:

1.) NCS, CCl_4

2.) 1-CO_2H,2-CO_3H-C_6H_4

3.) t-BuO^-

(27%) ①

t-BuO^-, CCl_4

CH_2Ph SO_2 O $-CH_2Ph$ (69%) ②

C_5H_{11} $-SO_2PhCH_3$ SO_2 NaH-KH DMSO C_5H_{11} O C_5H_{11} (87%) ③

SO_2CH_2Cl H 20% NaOH Phase transfer catalyst $=CH_2$ (86%) ④

Cl SO_2 KOH H_2O, Δ (75%) ⑤

REACTION: REFORMATSKY REACTION

March's Advanced Organic Chemistry : 822- 24

GENERAL SCHEME:

$$>C{=}O + Br{-}\overset{|}{\underset{|}{C}}{-}COOR \xrightarrow[\text{2. Hyd.}]{\text{1. Zn}} HO{-}\overset{|}{\underset{|}{C}}{-}\overset{|}{\underset{|}{C}}{-}COOR$$

MECHANISM:

$$Br{-}\overset{|}{\underset{|}{C}}{-}\overset{O}{\overset{\|}{C}}{-}O{-}R + Zn \longrightarrow >C{=}\underset{O-ZnBr}{\underset{|}{C}}{-}O{-}R + >C{=}O \longrightarrow$$

$$\left[\text{Br-Zn, O, C-O-R, C, O=C- (cyclic transition state)}\right] \longrightarrow BrZn{-}O{-}\overset{|}{\underset{|}{C}}{-}\overset{|}{\underset{|}{C}}{-}\overset{O}{\overset{\|}{C}}{-}OR \xrightarrow{\text{Hyd.}} HO{-}\overset{|}{\underset{|}{C}}{-}\overset{|}{\underset{|}{C}}{-}COOR$$

NOTES:

1. The organozinc reagent is formed in the first step and the reagent is similar to the Grignard reagent; however, it is lower in reactivity. This allows for its formation in the presence of an ester.

2. A modification of the Reformatsky reaction using nitriles is called the **Blaise Reaction.**

$$R{-}\overset{R}{\overset{|}{\underset{ZnBr}{\underset{|}{C}}}}{-}COOEt + R'{-}C{\equiv}N \longrightarrow \xrightarrow{H_2O} R'{-}\overset{O}{\overset{\|}{C}}{-}\overset{R}{\overset{|}{\underset{R}{\underset{|}{C}}}}{-}COOEt$$

3. After the hydrolysis step, dehydration can be a common observation if an alpha hydrogen is available.

REFERENCES:

1. L. Friedrich, N. DeVera and M. Hamilton, **Synth. Commun.**, (1980), 10, 637.

2. B.M. Trost, C.D. Shuey, F. DiNinno, Jr., and S.S. McElvain, **J. Am. Chem. Soc.**, (1979), 101, 1284.

3. H. Mattes and C. Benezra, **Tetrahedron Lett.**, (1985), 5697.

4. J. Wicha and K. Bal, **Chem. Commun.**, (1975), 968.

5. R. Heilmann and R. Glenat, **Bull. Soc. Chim. Fr.**, (1955), 1586.

EXAMPLES:

$Ph_2C{=}O$ + $BrCH_2C{\equiv}CH$ —(Zn, THF)→ $Ph_2C(OH)CH_2C{\equiv}CH$ (NYA) ①

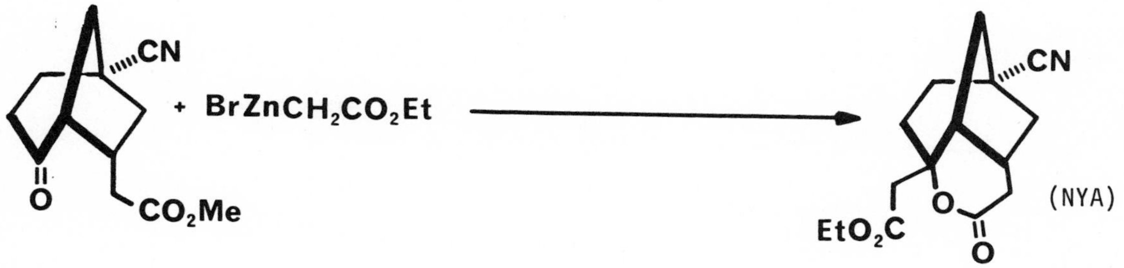

CN, CO_2Me, O + $BrZnCH_2CO_2Et$ → CN, EtO_2C, O, O (NYA) ②

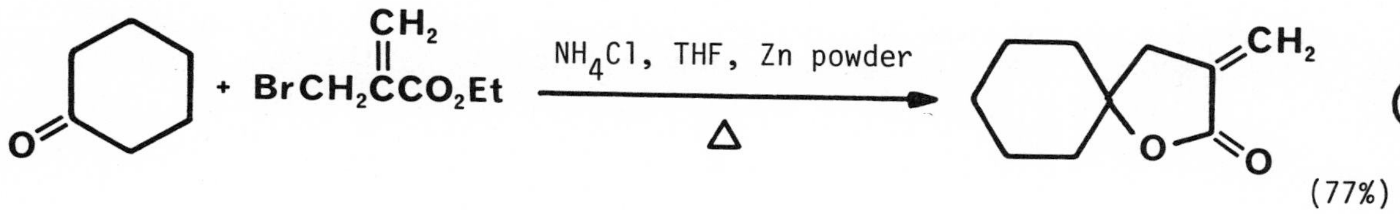

O + $BrCH_2C(=CH_2)CO_2Et$ —(NH_4Cl, THF, Zn powder, Δ)→ CH_2, O, O (77%) ③

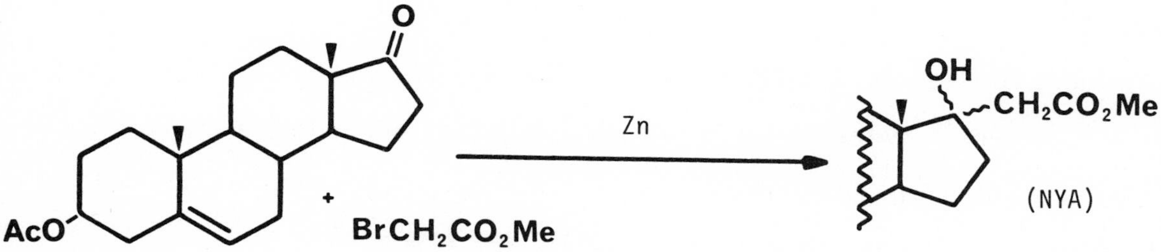

O, AcO + $BrCH_2CO_2Me$ —(Zn)→ OH, CH_2CO_2Me (NYA) ④

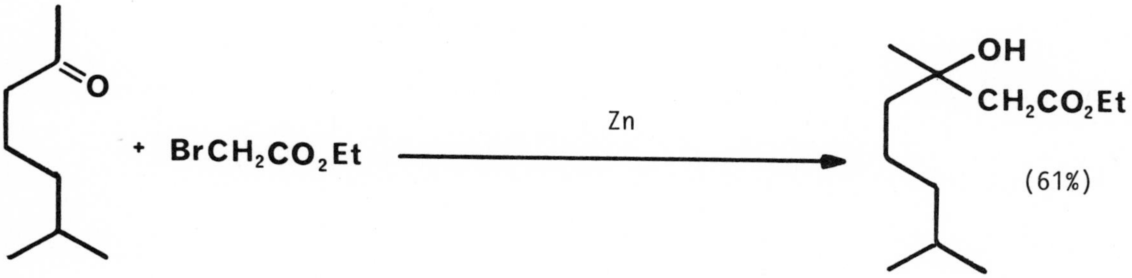

O + $BrCH_2CO_2Et$ —(Zn)→ OH, CH_2CO_2Et (61%) ⑤

REACTION: RITTER REACTION

March's <u>Advanced Organic Chemistry</u> : 860

GENERAL SCHEME:

$$ROH + R'CN \xrightarrow{H^{\oplus}} RNHCR' \text{ (with } C{=}O\text{)}$$

MECHANISM:

$$ROH \xrightarrow{H^{\oplus}} R^{\oplus} + R'CN \longrightarrow R{-}N{=}\overset{\oplus}{C}{-}R' \xrightarrow{H_2O} R{-}N{=}\overset{OH}{C}{-}R'$$

$$\xrightarrow{\text{Taut.}} R{-}\underset{H}{N}{-}\underset{O}{\overset{}{C}}{-}R'$$

NOTES:

1. The **Meyers Reagent,** formed by this procedure, has found a variety of useful synthetic applications. See, for example: **Meyers Aldehyde Synthesis.**

REFERENCES:

1. S. Top and G. Jaoven, **J. Org. Chem.**, (1981), <u>46</u>, 78.

2. See: A.I. Meyers in **Heterocycles in Organic Synthesis**, Wiley, New York (1974).

3. T. Gajda, A. Kuziara, S. Zawadski and A. Zwierzak, **Synthesis**, (1979), 549.

4. R.J. Ryan and S. Julia, **Tetrahedron,** (1973), <u>29</u>, 3649.

5. R.A. Wohl, **J. Org. Chem.**, (1973), <u>38</u>, 3099.

EXAMPLES:

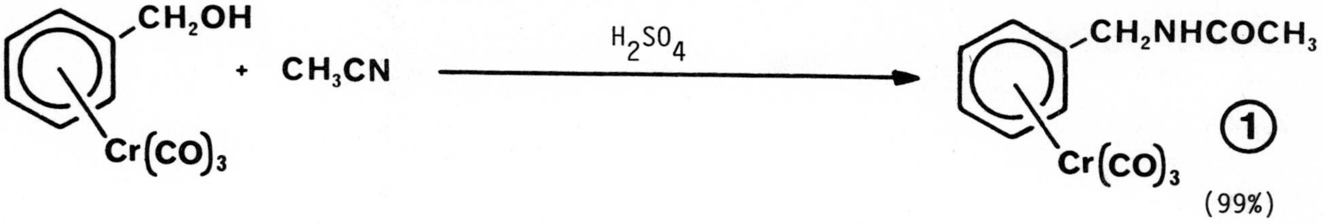

(Cr(CO)$_3$-complexed benzyl alcohol, ring–CH_2OH) + CH_3CN —H_2SO_4→ (Cr(CO)$_3$-complexed arene, ring–$CH_2NHCOCH_3$) ① (99%)

(diol with OH groups) —CH_3CN, H^+→ (dihydro-oxazine, O, N=) (NYA) ② △1

$$\mathrm{Me_3C{-}OH} \xrightarrow[\text{2.) } H_2O,\ K_2CO_3]{\text{1.) } KCN,\ H_2SO_4,\ HOAc} \mathrm{Me_3C{-}NH{-}CHO}$$ (70%) ③

(HO-substituted epoxy ring system) + CH_3CN —MeCN, $HClO_4$→ (HO, HO, $NHCCH_3$ with C=O) (49%) ④

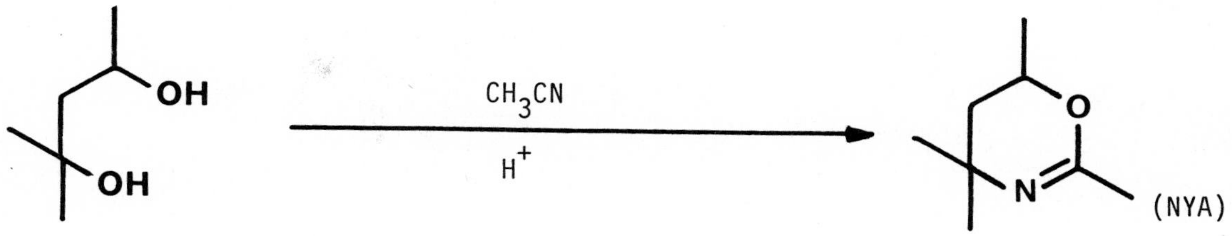

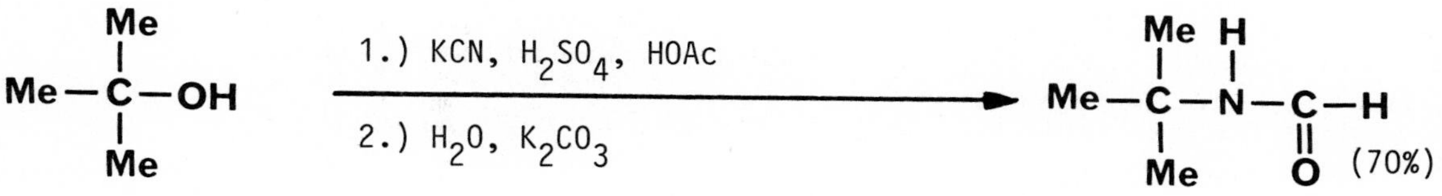

$CH_3CH(OH)CH(Br)CH_3$ (Threo) + Ph–C≡N —PhCN, H_2SO_4→ $CH_3CH(NH{-}C(=O){-}Ph)CH(Br)CH_3$ (Threo) (81%) ⑤

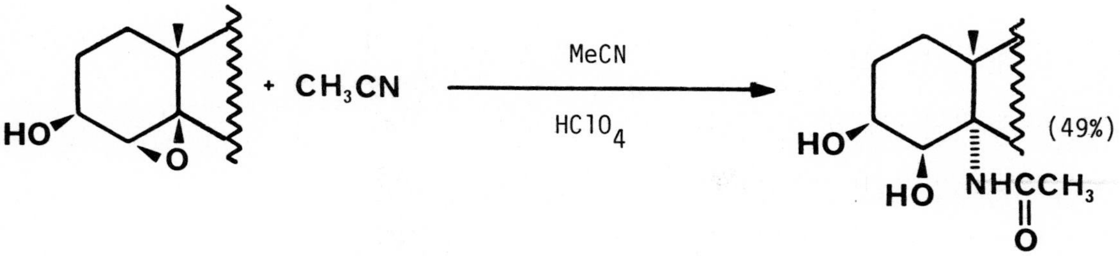

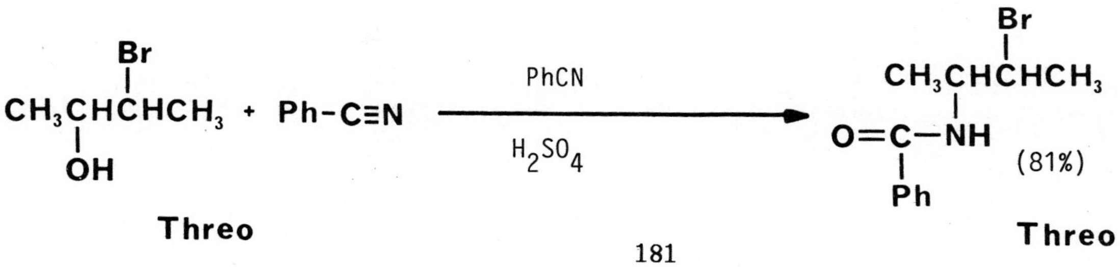

REACTION: ROBINSON ANNULATION REACTION

March's Advanced Organic Chemistry : 834

GENERAL SCHEME:

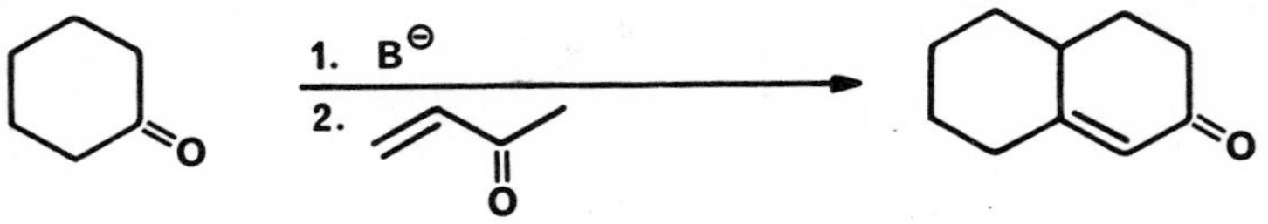

1. $B^{\ominus}$
2.

MECHANISM:

$-H_2O$

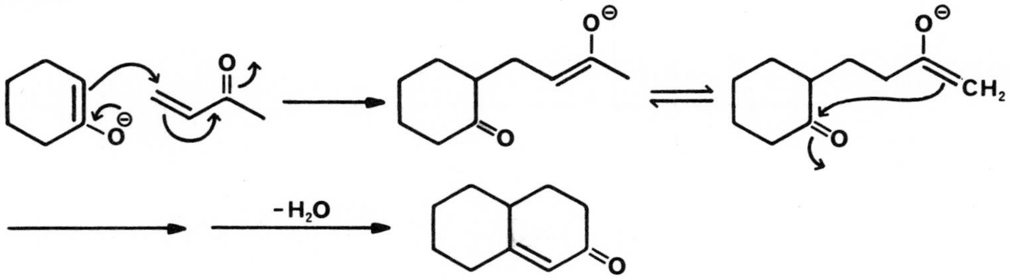

NOTES:

1. This is an example of an **Aldol Condensation** applied to ring formation.

2. Methyl 5-methoxy-3-oxopentanoate is known as **Nazarov's Reagent**.

3. Via:

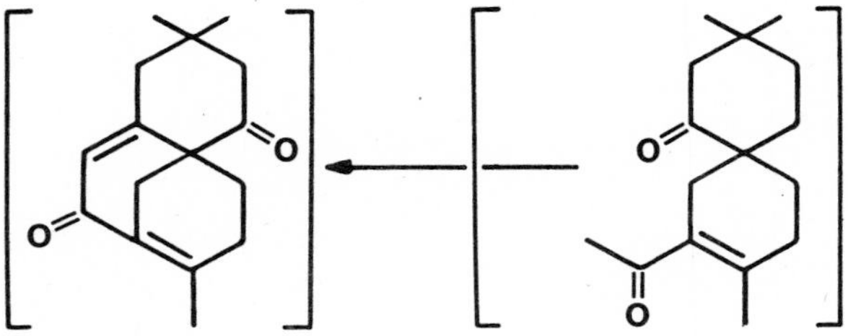

REFERENCES:

1. J. Ellis, J. Dutcher and C. Heathcock, **Synth. Commun.**, (1974), 4, 71.

2. W. Oppolzer, K. Battig and T. Hudlicky, **Helv.**, (1979), 62, 1493.

3. R.E. Ireland, C.A. Lipinski, C.J. Kowalski, J.W. Tilley and D.M. Walba, **J. Am. Chem. Soc.**, (1974), 96, 3333.

4. T. Honda, T. Murae, S. Ohta, Y. Kurata, H. Kawai, T. Takahashi, A. Itai and Y. Iitaka, **Chem. Letters**, (1981), 299.

5. Y. Houbrechts, P. Laszlo and P. Pennetreau, **Tetrahedron Lett.**, (1986), 705.

EXAMPLES:

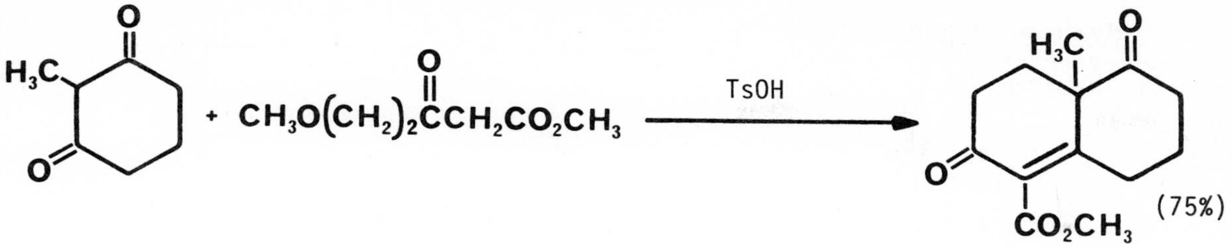

1

2

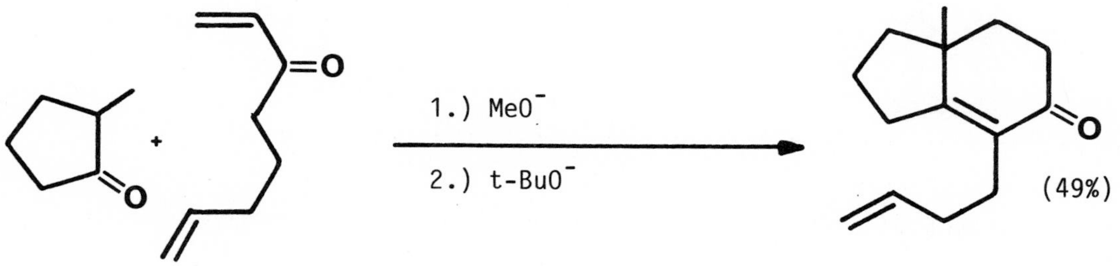

2

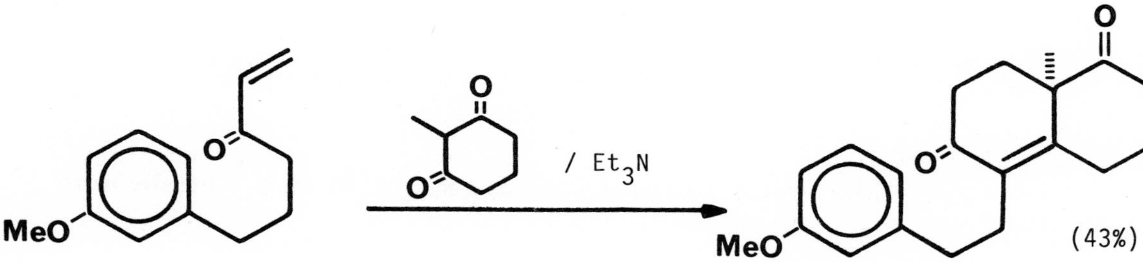

3

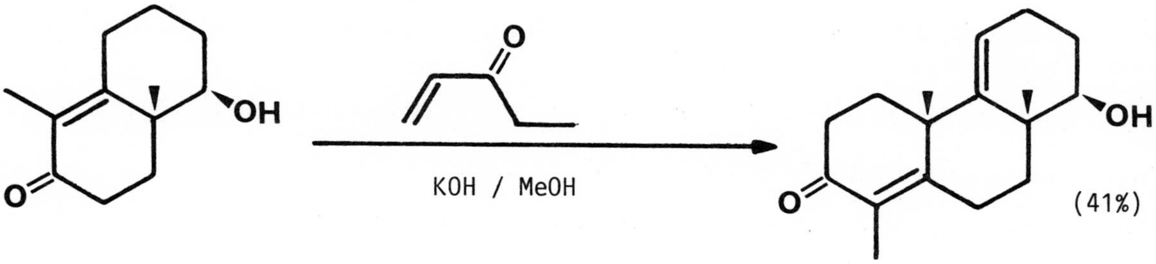

4

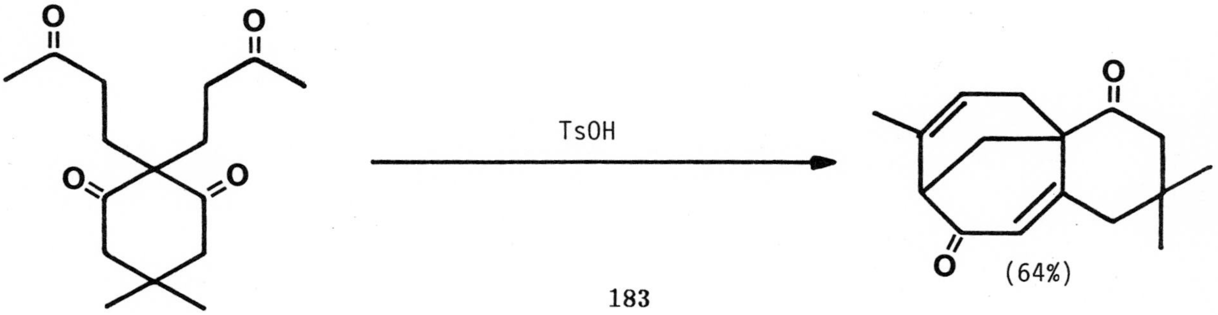

5

3

REACTION: ROSENMUND REDUCTION

March's Advanced Organic Chemistry : 396

GENERAL SCHEME:

⚠1

$$RC(=O)Cl \xrightarrow{H_2,\ Pd\text{-}BaSO_4} RC(=O)H$$

MECHANISM:

NOTES:

1. The use of 3-methyl-1-phenyl-2-phospholine allows an alternative for the reduction of aromatic acid chlorides:

$$\text{3-methyl-1-phenyl-2-phospholine (P-Ph)} + Ar\text{-}C(=O)\text{-}Cl \longrightarrow \text{phospholium } (\oplus P(Ph)C(=O)Ar)\ Cl^{\ominus} \xrightarrow{H_2O} Ar\text{-}C(=O)\text{-}H$$

D.G. Smith and D.J.H. Smith, **Chem. Commun.**, (1975), 459; F. Mathey, **Tetrahedron**, (1973), 29, 707.

REFERENCES:

1. A.W. Burgstahler, L.O. Weigel and C.G. Shafer, **Synthesis**, (1976), 767.

2. S. Danishefsky, M. Hinama, K. Gombatz, T. Harayama, E. Berman and P. Schirda, **J. Am. Chem. Soc.**, (1979), 101, 7020.

3. A.I. Rachlin, H. Gurien and D.P. Wagner, **Org. Synthesis**, (1971), 51, 8.

4. K.W. Rosenmund, **Ber.**, (1918), 51, 585.

5. See E. Mosettig and R. Mozingo in **Org. Reactions**, (1948), IV, 362.

EXAMPLES:

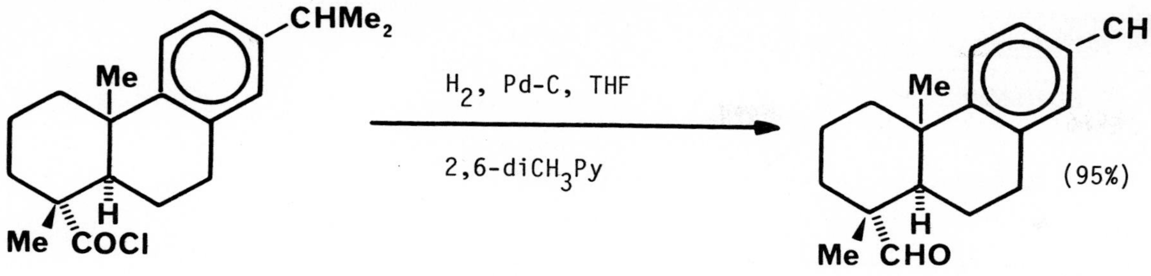
CHMe$_2$
Me
H
Me COCl
H$_2$, Pd-C, THF
2,6-diCH$_3$Py
CHMe$_2$
Me
H
Me CHO
(95%)
①

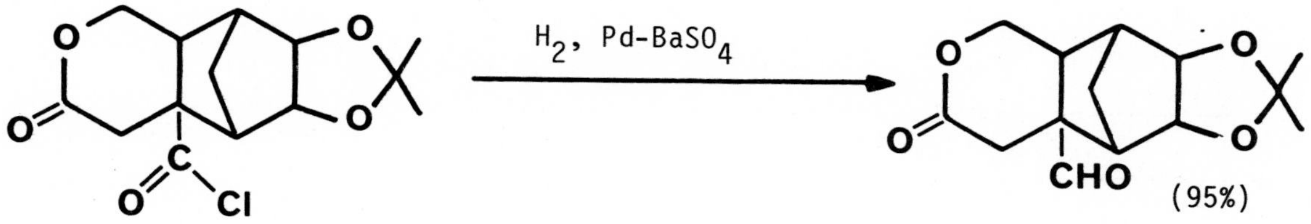
H$_2$, Pd-BaSO$_4$
CHO
(95%)
②

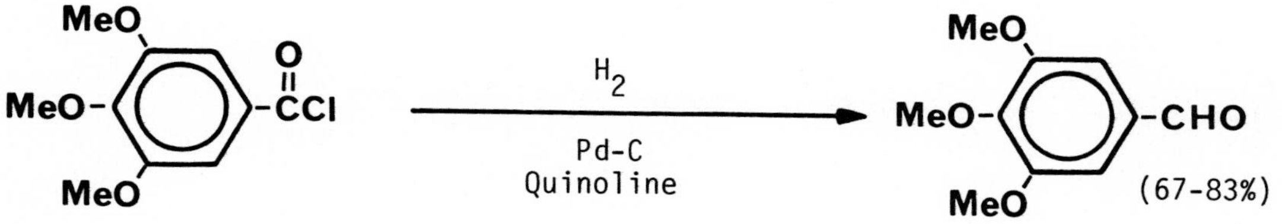
MeO
MeO
MeO
CCl
H$_2$
Pd-C
Quinoline
MeO
MeO
MeO
CHO
(67-83%)
③

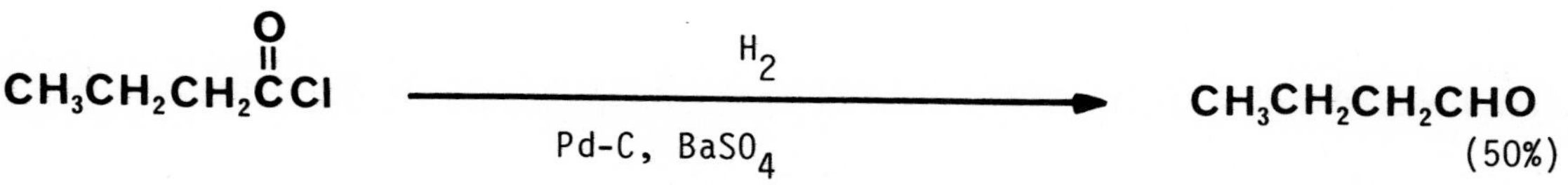
CH$_3$CH$_2$CH$_2$CCl
H$_2$
Pd-C, BaSO$_4$
CH$_3$CH$_2$CH$_2$CHO
(50%)
④

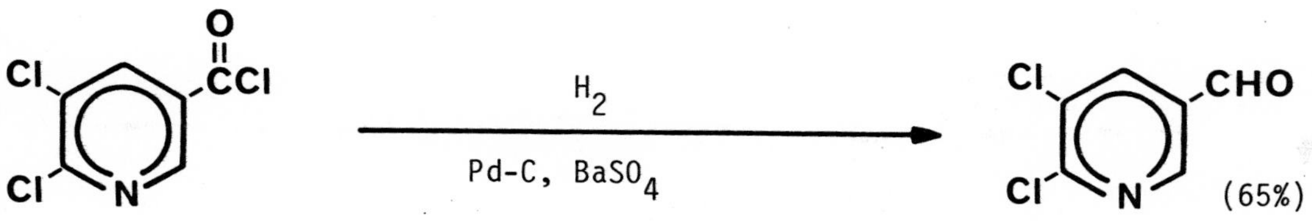
Cl
Cl
N
CCl
H$_2$
Pd-C, BaSO$_4$
Cl
Cl
N
CHO
(65%)
⑤

REACTION: RUPE REARRANGEMENT

===

March's **Advanced Organic Chemistry :** Not indexed

GENERAL SCHEME:

$$R-\overset{OH}{\underset{R}{C}}-C\equiv C-R' \xrightarrow{HCOOH} R-\underset{R}{C}=CH-\overset{O}{\overset{\|}{C}}-R'$$

MECHANISM:

$$R-\overset{OH}{\underset{R}{C}}-C\equiv C-R' \xrightarrow{H^{\oplus}} R-\overset{\oplus OH_2}{\underset{R}{C}}-C\equiv C-R' \longrightarrow \begin{matrix}R\\R\end{matrix}{>}C=C=\overset{\oplus}{C}-R'$$

$$\xrightarrow{H_2O} \begin{matrix}R\\R\end{matrix}{>}C=C=C{<}\begin{matrix}R'\\OH\end{matrix} \rightleftharpoons \begin{matrix}R\\R\end{matrix}{>}C=CH-\overset{O}{\overset{\|}{C}}-R'$$

NOTES:

1. The **Meyer- Schuster rearrangement** is the base- catalyzed S_Ni' reaction of similar alkynols.

$$R-C\equiv C-\underset{OH}{\overset{|}{C}}- \quad {}^{\ominus}OH \longrightarrow R-\overset{OH}{\overset{|}{C}}=C=\overset{|}{C}- \rightleftharpoons R-\overset{O}{\overset{\|}{C}}-CH=C{<}$$

REFERENCES:

1. J.D. Chanley, **J. Am. Chem. Soc.**, (1948), 70, 240.

2. W.S. Johnson, S.L. Gray, J.K. Crandall and D.M. Bailey, **J. Am. Chem. Soc.**, (1964), 86, 1966.

3. R.W. Hasbrouck and A.D.A. Kiessling, **J. Org. Chem.**, (1973), 38, 2103.

4. S.W. Pelletier and S. Prabhakar, **J. Am. Chem. Soc**, (1968), 90, 5318, and earlier references.

5. S.W. Pelletier and N.V. Mody, **J. Org. Chem.**,(1976), 41, 1069.

EXAMPLES:

HCOOH, Δ → (50%) ①

OH, Me, C≡CH, Me; HCOOH, Δ → O, Me, Me, Me (81%) ②

$CH_3CH_2C(CH_3)(OH)C{\equiv}CH$ $\xrightarrow[\Delta]{HCOOH}$ $CH_3CH{=}C(CH_3){-}C(=O){-}CH_3$ (60%) ③

OH, AcO, OAc; HCOOH, Δ → O, AcO, OAc (48%) ④

C≡CH; 1.) 80% HOAc, Ag_2CO_3, 90° 2.) Hydrolysis → H, CHO (Apx. 90%, 1:1 isomer mixture) ⑤

REACTION: SANDMEYER REACTION

March's Advanced Organic Chemistry : 477, 647, 648

GENERAL SCHEME:

$$ArN_2^{\oplus} + CuX \longrightarrow ArX$$

MECHANISM:

$$ArN_2^{\oplus}X^{\ominus} + CuX \longrightarrow Ar\cdot + N_2 + CuX_2$$

$$Ar\cdot + CuX_2 \longrightarrow ArX + CuX$$

NOTES:

1. A number of similar reactions of aromatic diazonium salts are possible.

REFERENCES:

1. J. Hartwell, **Org. Synth., III**, (1955), 185.

2. P.J. Harrington and L.S. Hegedus, **J. Org. Chem.** , (1984), 49, 2657.

3. H.T. Clarke and R.R. Read, **Org. Synth.,**, (1941), 1, 514.

4. N. Kornblum, **Org. Synth.,III**, (1955), 295.

5. H. Heaney and I.T. Millar, **Org. Synth.**, (1960), 40, 105.

EXAMPLES:

NH_2, Cl (2-chloroaniline) — 1.) HONO; 2.) CuBr → Br, Cl (1) (89-95%)

NH_2, Me, NO_2 — 1.) $NaNO_2$, HBr; 2.) CuBr → Br, Me, NO_2 (2) (94%)

NH_2, Me — 1.) HONO; 2.) CuCN → CN, Me (3) (64-70%)

H_2N–, Me, Me, –NH_2 — 1.) HONO; 2.) H_3PO_2 → Me, Me (4) (76-82%)

Br, NH_2 — 1.) HONO; 2.) KI → Br, I (5) (72-83%)

REACTION: SCHMIDT REARRANGEMENT

March's Advanced Organic Chemistry : 945, 986- 87.

GENERAL SCHEME:

⚠1

$$\mathrm{RCOH} \; (\mathrm{C{=}O}) + \mathrm{HN_3} \xrightarrow[\mathrm{H_2O}]{\mathrm{H^{\oplus}}} \mathrm{RNH_2}$$

MECHANISM:

$$\mathrm{RCOH} \xrightarrow[\mathrm{H^{\oplus}}]{\mathrm{HN_3}} \mathrm{RC(=O){-}N(H){=}N^{\oplus}{=}N^{\ominus}} \xrightarrow{-\mathrm{N_2}} \mathrm{R{-}\overset{\oplus}{N}(H){=}C{=}O} \xrightarrow{\mathrm{H_2O}} \mathrm{RNH_2} + \mathrm{CO_2}$$

NOTES:

1. There are actually three reactions listed as **Schmidt reactions**: (1) the addition of hydrazoic acid to carboxylic acids, (2) the addition of hydrazoic acid to aldehydes and ketones, and (3) the addition of hydrazoic acid to alkenes and alcohols.

REFERENCES:

1. R. Palmere and R. Conley, **J. Org. Chem.**, (1970), 35, 2703.

2. E. Moriconi and M. Stemniski, **J. Org. Chem.**, (1972), 37, 2035.

3. T. Sasaki, S. Eguchi and T. Toru, **J. Org. Chem.**, (1970), 35, 4109.

4. L.E. Fikes and H. Schecter, **J. Org. Chem.**, (1979), 44, 741.

5. E.A. Vogler and J.M. Hayes, **J. Org. Chem.**, (1979), 44, 3682.

EXAMPLES:

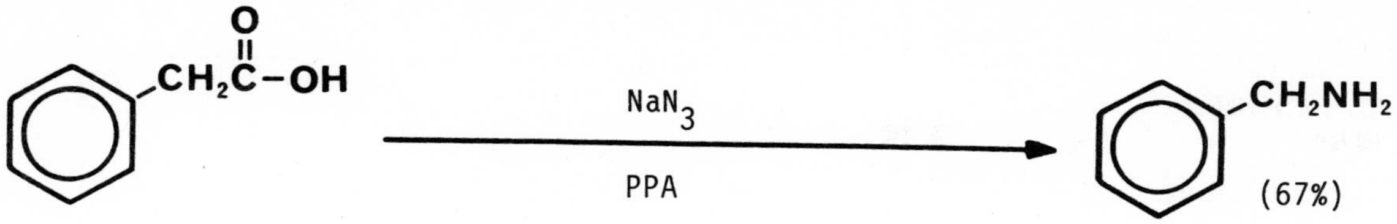

$C_6H_5CH_2COOH \xrightarrow[\text{PPA}]{NaN_3} C_6H_5CH_2NH_2$ (67%) ①

(Reaction scheme: Me O … –Me, H → NaN$_3$, H$_2$SO$_4$ / Benzene → Me O, NH, Me, H (30%)) ②

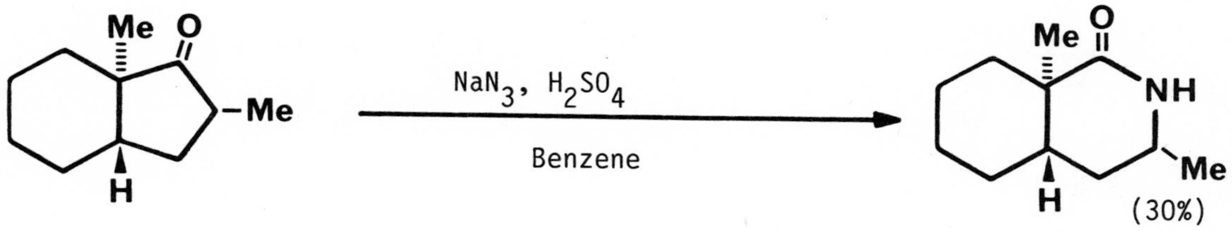

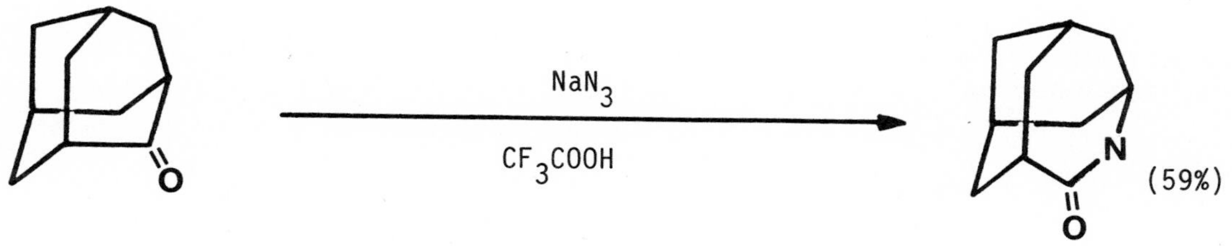

$\xrightarrow[CF_3COOH]{NaN_3}$ N, O (59%) ③

$EtCPh$ (C=O) $\xrightarrow{HN_3}$ $EtC(O)NHPh$ + $EtNHC(O)Ph$ (NYA) ④

$CH_3(CH_2)_6COOH \xrightarrow{HN_3} CH_3(CH_2)_5CH_2NH_2$ (NYA) ⑤

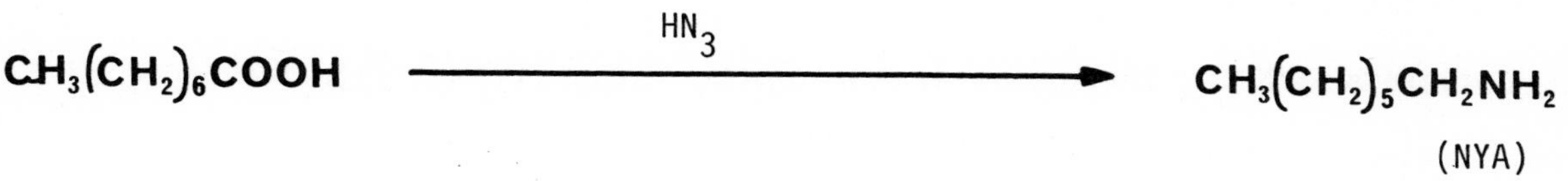

REACTION: SHARPLESS EPOXIDATION

March's Advanced Organic Chemistry : 736

GENERAL SCHEME:

$$HO(CH_2)_nC(R){=}C(R){-}R \xrightarrow[VO(acac)_2 \text{ or } Mo(CO)_6]{t\text{-BuOOH}} HO(CH_2)_nC(R)\overset{O}{—}C(R){-}R$$

MECHANISM:

$$\underset{\text{HO}}{\text{C=C}} \xrightarrow[VO(acac)_2]{t\text{-BuOOH}}$$

NOTES:

1. The use of tartrate esters results in enantioselective epoxidation of the alkene. This can be formulated as:

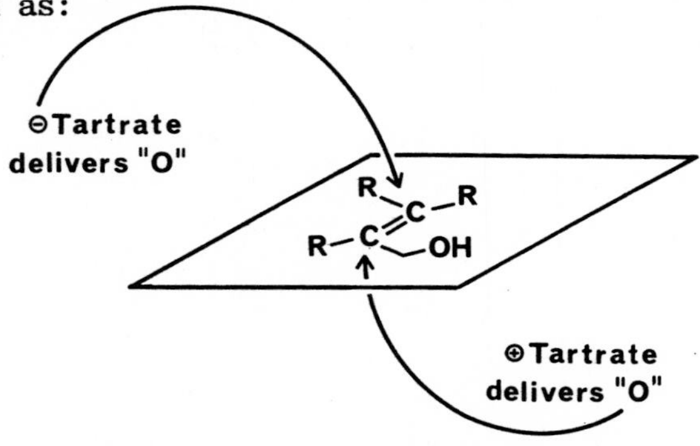

REFERENCES:

1. S. Danishefsky, R. Zamboni, M. Kahn, and S.J. Etheredge, **J. Am. Chem. Soc.**, (1981), 103, 3460.

2. E.J. Corey, A. Marfat, J.R. Falck and J.O. Albright, **J. Am. Chem. Soc.**, (1980), 102, 1433.

3. T. Nakata, G. Schmid, B. Vranesis, M. Okigawa, T. Smith-Palmer and Y. Kishi, **J. Am. Chem. Soc.**, (1978), 100, 2933.

4. T. Katsuki and K.B. Sharpless, **J. Am. Chem. Soc.**, (1980), 102, 5974.

5. B.E. Rossiter, T. Katsuki and K.B. Sharpless, **J. Am. Chem. Soc.**, (1981), 103, 464.

EXAMPLES:

t-BuOOH / VO(acac)$_2$ ①

(65%)

t-BuOOH / VO(acac)$_2$

Benzene ②

(66%)

1.) t-BuOOH / VO(acac)$_2$

2.) AcOH ③

(NYA)

Me$_3$COOH, Ti(O-iPr)$_4$

CH_2Cl_2, -20° ④

(77%)

(+) Diethyltartrate

"Standard conditions" ⑤ △

(80%, 91% ee)

REACTION: SIMMONS- SMITH REACTION

===

March's Advanced Organic Chemistry : 772- 73

GENERAL SCHEME:

$$>C=C< + CH_2I_2 \xrightarrow{Zn-Cu} >C(CH_2)C<$$

MECHANISM:

$$CH_2I_2 + ZnCu \longrightarrow ICH_2ZnI$$

$$>C=C< \xrightarrow{ICH_2ZnI} \left[-\overset{I}{C}-\overset{CH_2ZnI}{C}- \right] \longrightarrow >C(CH_2)C<$$

NOTES:

1. The **Simmons-Smith reaction** is generally subject to steric effects, and the cyclopropanation will generally take place from the less-hindered side. If there is a neighboring hydroxyl group, this will both speed up the reaction as well as place the cyclopropane ring syn to the hydroxyl group.

REFERENCES:

1. E. Wenkert, R.A. Mueller, E.J. Reardon, Jr., S.S. Sathe, D.J. Scharf ad G. Tosi, **J. Am. Chem. Soc.**, (1970), 92, 7428.

2. L.K. Bee, J. Beeby, J.W. Everett and P.J. Garratt, **J. Org. Chem.**, (1975), 40, 2212.

3. C. Filliatre and C. Gueraud, **C.R. Acad. Sci, C**, (1971), 273, 1186.

4. W.G. Dauben and A.C. Ashcroft, **J. Am. Chem. Soc.**, (1963), 85, 3673.

5. R.J. Rawson and I.T. Harrison, **J. Org. Chem.**, (1970), 35, 2057.

EXAMPLES:

OMe — CH_2I_2, I_2 / [Zn-Cu] → OMe (91%) ①

CH_2I_2 / [Zn-Cu] → (35%) ②

CH_2, OH — CH_2I_2 / [Zn-Cu] → OH (62%) ③

OH — CH_2I_2 / [Zn-Cu] → OH (23%) ④

CH_2I_2 / Zn, CuCl → (92%) ⑤

REACTION: SKRAUP QUINOLINE SYNTHESIS

March's Advanced Organic Chemistry : Not indexed

GENERAL SCHEME:

MECHANISM:

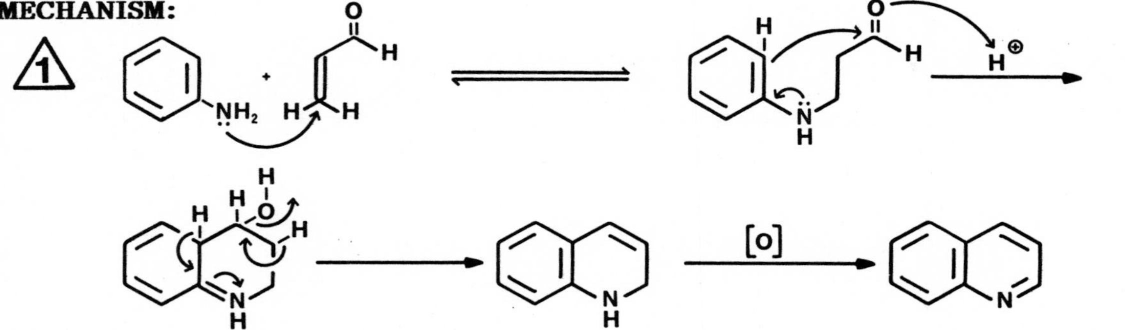

NOTES:

1. By running the reaction in nitrobenzene there seems to be sufficient oxidizing ability to complete the aromatization of the heterocyclic ring.

2. In the original preparation, acrolein was prepared, in situ, from the dehydration of glycerol.

3. **PPE** is formed from polyphosphoric acid and ethanol.

4. This reaction is similar to the **Dobner von Miller** (1) and the **Conrad-Limpach** (2) quinoline syntheses.

REFERENCES:

1. H. Rapoport and A.D. Batcho, **J. Org. Chem.**, (1963), 28, 1753.

2. W.P. Utermohlen, **J. Org. Chem.**, (1943), 8, 544.

3. E.B. Mullock, R. Searby and H. Suschitzky, **J. Chem. Soc.**, C, (1970), 829.

4. J.-C. Perche, G. Saint-Ruf and N.P. Bun-Hoi, **J. Chem. Soc.**, Perkin I, (1972), 261.

EXAMPLES:

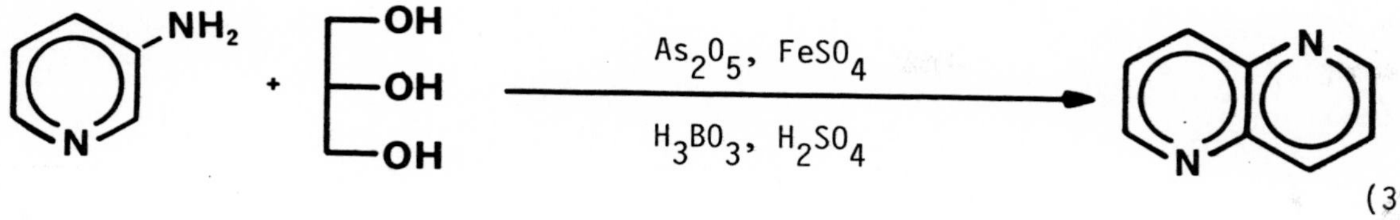

3-aminopyridine + glycerol ($HOCH_2CH(OH)CH_2OH$) —[As_2O_5, $FeSO_4$ / H_3BO_3, H_2SO_4]→ 1,5-naphthyridine (31%) ① △2

aniline + $CH_2=C(CHO)-Et$ —[H_2SO_4 / m-Nitrosulfonate]→ 3-ethylquinoline (42%) ②

amino-tetrahydropyrido[1,2-a]benzimidazole (NH_2) —[$CH_3C(O)CH_2(Me)CO_2Et$ / PPE]→ OH, Me, Me-substituted fused quinoline (80%) ③ △3

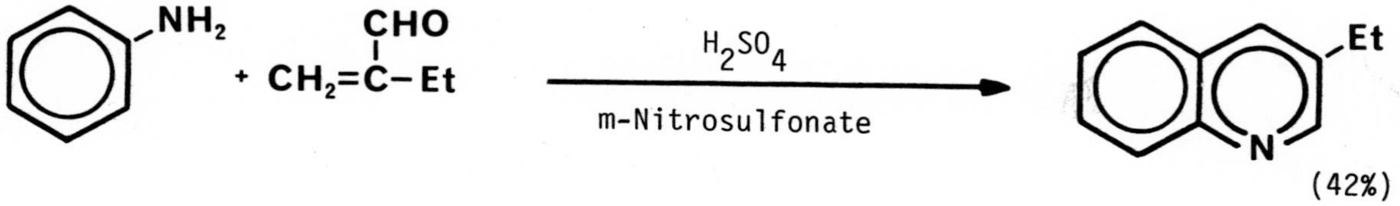

Me-substituted N-Et carbazole amine (NH_2) + glycerol —[H_2SO_4 / $PhNO_2$, △]→ Me, N-Et pyridocarbazole (Yield described as "excellent") △2 ④

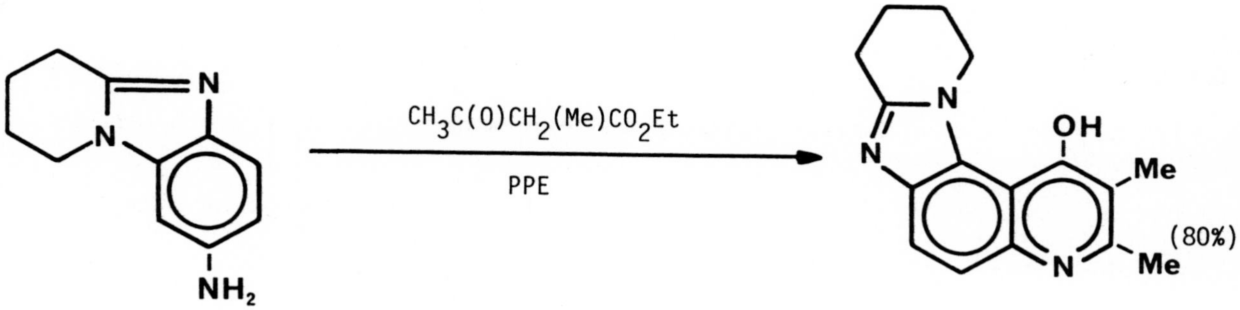

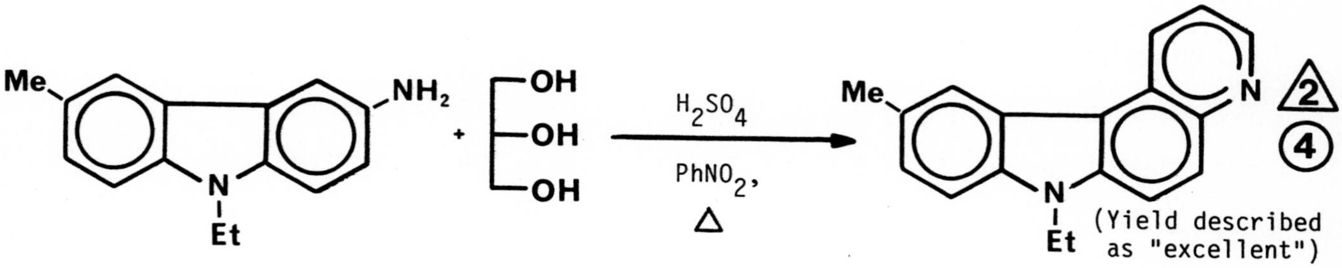

REACTION: SOMMELET- HAUSER REARRANGEMENT

March's Advanced Organic Chemistry : 605

GENERAL SCHEME:

$C_6H_5CH_2\overset{\oplus}{N}CH_3R_2 \xrightarrow{NH_2^{\ominus}}$ o-$CH_3C_6H_4CH_2NR_2$

MECHANISM:

$C_6H_5CH_2-\overset{\oplus}{N}R_2-CH_3 \xrightarrow{NH_2^{\ominus}} C_6H_5CH_2-\overset{\oplus}{N}R_2-\overset{\ominus}{C}H_2 \longrightarrow$

(exo-$=CH_2$, H, CH_2NR_2 cyclohexadiene) $\longrightarrow$ o-$CH_3C_6H_4CH_2NR_2$

NOTES:

1. The product can be further alkylated and rearranged until there are no ortho positions left.

2. The reaction may take place with sulfonium salts instead of ammonium salts.

3. **n-BuLi- TMDEA** = n-Butyllithium- N,N,N',N'-tetramethylenediamine.

4. The reaction may compete with the **Stevens Rearrangement**.

REFERENCES:

1. J. Biellmann and J. Schmitt, **Tetrahedron Lett.**, (1973), 4615.

2. G.C. Jones and C.R. Hauser, **J. Org. Chem.**, (1962), 27, 3572.

3. S.W. Kantor ad C.R. Hauser, **J. Am. Chem. Soc.**, (1951), 73, 4122.

4. C.R. Hauser and A.J. Weinheimer, **J. Am. Chem. Soc.**, (1954), 76, 1264.

5. W.Q. Beard, Jr., and C.R. Hauser, **J. Org. Chem.**, (1960), 25, 334.

EXAMPLES:

$C_6H_5-CH_2SCH_2-C_6H_5$ → 1.) n-BuLi-TMEDA, HMPT, -78° 2.) CH_3I → o-$CH_3C_6H_4-CH(SCH_3)-C_6H_5$ (57%) ① △3

2-phenyl-1,1-dimethylpiperidinium ($N^{\oplus}$, H_3C, CH_3) → $NaNH_2$ / NH_3 → N-CH_3 benzo-fused azacycle (83%) ②

$C_6H_5CH_2-\overset{\oplus}{N}(CH_3)_2-CH_2C_6H_5$ → 2eq. $NaNH_2$ / NH_3 → o-$H_3C-C_6H_4-CH(N(CH_3)_2)-C_6H_5$ (95%) ③

$C_6H_5CH_2\overset{\oplus}{N}Me_3$ → $NaNH_2$ / NH_3 → o-$Me-C_6H_4-CH_2NMe_2$ (97%) ④

p-$MeO-C_6H_4-CH_2\overset{\oplus}{N}Me_3$ → $NaNH_2$ / NH_3 → MeO, Me, CH_2NMe_2 substituted benzene (93%) ⑤

REACTION: STEPHEN REDUCTION

March's Advanced Organic Chemistry : 816

GENERAL SCHEME:

$$R-C\equiv N \xrightarrow[\text{2. Hyd}]{\text{1. HCl, SnCl}_2} R-\overset{O}{\overset{\|}{C}}-H$$

MECHANISM:

$$R-C\equiv N \cdot HCl \longrightarrow R-\overset{Cl}{\overset{|}{C}}=\overset{\oplus}{N}H_2\ Cl^{\ominus} \longrightarrow R-CH=NH \longrightarrow R-CHO$$

NOTES:

1. The reaction is most useful when **R-** is an aromatic group. The reaction tends to lose its usefulness for **R-** = alkyl.

2. The R-CH=NH precipitates out as a $SnCl_4$ salt.

3. The **Sonn-Muller Reaction** is a variation of this same process.

$$C_6H_5-\overset{O}{\overset{\|}{C}}-NHPh \xrightarrow{PCl_5} C_6H_5-\overset{Cl}{\overset{|}{C}}=\overset{\oplus}{N}HPh$$

4. Overreduction to the amine is not uncommon.

REFERENCES:

1. J. Williams, **Org. Synth.**, Coll. Vol. 3, (1955), 626.

2. J. Williams, C. Witten and J. Krynitsky, ibid, 818.

3. H. Stephen, **J. Chem. Soc.**, (1925), 1874.

4. C.G. Stuckwisch, **J. Org. Chem.**, (1972), 37, 318.

5. T.S. Gardner, F.A. Smith, E. Wenis and J. Lee, **J. Org. Chem.**, (1951), 16, 1121.

EXAMPLES:

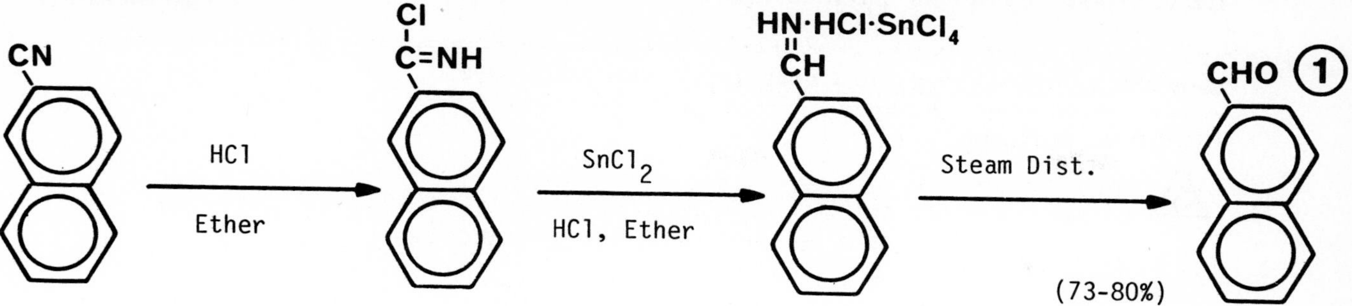

O
C–NHPh
Me
PCl_5
C_6H_6
Cl
C=NPh
Me
$SnCl_2$
HCl, Ether
CH=NPh
Me
Steam Dist.
CHO
Me
2
(62-70%)

CH_2CN
Me
1.) $SnCl_2$
2.) HCl, Hyd.
CH_2CHO
Me
3
("nearly Quant.")

O
CN
1.) $SnCl_2$, HCl, Ether
2.) H_2S, H_2O
O
$CH_2\overset{\oplus}{N}H_3\overset{\ominus}{Cl}$
4
(67%)

CN
N
$SnCl_2$ / Thiosemicarbazide
HCl, Ether
O
C–NH–NHCNH$_2$
SO
N
(86%)
5

REACTION: STEVENS REARRANGEMENT

==

March's Advanced Organic Chemistry : 605, 992- 994

GENERAL SCHEME:

$$\text{EWG-CH}_2\text{-}\overset{\text{R}}{\underset{\text{R}''}{\text{N}^{\oplus}}}\text{-R}' \xrightarrow{\text{Base}} \text{EWG-}\overset{\text{R}''}{\text{CH}}\text{-N}\langle^{\text{R}}_{\text{R}'}$$

MECHANISM:

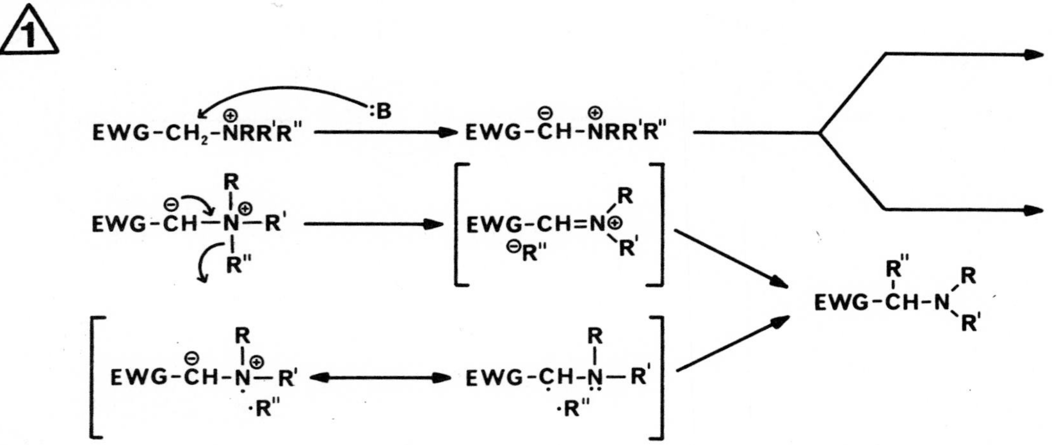

NOTES:

1. Both mechanisms have some acceptance.

2. CH_3-SO-CH_2^- Na^+ = Sodium methylsulfinylmethylide = **"Dimsyl sodium"**.

3. **Phenacyl > propargyl > allyl > benzyl > alkyl** as the migration terminus.

4. Evidence for the intramolecular character of the reaction.

REFERENCES:

1. S. Kano, T. Yokomatsu, E. Komiyama, S. Tokita, Y. Takahagi and S. Shibuya, **Chem. Pharm. Bull. Japan**, (1975), 23, 1171.

2. E. Grovenstein, Jr. and G. Wentworth, **J. Chem. Soc.**, (1963), 3397.

3. P.C. Anderson and B. Staskun, **J. Org. Chem.**, (1965), 9, 3033.

4. C.R. Hauser, R.M. Manyik, W.R. Brasen and P.L. Bayless, **J. Org. Chem.**, (1955), 20, 1119.

5. W.H. Puterbaugh and C.R. Hauser, **J. Am. Chem. Soc.**, (1964), 86, 1394.

EXAMPLES:

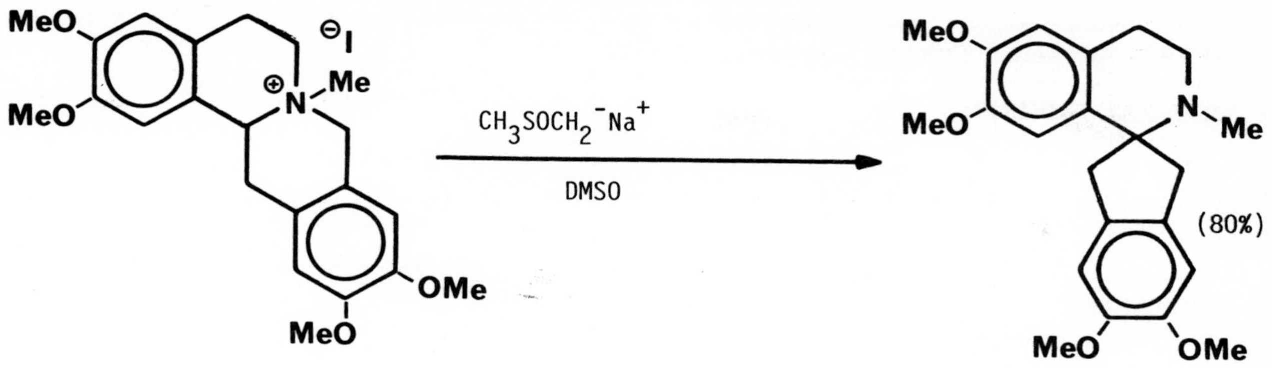

MeO, MeO, $\overset{\ominus}{I}$, $\overset{\oplus}{N}$–Me, OMe, MeO $\xrightarrow[\text{DMSO}]{CH_3SOCH_2^-Na^+}$ MeO, MeO, N–Me, (80%), MeO, OMe ① △2

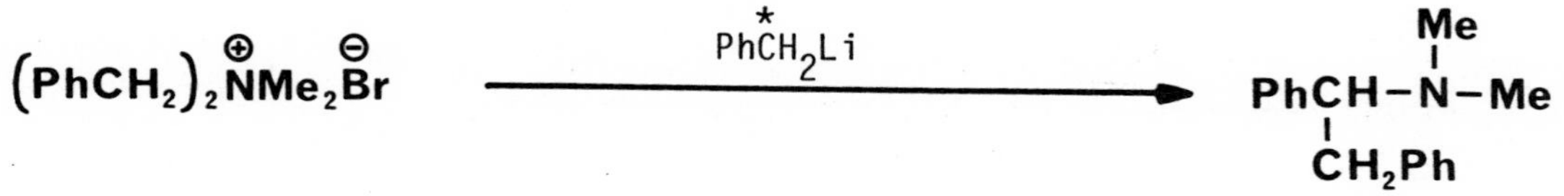

$(PhCH_2)_2\overset{\oplus}{N}Me_2\overset{\ominus}{Br}$ $\xrightarrow{Ph\overset{*}{C}H_2Li}$ PhCH(CH₂Ph)–N(Me)–Me ② △4

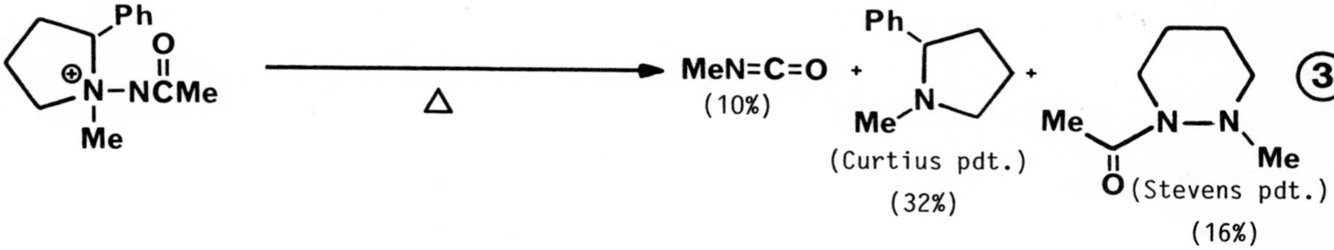

Ph, $\overset{\oplus}{N}$–NC(=O)Me, Me $\xrightarrow{\Delta}$ MeN=C=O (10%) + Ph, Me–N (Curtius pdt.) (32%) + Me–C(=O)–N—N–Me (Stevens pdt.) (16%) ③

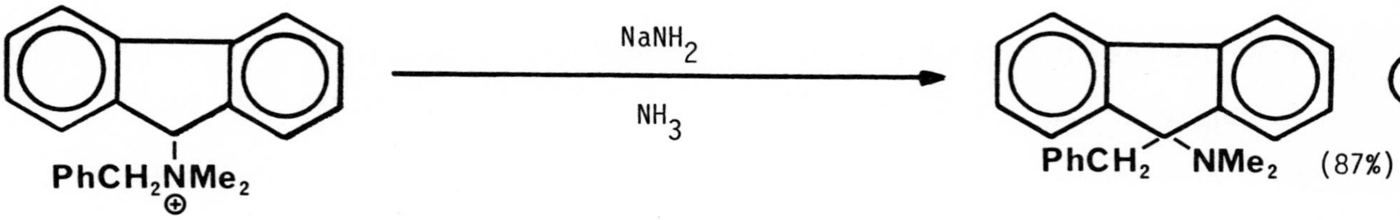

$PhCH_2\underset{\oplus}{N}Me_2$ $\xrightarrow[NH_3]{NaNH_2}$ $PhCH_2$ NMe_2 (87%) ④

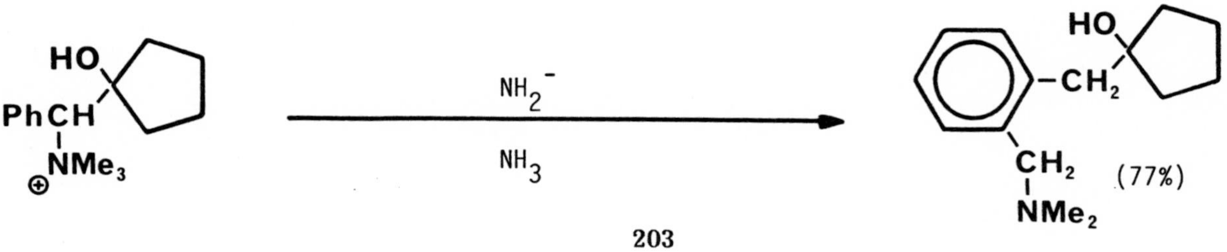

HO, PhCH, $\underset{\oplus}{N}Me_3$ $\xrightarrow[NH_3]{NH_2^-}$ HO, –CH_2, CH_2, NMe_2 (77%) ⑤

REACTION: STOBBE CONDENSATION

==

March's Advanced Organic Chemistry: 835

GENERAL SCHEME:

1

$$EtO\overset{O}{\overset{\|}{C}}CH_2CH_2\overset{O}{\overset{\|}{C}}OEt \xrightarrow[:B]{R\overset{O}{\overset{\|}{C}}R'} (R)(R')C{=}C(CO_2Et)(CH_2CO_2Et)$$

MECHANISM:

$$EtO_2CCH_2CH_2CO_2Et \xrightarrow{:B} EtO_2C\overset{\ominus}{C}HCH_2CO_2Et + R\overset{O}{\overset{\|}{C}}R' \longrightarrow$$

$$EtO_2C\text{–}CH(CH_2COEt{=}O)\text{–}C(R)(R')\text{–}O^{\ominus} \longrightarrow \text{lactone} \xrightarrow[H_3O^{\oplus}]{:B} (R)(R')C{=}C(CO_2Et)(CH_2CO_2H)$$

NOTES:

1. The **Stobbe Condensation** has the limitation that α-keto esters cannot be used because of their sensitivity to base. This can be remedied by using diethyl lithiosuccinate, generated from the ethyl succinate and **LDA** at -78°.

REFERENCES:

1. R. Baker, P.H. Briner and D.A. Evans, **Chem. Commun.**, (1978), 410.

2. W.S. Johnson, J.W. Peterson and C.D. Gutsche, **J. Am. Chem. Soc.**,(1947), 69, 2942.

3. V.B. Bachos, F.H. Nasr and M. Gindy, **Helv.**, (1979), 62, 90.

4. G. Bagavant and K.B.V. Swaminathan, **Current Science**, (1975), 44, 661.

5. M.F. El-Newaihy, M.R. Salem and F.A. El-Bassiony, **Aust. J. Chem.**, (1976), 29, 223.

EXAMPLES:

① + CO_2Et / CO_2Et —(t-BuO⁻)→ CO_2Et, CO_2Et (68%)

CHO

t-BuO$^-$

② Me, CN, O, MeO + $(CH_2CO_2Me)_2$ —(t-BuOK, t-BuOH)→ Me, O, CO_2Me, MeO (79%)

t-BuOK, t-BuOH

③ OMe, O, Cl + CH_2CO_2Et / CH_2CO_2Et

1.) t-BuOK, t-BuOH

2.) t-BuOK, t-BuOH
Reflux, 45 min

OMe, Cl, CH_2CO_2H, CO_2Et (90%)

④ CHO, OMe + $CH_2CH_2CO_2Et$ / C=O / $CH_2CH_2CO_2Et$

K^+ / t-BuOH

O=C-$CH_2CH_2CO_2Et$
HC=C-CH_2CO_2Et
OMe (90%)

⑤ CHO, OMe + CH_2CO_2Et / O / CH_2CO_2Et

1.) NaH, C_6H_6

2.) t-BuOK / t-BuOH
Reflux

CO_2Et
HC=C-OCH_2CO_2H
OMe (80%)

REACTION: STORK ENAMINE SYNTHESIS

March's Advanced Organic Chemistry: 540- 42

GENERAL SCHEME:

1. RX
2. $H_3O^{\oplus}$

MECHANISM:

R–X

$H_3O^{\oplus}$

NOTES:

1. Water can be removed using an azeotrope and water separator or can be removed by allowing the reaction mixture to stand over molecular sieves.

2. This is an equilibrium process, thus addition of excess water in acid will convert the enamine to the starting carbonyl compound.

REFERENCES:

1. J. Gore and A. Doutheau, **Tetrahedron Lett.**, (1972), 4545.

2. A. Doutheau and J. Gore, **Tetrahedron Lett.**, (1976), 2703.

3. G. Stork, A. Brizzolara, H. Landesman, J. Szmuskovicz and R. Terrell, **J. Am. Chem. Soc.**, (1963), 85, 207.

4. E. Demole and M. Winter, **Helv.**,(1962), 45, 1256.

5. H.J. Davis and B.G. Main, **J. Chem. Soc.**,C, (1970), 327.

EXAMPLES:

+ $ClCH_2CH{=}C(CH_3)C{\equiv}CH$ —(H_3O^+)→ (96%) ①

+ $BrCH_2CH{=}C(Me)C{\equiv}CH$ —(HCl)→ CO_2Et (100%) ②

+ $ClCO_2Et$ —(Dry benzene / N_2)→ CO_2Et (62%) ③

+ $BrCH_2C{\equiv}CCH_2CH_3$ —(H_3O^+)→ $CH_2C{\equiv}CCH_2CH_3$ (83%) ④

+ $BrCH_2CH{=}CH_2$ —(Dry acetonitrile, Reflux)→ $CH_2CH{=}CH_2$ (76%) ⑤

REACTION: THORPE (ZIEGLER) REACTION

March's Advanced Organic Chemistry: 834, 1116.

GENERAL SCHEME:

MECHANISM:

NOTES:

1. The enamine form of the product has been shown to be more stable than the imine form.

2. When the reaction is run intermolecularly it is known as the **Thorpe** reaction, when it is run intramolecularly it is called the **Thorpe-Ziegler** reaction.

3. This reaction is very similar to the **Dieckmann Cyclization.** It generally allows for higher yields of products.

4. A comparison of ring-size and yields (from Concepts of Organic Synthesis, B.P. Mundy, Marcel Dekker, Inc., New York, N.Y., 1979, page 55).

REFERENCES:

1. T. Doornbos and J. Strating, **Synth. Commun.**, (1971), 1, 193.

2. J.J. Bloomfield and P.V. Fennessey, **Tetrahedron Lett.**, (1964), 2273.

EXAMPLES:

Na^+PhN^-Me

(16%)

①

1.) $Me\text{-}SO\text{-}CH_2^-$

2.) H_2O

(92%)

②

Ring Size	Thorpe-Ziegler	Dieckmann	Acyloin
	Percent Yield		
6	95	---	58
7	96	47	52
8	89	15	36
9	0	0	38
10	0	0	52
11	2	0.5	53
12	---	0.5	80
13	14	24	80
14	57	32	85

REACTION: TIFFENEAU- DEMJANOV REARRANGEMENT 

March's Advanced Organic Chemistry : 965

GENERAL SCHEME:

MECHANISM:

NOTES:

1. Sometimes **Demyanov**

2. This is closely related to the **Semi-pinacolonic Deamination**

3. If the reaction is only the carbonium ion rearrangement of a diazonium intermediate, it is called the **Demjanov** reaction.

REFERENCES:

1. R.B. Woodward, J. Gosteli, I. Ernest, R.J. Friary, G.N. Nestler, H. Raman, R. Sitrin, C. Suter and J.K. Whitesell, **J. Am. Chem. Soc.**, (1973), 95, 6853.

2. F. Blicke, J. Azuara, N. Dorrenbos and E. Hotelling, **J. Am. Chem. Soc.**, (1953), 75, 5418.

3. A. Nickon and G. Stern, **Tetrahedron Lett.**, (1985), 5915.

4. J.D. Rogers and W.F. Gorham, **J. Am. Chem. Soc.**, (1952), 74, 2278.

5. R.G. Murray,Jr. and T.M. Ford, **J. Org. Chem.**, (1979), 44, 3504.

EXAMPLES:

OMe

NH_2 MeOH H^+ HO OH $NH_3^{\oplus}$ OH

HONO OMe ① HO CHO (80%)

CH_2NH_2 OH HONO =O (61%) ②

HONO ③ OH CH_2NH_2 (98%) O

OH CH_2NH_2 HONO O (47%) + =O (8%) ④

OH CH_2NH_2 $NaNO_2$ HOAc =O (70%) ⑤

REACTION: ULLMANN REACTION

March's Advanced Organic Chemistry : 597- 98

GENERAL SCHEME:

$$2 \cdot ArX \xrightarrow[\Delta]{Cu} Ar{-}Ar$$

MECHANISM:

$$ArX + Cu \longrightarrow Ar\cdot \longrightarrow ArCu \xrightarrow{ArX} Ar{-}Ar$$

NOTES:

1. Aryl halides react in the order: **Ar-I > Ar-Br > Ar-Cl.**

2. Nickel has been used in place of copper.

REFERENCES:

1. M. Semmelhack, P. Helqist, L. Jones, L. Keller, L. Mendelson, L. Ryono, J. Smith ad R. Stauffer, **J. Am. Chem. Soc.**, (1981), 103, 6460.

2. J. Cornforth, A. Sierakowski and T. Wallace, **J.C.S., Chem. Commun.**, (1979), 294.

3. R.C. Fuson and E.A. Cleveland, **Org. Synth., III**, (1955), 339.

4. Reported in a Review: M. Tashiro, **Synthesis**, (1979), 933.

5. P.M. Everitt, D.M. Hall and E.E. Turner, **J. Chem .Soc.**, (1956), 2286.

EXAMPLES:

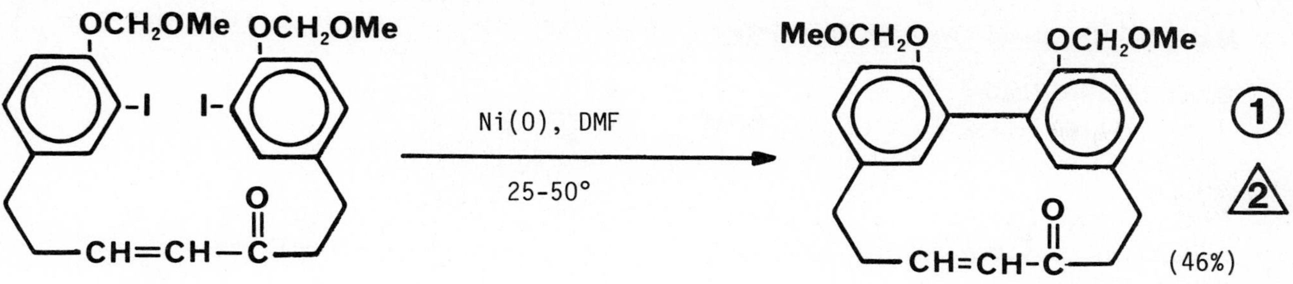

O_2N ... NO_2 + OMe ... I

Cu-O-CMe_3

Pyridine, DME

NO_2 ... NO_2 ... OMe (95%) ②

NO_2, Cl

Copper-Bronze

215-225°

O_2N ... NO_2 (52-61%) ③

I, Me, Me, *t*-Bu

Cu

Me, Me, *t*-Bu, *t*-Bu, Me, Me (NYA) ④

Et, I

Cu-Bronze

Et, Et (60%) ⑤

REACTION: VILSMEIER REACTION 

March's Advanced Organic Chemistry : 539, 732

GENERAL SCHEME:

$R_2\overset{\oplus}{N}=C(R')(OPOCl_2) + {}^{\ominus}Cl \longrightarrow R_2\overset{\oplus}{N}=C(R')Cl + {}^{\ominus}OP(O)Cl_2$

MECHANISM:

$(H)RC(=O)-NR'R'' \xrightarrow{POCl_3} (H)RC(OPOCl_2)=\overset{\oplus}{N}R'R''\ Cl^{\ominus} \longrightarrow (H)R-C(OPOCl_2)(Ar)-NR'R''-H$

$\longrightarrow (H)RC(=O)-Ar + H_2\overset{\oplus}{N}R'R''$

NOTES:

1. Sometimes referred to as the **Vilsmeier-Haack Reaction.**
2. DMF- $POCl_3$ is known as the **Vilsmeier Reagent**.
3. The reaction gives best results when the aromatic ring is activated.
4. Via:

$\left[Me_2N-C(Cl)(H)-N\text{(isoxazolone, Me, Ph)} \longrightarrow Me_2N-C(Cl)=N-C(Me)=C(Ph)-CO_2H \right]$

REFERENCES:

1. D. Reid, R. Webster and S. McKenzie, **J. Chem. Soc.,Perkin I**, (1979), 2334.
2. C.P. Traas, H.J. Takken and H. Boelens, **Tetrahedron Lett.**, (1977), 2129.
3. E.M. Becalli, A. Marchesini and H. Molinari, **Tetrahedron. Lett.**, (1986), 627.
4. G.F. Smith, **J. Chem. Soc.**, (1954), 3842.
5. T. Shono, Y. Matsumura, K. Tsubata and Y. Sugihara, **Tetrahedron Lett.**, (1975), 3391.

EXAMPLES:

+ Me_2N-CHO + NaHSe —($POCl_3$)→ (46%) ①

—($POCl_3$, DMF)→ CHO ② △2

1.) DMF-$POCl_3$, CCl_4, △ 2.) -HCl

Me, Ph, HN, O, O → Me, N, Ph, Me_2N, O, O (60-85%) ③ △4

—(DMF, $POCl_3$)→ CHO, N H (83%) ④

AcO —($POCl_3$, DMF)→ AcO, O, CHO ⑤

REACTION: VINYLCYCLOPROPANE- CYCLOPENTENE REARRANGEMENT

==

March's Advanced Organic Chemistry : 1019

GENERAL SCHEME:

$$\text{vinylcyclopropane} \xrightarrow{\Delta} \text{cyclopentene}$$

MECHANISM:

1

D D D D D D

NOTES:

1. There is still controversy over the mechanism of this reaction, and both a diradical and a concerted [1,3] sigmatropic migration have been considered.

REFERENCES:

1. E.N. Marvel and C. Lin, **J. Am. Chem. Soc.**, (1978), 100, 877.

2. L.A. Paquette, G.J. Wells, K.A. Horn and T.H. Yan, **Tetrahedron Lett.**, (1982), 263.

3. T. Hudlicky, T.M. Kutchan, S.R. Wilson and D.T. Mao, **J. Am. Chem. Soc.**, (1980), 102, 6351.

4. T.C. Shields, W.E. Billups and A.R. Lepler, **J. Am. Chem. Soc.**, (1968), 90, 4749.

5. T. Hudlicky, T.M. Kutchan and S.M. Naqvi, **Org. Reactions**, (1985), 33, 247. This is a useful review of the reaction.

EXAMPLES:

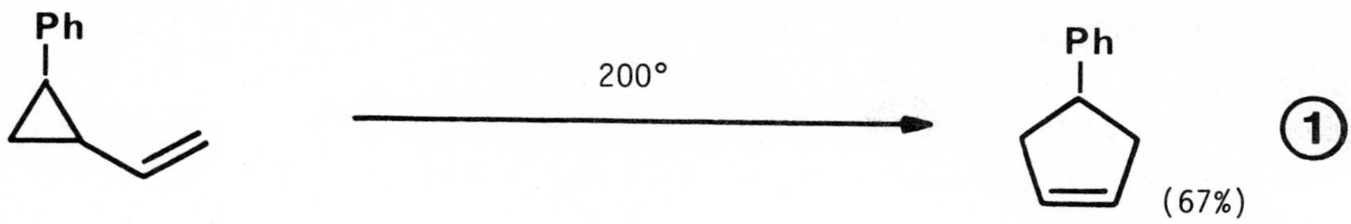

O

570°

O

SiMe$_3$

(70-82%)

SiMe$_3$

2

O

600°

H

H

O

(68%)

3

80°

(90%)

4

O

70°, then to 90°

O

(55%)

5

REACTION: VON BRAUN REACTION

March's Advanced Organic Chemistry : 387

GENERAL SCHEME:

$$BrCN + R_3N \longrightarrow RBr + R_2NCN$$

MECHANISM:

$$N{\equiv}C{-}Br + \ddot{N}R_3 \longrightarrow N{\equiv}C{-}\overset{\oplus}{N}R_3 + Br^{\ominus} \longrightarrow NC{-}NR_2 + RBr$$

NOTES:

1. Reactions run on secondary amines seem to have lower yields.

2. This reaction should not be confused with the **von Braun Amide Degradation** of N-alkyl-substituted amides:

$$RC(=O)NHR' \xrightarrow{PCl_5} RCl + R'CN$$

3. See B.A. Phillips, G. Fodor, J. Gal, F. Letourneau and J.J. Ryan, **Tetrahedron,** (1973), 29, 3309 for a discussion of the mechanism.

4. Thioether cleavage takes place by a similar mechanism.

REFERENCES:

1. G. Werner and R. Schickfluss, **Ann,** (1969), 729, 152.

2. See H.A. Hageman, **Org. Reactions,** (1953), VII, 213.

3. H. Rapoport, C.H. Lovell, H.R. Reist and M.E. Warren, Jr., **J. Am. Chem. Soc.,** (1967), 89, 1942.

4. J. von Braun and P. Engelbertz, **Ber.,** (1923), 56, 1573.

5. Ibid.

EXAMPLES:

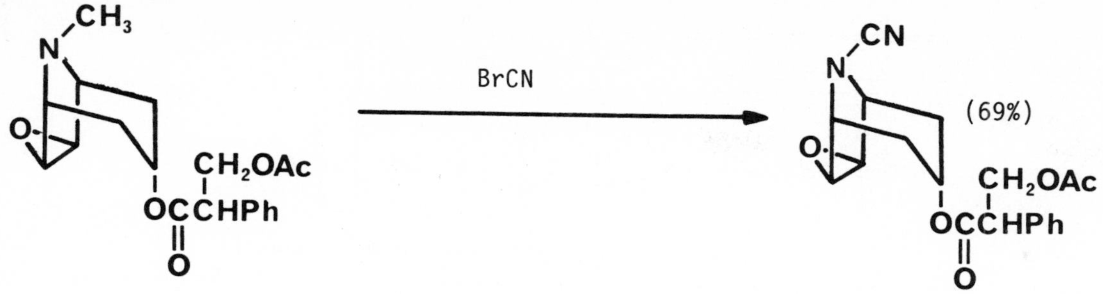

BrCN (69%) ①

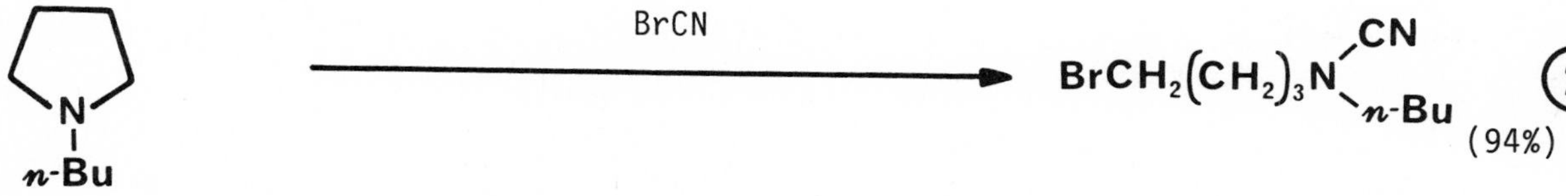

BrCN $BrCH_2(CH_2)_3N(CN)$-*n*-Bu (94%) ②

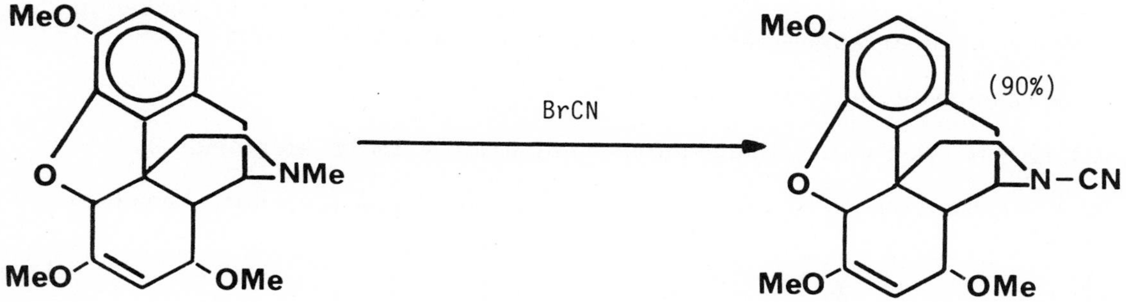

BrCN (90%) ③

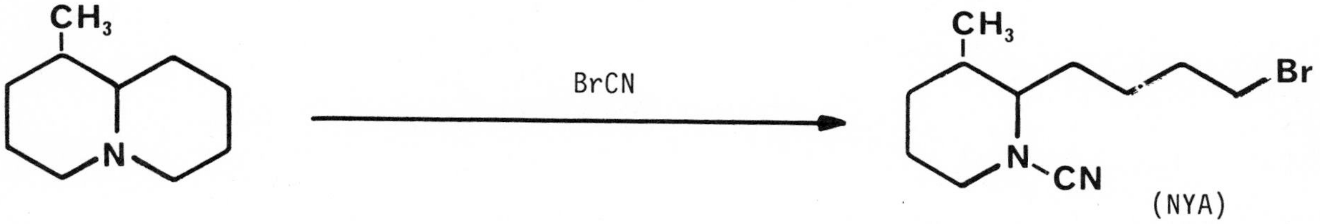

BrCN (NYA) ④

$CH_3CH_2SCH_3 \xrightarrow{BrCN} CH_3CH_2SCN + CH_3Br$ ⑤

(NYA)

△4

March's Advanced Organic Chemistry : 945, 949, 958- 963

GENERAL SCHEME:

$$\mathrm{R{-}\underset{R'''}{\overset{R'}{C}}{-}\underset{R''''}{\overset{R''}{C}}{-}OH \xrightarrow{H^{\oplus}} \underset{R'''}{\overset{R}{}}\!\!>C{=}C\!\!<\underset{R''''}{\overset{R'}{}} \quad or \quad R'''{-}\overset{R}{\overset{\|}{C}}{-}\underset{R''''}{\overset{R''}{C}}{-}R'}$$

if R''=H — if R has an αH

MECHANISM:

$$\mathrm{-\underset{H}{C}{-}\underset{R''}{\overset{R}{C}}{-}\underset{H}{\overset{R'}{C}}{-}OH \xrightarrow{H^{\oplus}} -\underset{H}{C}{-}\underset{R''}{\overset{R}{C}}{-}\underset{H}{\overset{\oplus}{C}}{-}R' \longrightarrow -\underset{H}{C}{-}\underset{R''}{\overset{\oplus}{C}}{-}\underset{H}{\overset{R}{C}}{-}R'}$$

$$\mathrm{\longrightarrow\ >C{=}\underset{R''}{C}{-}\underset{H}{\overset{R}{C}}{-}R' \quad or \quad -\underset{H}{C}{-}\underset{R''}{C}{=}C\!\!<\underset{R'}{\overset{R}{}}}$$

NOTES:

1. For a well-written, but dated , review of this rearrangement, see: Y. Pocker in DeMayo's, Molecular Rearrangements, Interscience, New York, (1963).

2. **TsOH** = p-toluenesulfonic acid (tosic acid)

3. **$(CF_3\text{-}SO_2)_2O$** = Trifluoromethanesulfonic anhydride (**triflic anhydride**).

4. The rearrangement of camphene hydrochloride has a special name, **Nametkin Rearrangement**.

Me, Cl, Me, Me —Δ→ Me, Me, CH_2

REFERENCES:

1. M. Pirrung, **J. Am. Chem. Soc.**, (1979), 101, 7130.

2. W. Dauben and J. Vinson, **J. Org. Chem.**, (1975), 40, 3756.

3. T.C.W. Mak, Y.C. Yip, and T.-Y. Luh, **Tetrahedron**, (1986), 42, 1981.

4. L.A. Paquette, L. Waykole, H. Jendralla and C.E. Cottrell, **J. Am. Chem. Soc.**, (1986), 108, 3739.

5. A.B. Smith,III, B.A. Wexler, C.-Y. Tu and J.P. Konopelski, **J. Am. Chem. Soc.**, (1985), 107, 1308. See, also: K. Kakiuchi, et. al., **J. Org. Chem.**,(1985), 50, 488.

EXAMPLES:

CH_2, CH_3, CH_3, H_3C — TsOH / C_6H_6 → CH_3, CH_3, CH_3, H_3C (98%) ① △2

CH_2OH — 1.) n-BuLi 2.) $(CF_3SO_2)_2O$ → CH_2OTf — CF_3CH_2OH / Et_3N → CF_3, CH_2, O, H (80%) ② △3

O, O, Br, CPh_2OH — HCl / MeOH → O, O, Br, MeO, Ph, Ph (91%) ③

HO — $Et_3N^+SO_2N^-CO_2Me$ / C_6H_6, → (57%) ④

OH — H_2SO_4 / THF → OH (40%, 2.5:1) — H^+ → OH ⑤

REACTION: WHARTON REACTION

March's Advanced Organic Chemistry : Not Indexed

GENERAL SCHEME:

1

H_2NNH_2

Δ

OH

MECHANISM:

Δ

NNH_2

H

N

N—H

O⊖

H⊕

OH

NOTES:

1. This reaction can be considered an eliminative **Wolff- Kishner** reduction.

2. Possible mechanism:

H

N—H

N

N—H

N

OH

OH

REFERENCES:

1. G.V. Nair and G.D. Pandit, **Tetrahedron Lett.**, (1966), 5097.

2. G. Stork and P.G. Williard, **J. Am. Chem. Soc.**, (1977), 99, 7067.

3. Ibid.

4. P.S. Wharton and D.H. Bohlen, **J. Org. Chem.**, (1961), 26, 3615.

5. P.S. Wharton, **J. Org. Chem.**, (1961), 26, 4781.

EXAMPLES:

CHO; O; N_2H_4; OH; (30-35%) ①

O; O; N_2H_4; OH; (85%) ② △2

O; O; N_2H_4; OH; (70%) ③

O; O; N_2H_4; OH; (75%) ④

Me; OCH_2Ph; O; O; N_2H_4; Me; OCH_2Ph; OH; (NYA) ⑤

REACTION: WILLIAMSON REACTION 

==

March's **Advanced Organic Chemistry**: 342

GENERAL SCHEME:

$$R'X + OR^{\ominus} \longrightarrow R'OR$$

MECHANISM:

$$R'{-}X + OR^{\ominus} \longrightarrow R'{-}O{-}R + X^{\ominus}$$

NOTES:

1. Also known as the **Williamson Ether Synthesis**.

2. Either classic S_N1 or S_N2 mechanistic considerations can be applied.

3. A variation using thallium (I) ethoxide. This method is best for substrates containing an additional oxygen function such as -OH, -COOR and -$CONR_2$.

4. With aromatic ethers, the reaction is most often run using a phenolate derivative with an alkyl halide. The **Ullmann Ether Synthesis** is used for diaryl ethers.

$$PhO^{\ominus} + PhX \xrightarrow[\text{salts}]{\text{Copper}} Ph{-}O{-}Ph$$

5. The use of benzyl ethers as protecting groups is very common, since the protecting group can be easily removed by catalytic reduction.

REFERENCES:

1. H. Kaunowski, G. Crass and D. Seebach, **Ber.**, (1981), 114, 477.

2. B.P. Mundy and D. Wilkening, **J. Org. Chem.**, (1984), 49, 3379.

3. G.S. Hiers and F.D. Hager, **Org. Synth.**, (1941), Coll. Vol. I, 58.

4. J.F. Norris and G.W. Rigby, **J. Am. Chem. Soc.**, (1932), 54, 2088.

5. R.C. Beier, B.P. Mundy and G.A. Strobel, **Carbohyd. Res.**, (1983), 121, 79.

EXAMPLES:

$EtO_2CCH(OH)CH(OH)CO_2Et$ —— 1.) 2 TlOEt; 2.) MeI ——> $EtO_2CCH(OMe)CH(OMe)CO_2Et$ (91%) ① △3

Bicyclic diol with two CH_2OH groups —— TsCl, Pyridine, △ ——> [CH_2OTs / CH_2OH intermediate] ——> cyclic ether ②

Phenol (OH) —— NaOH, CH_3-O-SO_2-O-CH_3 ——> anisole (OMe) (75%) ③

$O^{\ominus}Na^{\oplus}$ on $CH_3CHCH_2CH_3$ —— Et-Br ——> OCH_2CH_3 on $CH_3CHCH_2CH_3$ (NYA) ④

Disaccharide (HO-, CH_2OH, HO-, HO, -O-, CH_2OH, HO-, HO, -SPh) —— 1.) NaH / Dioxane; 2.) $PhCH_2Cl$ ——> A (BzO-, CH_2OBz, BzO-, BzO, -O-, CH_2OBz, BzO-, BzO, -SPh) (23%) △5 ⑤

REACTION: WITTIG REACTION

March's Advanced Organic Chemistry : 845- 854

GENERAL SCHEME:

△1 △3

$$RC(=O)R(H) + XCH(R'')(R') + Ph_3P \longrightarrow R(H)R\,C{=}C\,R''R'$$

MECHANISM:

$$Ph_3P + XCH(R'')(R') \longrightarrow X^{\ominus}\; Ph_3\overset{\oplus}{P}{-}CH(R'')(R') \longrightarrow \left[Ph_3\overset{\oplus}{P}{-}\overset{\ominus}{C}(R'')(R') \longleftrightarrow Ph_3P{=}C(R'')(R') \right]$$

(ylide)

$$\xrightarrow{RC(=O)R(H)} \text{R''-C(R')}\cdots\text{PPh}_3 \,/\, \text{R-C(R(H))}\cdots\text{O} \longrightarrow R''R'C{=}CR\,R(H)$$

NOTES:

1. See also: **Wittig- Horner Reaction, Horner- Emmons Reaction.** For a similar procedure, see: **Peterson Olefination Reaction.**

2. A **Wittig reaction,** and alkylation reaction.

3. Often, reactions with aldehydes give low yields; however, a new procedure finds much-improvement: Y. LeBigot, N. Hajjaji, I. Rico, A. Lattes, M. Delmas and A. Gasef, **Synth. Commun.**, (1985), 495.

$$(CH_3)_2C{=}O + Ph_3\overset{\oplus}{P}CH_2R \xrightarrow[\text{dioxane},\ HCNH_2\ (\text{C=O})]{K_2CO_3} (CH_3)_2C{=}CHR$$

4. Formally classified as a **Wittig- Horner reaction.**

5. Notice the disguised aldehyde as a hemiacetal.

REFERENCES:

1. H.-J. Altenbach, **Angew. Chem. Int. Ed.**, (1979), 18, 940.

2. K.C. Nicolaou, W.E. Barnette and P. Ma, **J. Org. Chem.**, (1980), 45, 1463.

3. W. Roush, D. Harris and B. Lesur, **Tetrahedron Lett.**, (1983), 2227.

4. J.H. Tumlinson, M.G. Klein, R.E. Doolittle, T.L. Ladd and A.T. Proveaux, **Science**, (1977), 197, 789.

5. R.D. Little and G.W. Muller, **J. Am. Chem. Soc.**, (1981), 103, 2744.

EXAMPLES:

$BrCH_2C(O)CH{=}PPh_3$ — NaH, DMF — (31%) ① ⚠2

$(Ph)_3P^+CH_2^-$ — CHO, SO_2 — (NYA) ②

$(i\text{-}PrO)_2P(O)CH_2COOEt$, $t\text{-}BuO^-$, THF, −78° — MeO, HC=O, O, Ph, EtO_2C — (Excess of 70%) ③ ⚠4

$n\text{-}C_8H_{17}CH^-P(Ph)_3^+$ — (NYA) ④

$(Ph)_3P^{+-}CHCO_2Me$ — OH, CH_2OH, CO_2Me — (90%) ⑤ ⚠5

REACTION: WITTIG REARRANGEMENT

March's Advanced Organic Chemistry : 994- 995

GENERAL SCHEME:

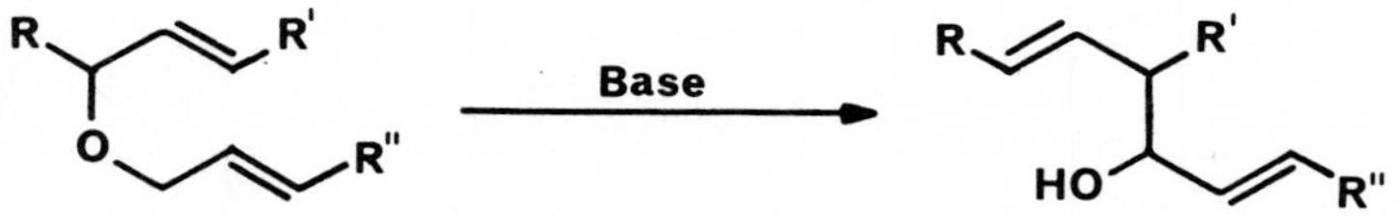

MECHANISM:

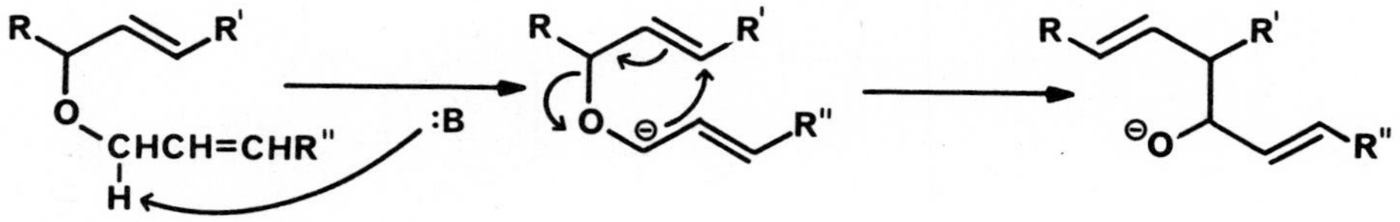

NOTES:

1. Formally this reaction can be considered a [2,3]-sigmatropic shift.

2. This is a combination of a **Wittig rearrangement** followed by an **Oxy-Cope rearrangement**:

Via:

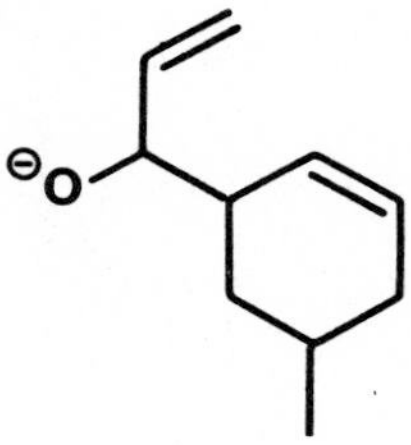

REFERENCES:

1. N. Sayo, Y. Kimura and T. Naka, **Tetrahedron Lett.**, (1982), 3931.

2. W. Still and A. Mitra, **J. Am. Chem. Soc.**, (1978), 100, 1927.

3. N. Sayo, E. Nakai and T. Nakai, **Chem. Letters**, (1985), 1723.

4. E. Nakai, E. Kitahara, N. Sayo, Y. Ueno and T. Nakai, **Chem. Letters**, (1985), 1725.

5. K. Mikami, O. Takahashi, T. Kasuga and T. Nakai, **Chem. Letters**, (1985), 1729.

EXAMPLES:

1.) BuLi

2.) KH

OHC

(14%)

Me

① ⚠2

n-Bu–CH–C=CH$_2$ (CH$_3$; O–CH$_2$Sn(n-Bu)$_3$)

n-BuLi

-78°

n-Bu, H, C=C, CH$_2$CH$_2$OH, CH$_3$ (95%)

②

Me$_3$Si—≡—

n-Bu

Me

n-BuLi

OH (96%)

③

Me

BuLi

OH

Me Me

(96%)

④

CH$_3$

O

OH

O=C N

1.) LDA

2.) H$_3$O$^+$

3.) CH$_2$N$_2$

CH$_3$

HO COOCH$_3$

(55% ee, 95% erythro)

⑤

REACTION: WOLFF- KISHNER REDUCTION

==

March's **Advanced Organic Chemistry :** 1119- 21

GENERAL SCHEME:

$$RC(=O)R \longrightarrow RCH_2R$$

MECHANISM:

$$R_2C{=}O \longrightarrow R_2C{=}N{-}NH_2 \overset{\ominus:B}{\rightleftharpoons} R_2CH{-}N{=}NH$$

$$R_2C{=}N{-}NH_2 \longrightarrow R_2C{=}N{-}\overset{\ominus}{N}H \longrightarrow R_2\overset{\ominus}{C}{-}N{=}NH \xrightarrow{H\text{-}sol} R_2CH{-}N{=}NH$$

$$\longrightarrow R_2CH{-}N{=}\overset{\ominus}{N} \xrightarrow{-N_2} R_2\overset{\ominus}{C}H \xrightarrow{H\text{-}sol} R_2CH_2 + \ominus{:}B$$

NOTES:

1. In the original procedure, the carbonyl was heated with hydrazine hydrate and base. This method has been almost completely replaced by the **Huang- Minlon Modification,** where the reaction is carried out in refluxing diethylene glycol.

REFERENCES:

1. R. Cory, D. Chan, Y. Naquib, M. Rastall and R. Renneboog, **J. Org. Chem.,** (1980), 45, 1852.

2. B.M. Trost and D.E. Keeley, **J. Org. Chem.,** (1975), 40, 2013.

3. K. Mori, **Tetrahedron,** (1977), 33, 289.

4. K. Hayano, Y. Ohfune, H. Shirahama and T. Matsumoto, **Tetrahedron Letters,** (1978), 1991.

5. W.G. Dauben and D.M. Walker, **J. Org. Chem.,** (1981), 46, 1103.

EXAMPLES:

Sodium diethylene glycolate / Diethylene glycol, 140° N_2H_4 (30%) ①

N_2H_4 KOH, 210° (85%) ②

N_2H_4 KOH, 210° (67%) ③

1.) H_2NNHTs 2.) $NaBH_3CN$ (95%) ④

N_2H_4 / KOH 205°, H^+ (69%) ⑤

REACTION: WOLFF REARRANGEMENT

March's Advanced Organic Chemistry : 176, 974

GENERAL SCHEME:

$$RC(=O)CHN_2 \longrightarrow RCH_2C(=O)OH$$

MECHANISM:

$$RC(=O)-CH=\overset{\oplus}{N}=N^{\ominus} \xrightarrow{-N_2} R-C(=O)-\ddot{C}H \longrightarrow RCH{=}C{=}O \xrightarrow{H_2O} RCH_2C(=O)OH$$

NOTES:

1. This is the second step in the **Arndt- Eistert** reaction; where the first step is the formation of the diazoketone:

$$RC(=O)Cl + CH_2N_2 \longrightarrow RC(=O)CHN_2 + HCl$$

2. An example of the **Photo- Wolff** rearrangement.

hν

REFERENCES:

1. V. Georgian, S. Boyer and B. Edwards, **J. Org. Chem.**, (1980), 45, 1686.

2. E.D. Bergmann and E. Hoffmann, **J. Org. Chem.**, (1961), 26, 3555.

3. J. Meinwald and P.G. Gassman, **J. Am. Chem. Soc.**, (1960), 82, 2857.

4. Ibid.

5. E.W. Della and M. Kendall, **J. Chem. Soc.**, Perkin I, (1973), 2729.

EXAMPLES:

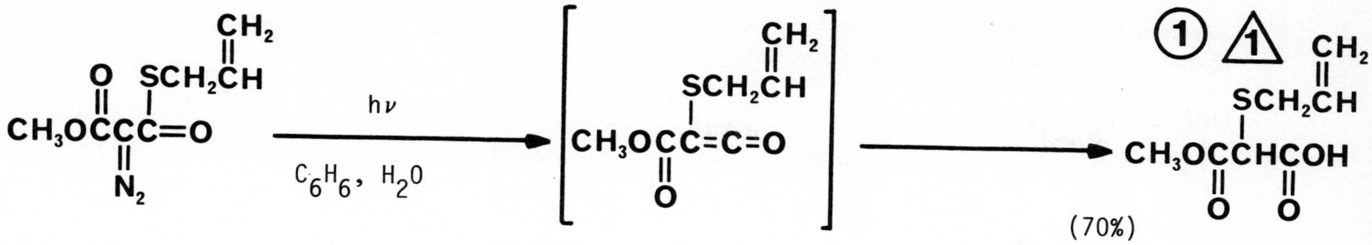

$CH_3OC(=O)C(=N_2)C(=O)SCH_2CH=CH_2$ $\xrightarrow[C_6H_6,\ H_2O]{h\nu}$ [$CH_3OC(=O)C(SCH_2CH=CH_2)=C=O$] $\longrightarrow$ $CH_3OC(=O)CH(SCH_2CH=CH_2)C(=O)OH$ (70%) ① △1

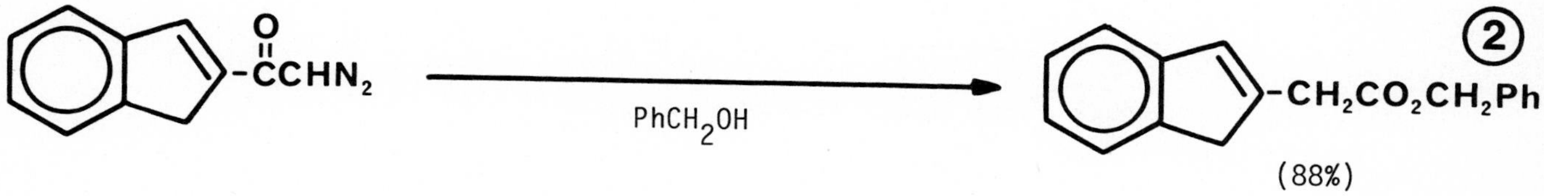

Indenyl–$C(=O)CHN_2$ $\xrightarrow{PhCH_2OH}$ Indenyl–$CH_2CO_2CH_2Ph$ (88%) ②

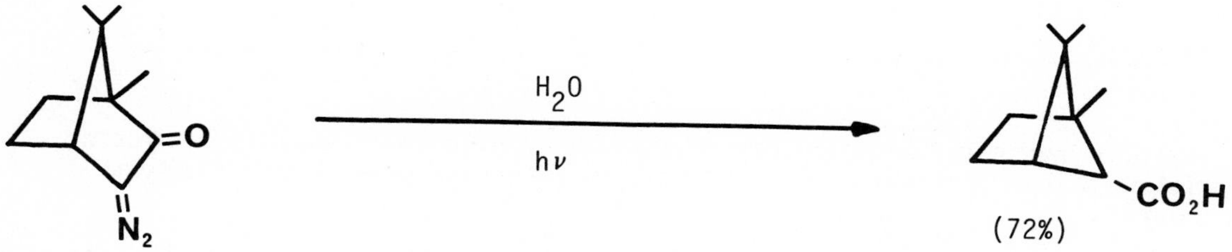

$\xrightarrow[h\nu]{H_2O}$ CO_2H (72%) ③ △2

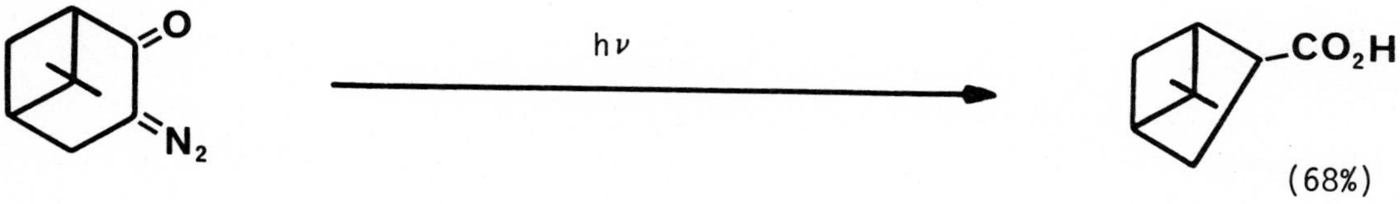

$\xrightarrow{h\nu}$ CO_2H (68%) ④ △2

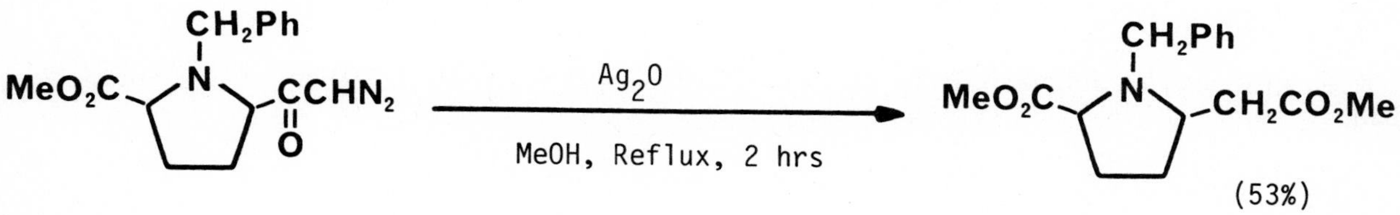

MeO_2C–pyrrolidine(N–CH_2Ph)–$C(=O)CHN_2$ $\xrightarrow[MeOH,\ Reflux,\ 2\ hrs]{Ag_2O}$ MeO_2C–pyrrolidine(N–CH_2Ph)–CH_2CO_2Me (53%) ⑤

REACTION: WURTZ REACTION

March's Advanced Organic Chemistry : 399

GENERAL SCHEME:

$$RX + R'X(M) \longrightarrow R{-}R'$$

MECHANISM:

This reaction could have two possible mechanisms:

Nucleophilic Displacement

$$RX + 2\,Na \longrightarrow R^{\ominus}Na^{\oplus} + NaX$$

$$R^{\ominus}Na^{\oplus} + R'{-}X \longrightarrow R{-}R' + NaX$$

Radical

$$RX + R'M \longrightarrow [R\cdot + R'\cdot + MX] \longrightarrow R{-}R'$$

NOTES:

1. The reaction has had limited utility due to many side reactions. The **Wurtz-Fittig** variation is the reaction of an aryl and an alkyl halide to form an alkylated aromatic ring.

2. Some intramolecular Wurtz reactions have been shown to give reasonable yields of products.

REFERENCES:

1. G.M. Whitesides and F.D. Gutowski, **J. Org. Chem.**, (1976), 41, 2882.

2. H. Nozaki, T. Shirafuji and Y. Yamamoto, **Tetrahedron**, (1969), 25, 3461.

3. G.M. Lampman and J.C. Aumiller, **Org. Synth.**, (1971), 51, 55.

4. R.R. Read, L.S. Foster, A. Russel and V.L. Simril, **Org. Synth.**, Coll. Vol. III, (1955), 157.

5. R.E. Pincock, J. Schmidt, W.B. Scott and E.J. Torupka, **Can. J. Chem.**, (1972), 50, 3958.

EXAMPLES:

1.) Mg / THF

2.) AgTf

(57%) ① △2

$PhCHCl_2$ —— $CuCl_2$ / DMSO ——→ $PhCH(Cl)CH(Cl)Ph$ (60%) ②

Br, Cl —— Na ——→ (78-94%) ③ △2

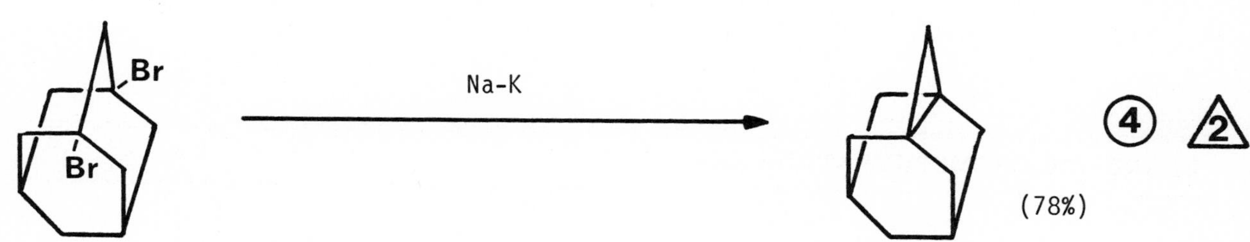

Br + *n*-BuBr —— Na / Ether ——→ *n*-Bu (65%) ⑤

BAKER- VENKATARAMAN REARRANGEMENT

$$\text{2-(PhCO-O)-C}_6\text{H}_4\text{-CO-CH}_2\text{Ph} \xrightarrow[t\text{-BuOH}]{t\text{-BuO}^{\ominus}} \text{2-HO-C}_6\text{H}_4\text{-CO-CH(Ph)-CO-Ph}$$

BETTI REACTION

$$\text{PhCHO} + \text{PhNH}_2 + \text{C}_6\text{H}_5\text{OH} \longrightarrow \text{2-HO-C}_6\text{H}_4\text{-CH(Ph)-NH-Ph}$$

BLANC CHLOROMETHYLATION REACTION

$$\text{Ph-H} + \text{CH}_2\text{O} + \text{HCl} \xrightarrow{\text{ZnCl}_2} \text{PhCH}_2\text{Cl}$$

BORSCHE CINNOLINE SYNTHESIS

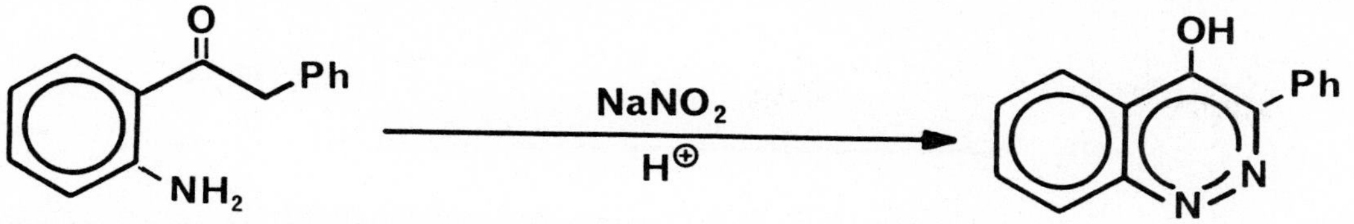

O, Ph, NH_2 — $NaNO_2$, $H^{\oplus}$ → OH, Ph, N, N

BRADSHER REACTION

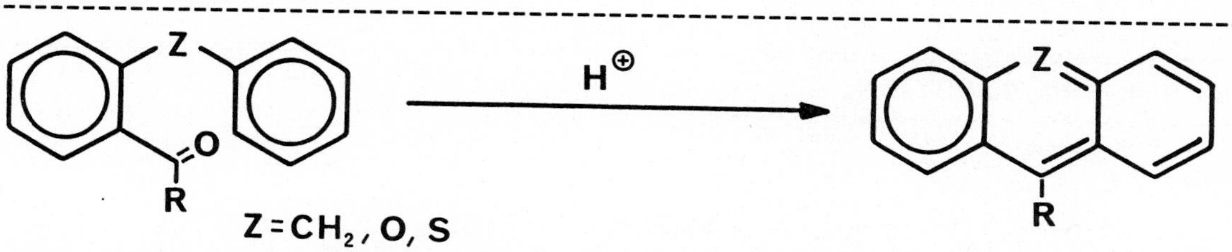

Z, O, R — $H^{\oplus}$ → Z, R

Z = CH_2, O, S

BUCHERER CARBAZOLE SYNTHESIS

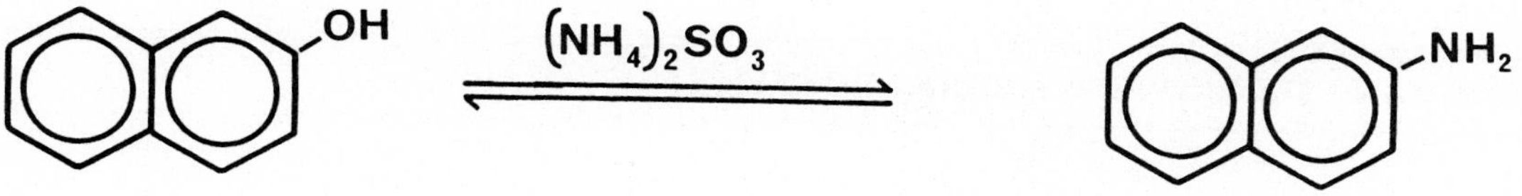

OH — $(NH_4)_2SO_3$ ⇌ NH_2

BUCHERER REACTION

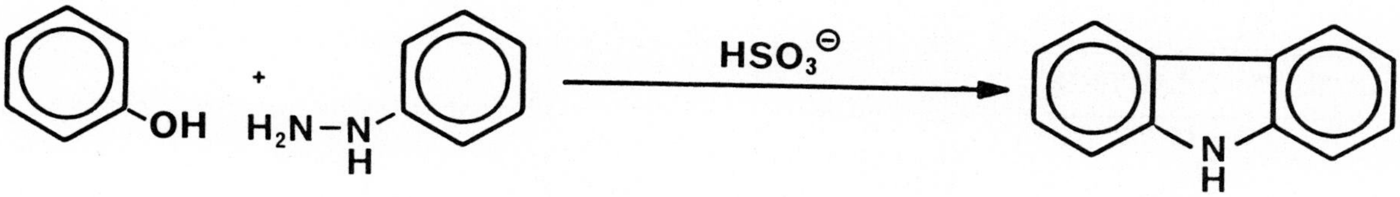

OH + H_2N-NH — $HSO_3^{\ominus}$ → N H

CAMPS QUINOLINE SYNTHESIS

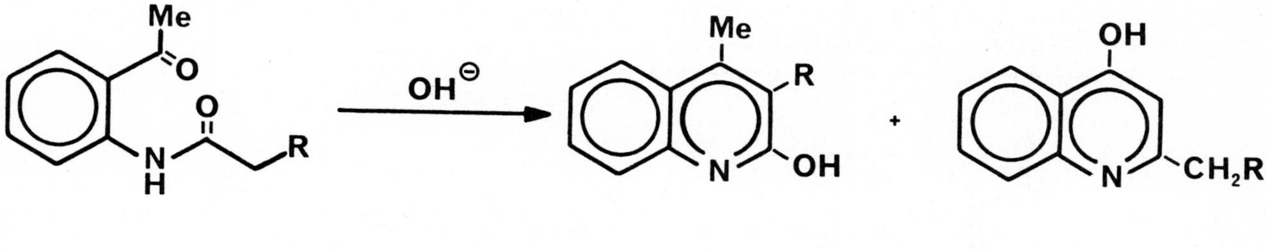

Me, O, O, N H, R — $OH^{\ominus}$ → Me, R, N, OH + OH, N, CH_2R

CHAPMAN REARRANGEMENT

Ph–C(O–Ph)=N–Ph —(Δ)→ Ph–C(=O)–NPh_2

CHICHIBABIN REACTION

Pyridine —($NaNH_2$, Δ)→ 2-aminopyridine (NH_2)

CIAMICIAN- DENNSTEDT REARRANGEMENT

Pyrrole (N–H) —($CHCl_3$, $OR^{\ominus}$)→ dichlorocyclopropane adduct (Cl, Cl; N–H) —($OR^{\ominus}$)→ 3-chloropyridine (Cl, N)

DE MAYO REACTION

R–C(=O)–CH=CH_2 + 2-butene —(hν)→ acetyl-dimethylcyclobutane

DIENONE- PHENOL REARRANGEMENT

4,4-R,R-cyclohexadienone (O) —($H^{\oplus}$)→ 3,4-R,R-phenol (OH, R, R)

ELBS PERSULFATE OXIDATION

Phenol $\xrightarrow[OH^{\ominus}]{K_2S_2O_8}$ hydroquinone (HO, OH)

ETARD REACTION

Toluene (CH_3) $\xrightarrow{CrO_2Cl_2}$ benzaldehyde (CHO)

FENTON REACTION

$$-\underset{\displaystyle OH}{\underset{|}{CH}}-CO_2H \xrightarrow[Fe^{+2}]{H_2O_2} -\underset{\displaystyle O}{\underset{\|}{C}}-CO_2H$$

$$-\underset{\displaystyle OH}{\underset{|}{CH}}-CH_2OH \xrightarrow[Fe^{+2}]{H_2O_2} -\underset{\displaystyle OH}{\underset{|}{CH}}-CHO$$

FIESSELMANN THIOPHENE SYNTHESIS

Me-C≡C-CO_2Et + HS-CH_2CO_2Me ⟶ Me(S-CH_2-CO_2Me)C=CH-CO_2Me $\xrightarrow{B^{\ominus}}$ 3-OH, 5-Me thiophene with 2-CO_2Me

FISCHER OXAZOLE SYNTHESIS

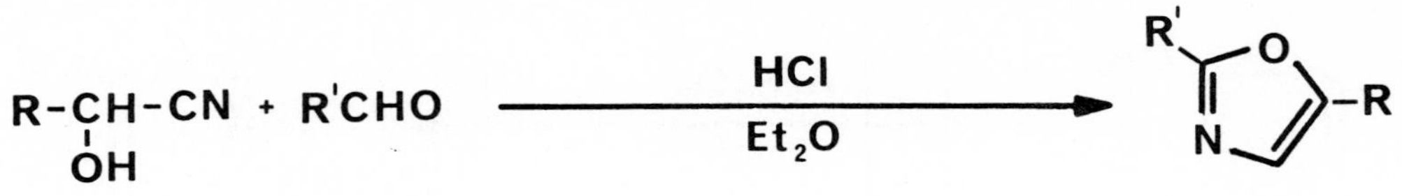

R-CH(OH)-CN + R'CHO → (HCl, Et_2O) 2-R'-5-R-oxazole

FORSTER REACTION

α-oximino ketone (C=O, C=N-OH) → (NH_2Cl) α-diazo ketone (C=O, C=N_2)

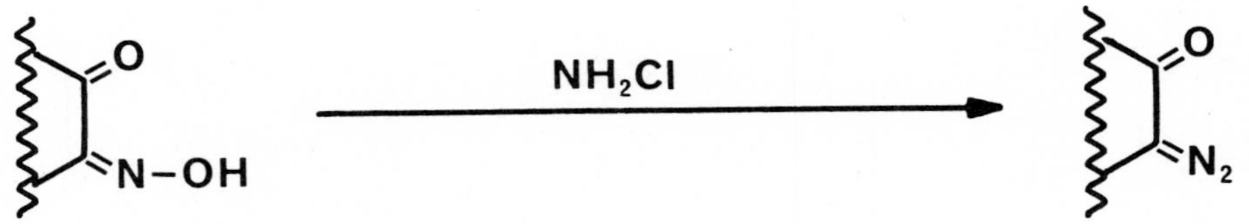

FUJIMOTO- BELLEAU REACTION

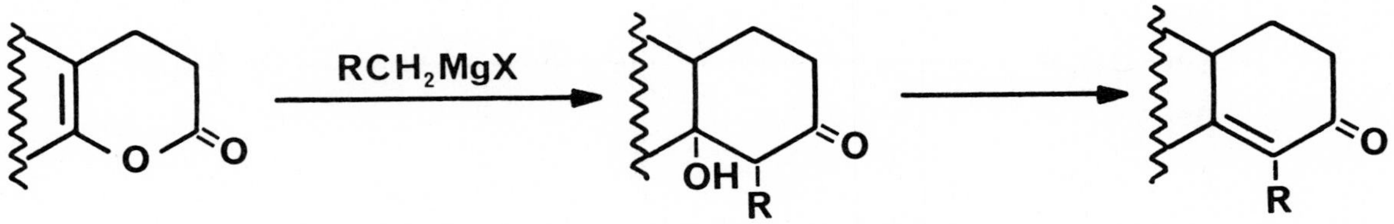

enol lactone → (RCH_2MgX) hydroxy ketone (OH, R) → enone (R)

GUARESCHI- THORPE CONDENSATION

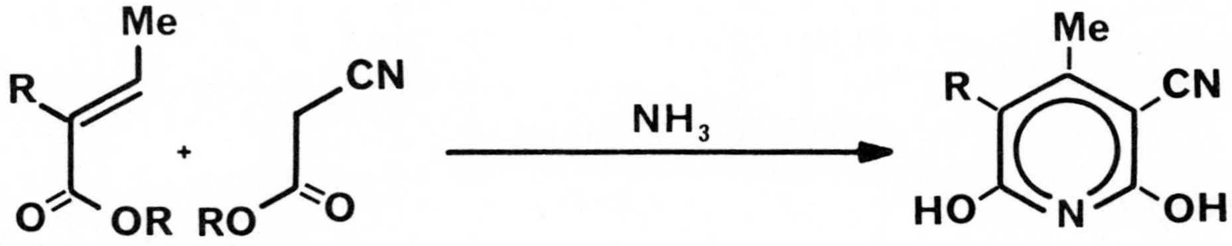

R(C=CHMe)COOR + NC-CH$_2$-COOR → (NH_3) 2,6-dihydroxy-4-methyl-5-R-pyridine-3-carbonitrile (HO, N, OH, Me, R, CN)

GIESE REACTION

alkene → (1. Hg(OAc)$_2$ / ROH; 2. $NaBH_4$, HC=C—EWG) product (OR, EWG)

HANTZSCH PYRROLE SYNTHESIS

Me, O, CH_2Cl + NH_3 + O, Me, CO_2Et → H, N, Me, Me, CO_2Et

HOUBEN- HOESCH REACTION

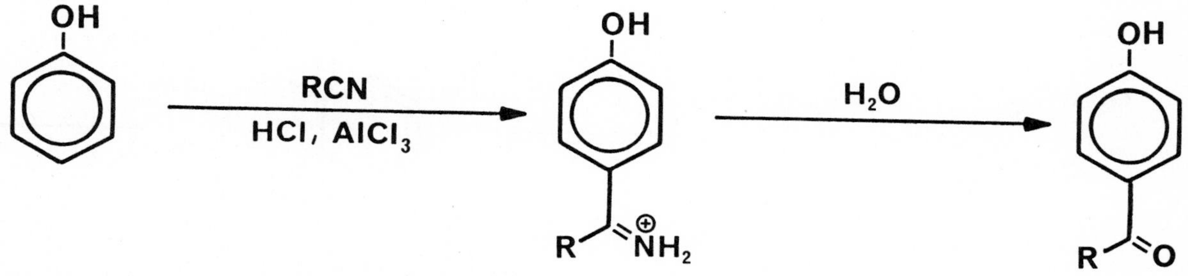

OH → (RCN; HCl, $AlCl_3$) → OH, R, $\overset{\oplus}{N}H_2$ → (H_2O) → OH, R, O

ISAY PTERIDINE SYNTHESIS

N, N, NH_2, NH_2 + O O, HC–CH → N, N, N, N

KOCHI REACTION

$R{-}CO_2H \xrightarrow[\text{2. LiCl, } \Delta]{\text{1. Pb(OAc)}_4} R{-}Cl + CO_2$

KOLBE- SCHMITT REACTION

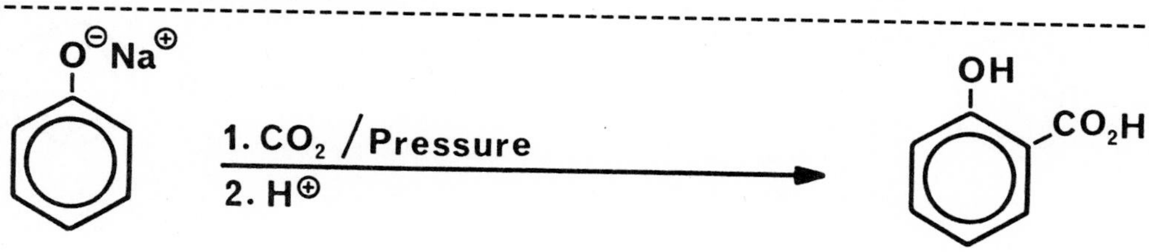

$O^{\ominus}Na^{\oplus}$ → (1. CO_2 / Pressure; 2. $H^{\oplus}$) → OH, CO_2H

MITSUNOBE REACTION

$$R\text{-}OH \xrightarrow[2.\ HX]{1.\ EtO_2CN{=}NCO_2Et,\ Ph_3P} R\text{-}X$$

NENITZESCU INDOLE SYNTHESIS

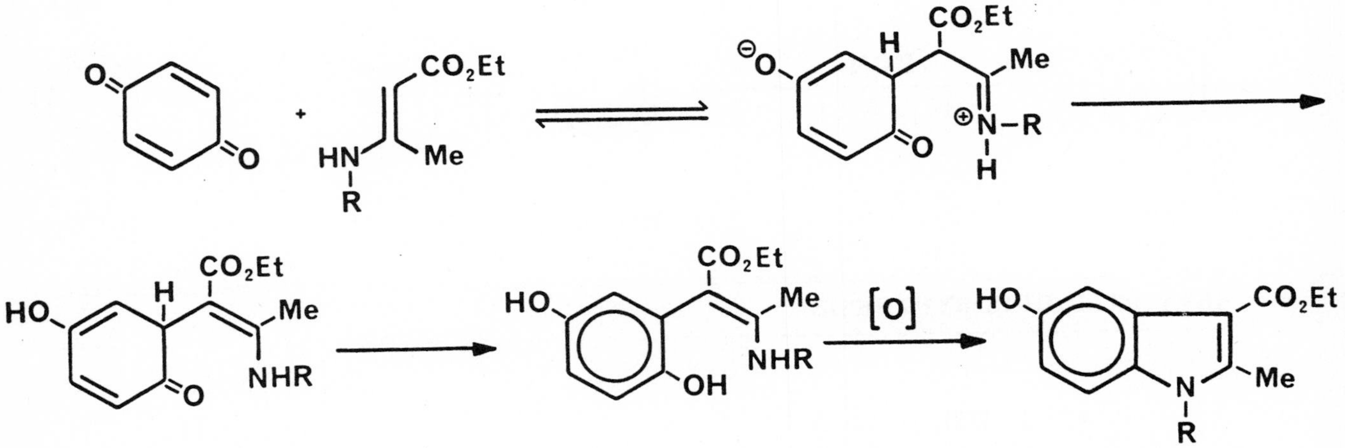

PATERNO- BUCHI REACTION

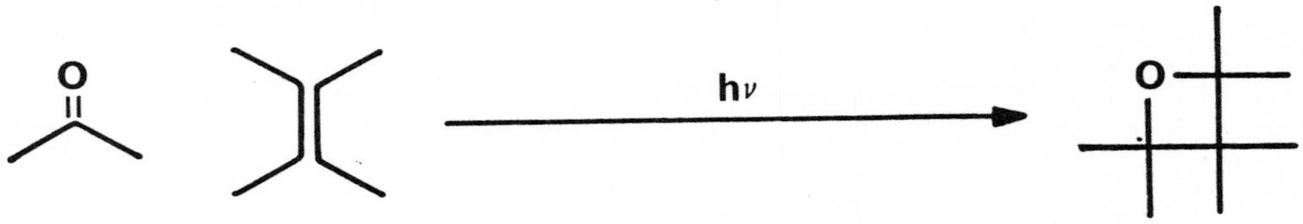

PICTET- GAMS ISOQUINOLINE SYNTHESIS

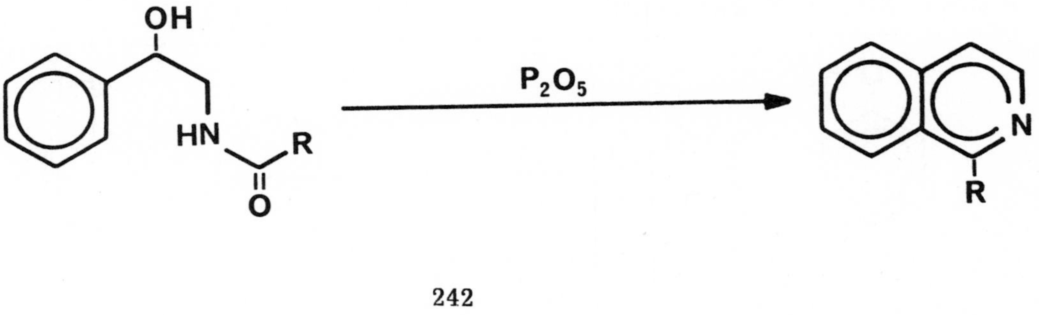

REIMER- TIEMANN REACTION

$CHCl_3$
KOH

REISSERT INDOLE SYNTHESIS

Me
NO_2
1. $EtO^{\ominus}$
2. EtO_2CCO_2Et
3. H_2O, Hyd.
CO_2H
NO_2
[H]
CO_2H
NH_2
CO_2H

VON RICHTER REACTION

NO_2
R
KCN
EtOH
CN
R

RILEY OXIDATION

SeO_2

ROBINSON- SCHOPF REACTION

CHO / CHO + $MeNH_2$ + O=C(CH_2CO_2H)$_2$ (HO_2C … HO_2C) ⟶ (N-Me bicyclic ketone, =O) + CO_2

RUZICKA SYNTHESIS

$(CH_2)_n(CO_2H)_2 \xrightarrow[\Delta]{Ba(OH)_2} (CH_2)_{n-1}C{=}O$

SARRETT OXIDATION

$$-\overset{H}{\underset{|}{\overset{|}{C}}}-OH \xrightarrow[\text{pyridine}]{CrO_3} -\overset{O}{\overset{\|}{C}}-$$

SEMMLER- WOLFF REACTION

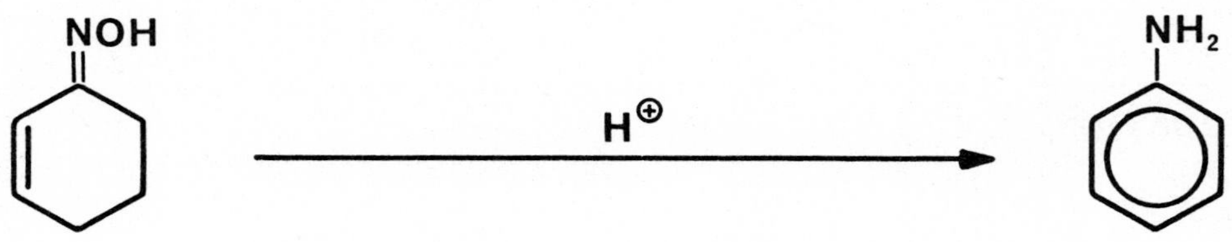

STRECKER AMINO ACID SYNTHESIS

$$R{-}CHO \xrightarrow[NH_3]{HCN} R{-}\overset{NH_2}{\overset{|}{C}}H{-}CN \longrightarrow R{-}\overset{NH_2}{\overset{|}{C}}H{-}CO_2H$$

SWERN OXIDATION

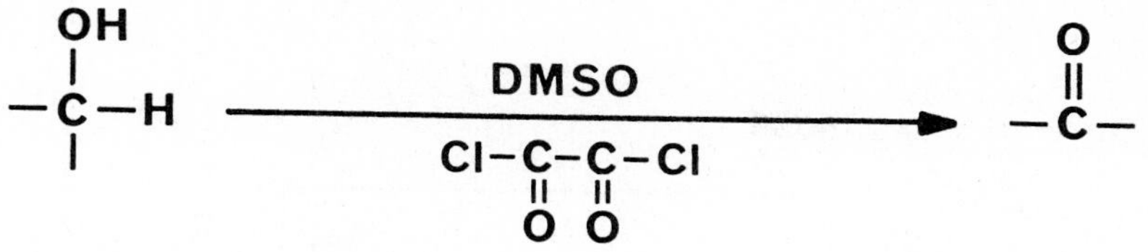

$$\begin{array}{c} OH \\ | \\ -C-H \\ | \end{array} \xrightarrow[\text{Cl-C(=O)-C(=O)-Cl}]{DMSO} \begin{array}{c} O \\ \| \\ -C- \end{array}$$

TRAUBE PURINE SYNTHESIS

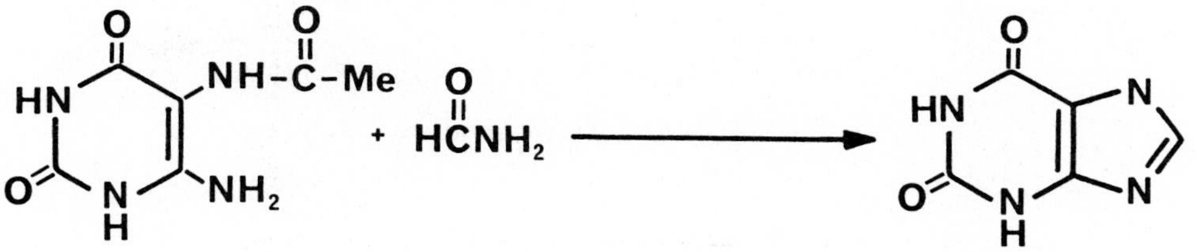

HN, NH-C(=O)-Me, O, N H, NH$_2$ + $HC(=O)NH_2$ → HN, O, N, N, N H

THIELE- WINTER REACTION

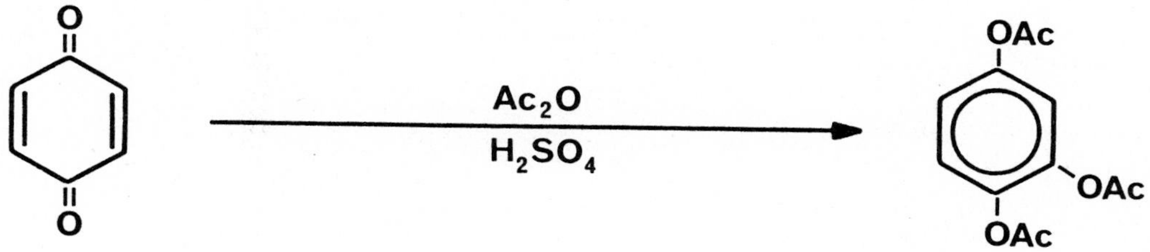

$$\xrightarrow[H_2SO_4]{Ac_2O}$$

OAc, OAc, OAc

WACKER OXIDATION

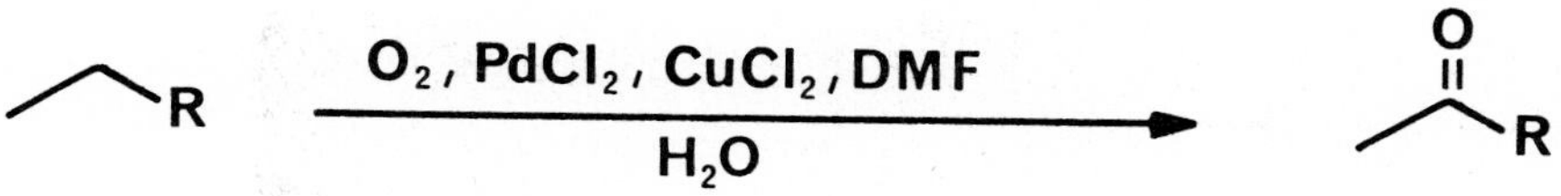

R $\xrightarrow[H_2O]{O_2,\ PdCl_2,\ CuCl_2,\ DMF}$ O, R

WICHTERLE REACTION

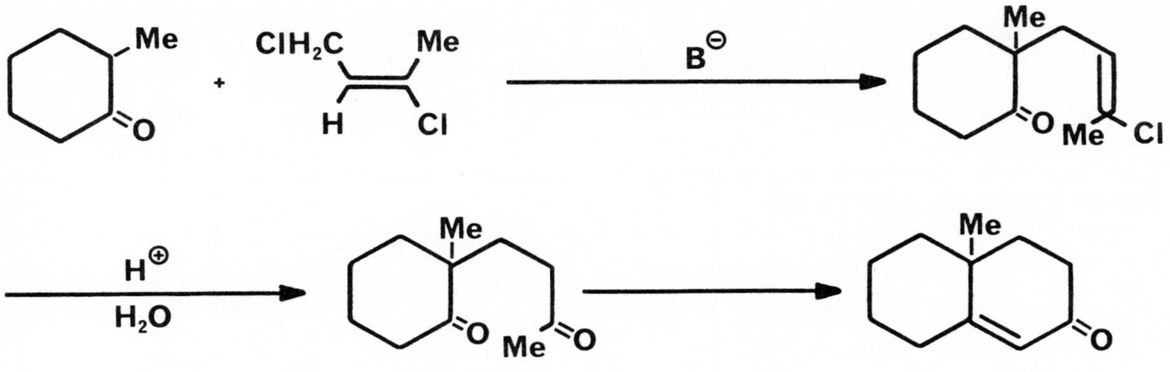

Me ClH$_2$C Me + H Cl $B^{\ominus}$ Me O Me Cl

$H^{\oplus}$ H_2O Me O Me O Me O

WIDMAN- STOERMER CINNOLINE SYNTHESIS

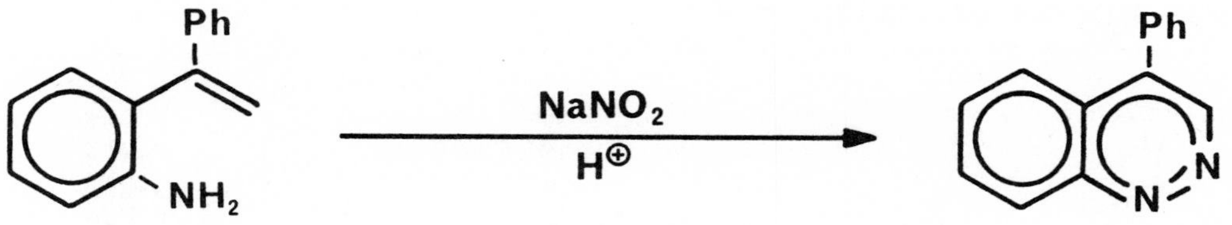

Ph NH$_2$ $NaNO_2$ $H^{\oplus}$ Ph N N

The following pages present a number of the most-commonly used reagents in organic chemistry. The left-facing page presents the name and structure of the reagent, some useful physical data, references to the **Reagents for Organic Synthesis** series and to March's book, some of the major uses, preparation, some of the precautions (**be sure to carefully read instructions on reagent use, and check source literature for special precautions**), notes and references to examples. The right-facing page gives several examples of the reagents use. The general format followed is:

==

Left-facing page:

==

REAGENT: REAGENT NAME

==

STRUCTURE:

Structure and physical properties

MW

BP **MP** **d**

<u>**Fieser's Reagents for Organic Synthesis**</u>**: Vol., pages**

References to the **Reagents** volumes

March's <u>**Advanced Organic Chemistry**</u>: References to March

----

MAJOR USES:

List of major uses. Any uses shown in the examples are indicated by □ next to the example.

PREPARATION:

A brief method of preparation and a comment on commercial availability.

PRECAUTION:

Brief commentary on some of the precautions. **Be sure to check additional sources.**

NOTES:

Pertinent notes. Those related to examples are seen by △ next to the example.

REFERENCES:

References to the examples

==

Right-facing Page:

==

EXAMPLES:

Representative examples of the reagent use.

REAGENT: ACETIC ANHYDRIDE ⚠1

STRUCTURE:

$(CH_3CO)_2O$

MW = 102.09
BP = 138- 40° **MP** = - 73° **d** = 1.082

Fieser's Reagents for Organic Synthesis: Vol., pages

1	3	**4**		**7**	1	**10**	
2	7- 10	**5**	3- 4	**8**	1- 2	**11**	1
3		**6**	1- 2	**9**	1	**12**	

March's Advanced Organic Chemistry: 354- 56, 687, 905, 930

MAJOR USES:

1. Esterification 2. **Perkin reaction** 3. Enol lactonization

---

PREPARATION:

Commercially available in high purity. The deuterated form also available.

PRECAUTION:

1. Corrosive and a lachrymator

2. If it has been on the lab shelf for prolonged periods, it should be tested for purity (**Reagents, Vol. 1, page 3**).

NOTES:

1. For a review, see: D. Kim, **J. Heterocyclic Chemistry**, (1976), 13, 179.

REFERENCES:

1. M. Watanabe, S. Nakamori, H. Hasegawa, K. Shirai, and T. Kumamoto, **Bull. Chem. Soc. Jpn.**, (1981), 54, 817.

2. I.K. Khanna and L.A. Mitscer, **Tetrahedron Lett.**, (1985), 691.

3. K. Takeda, Y. Shibata, Y. Sagawa, M. Urahata, K. Funaki, K. Hori, H. Sasahara and E. Yoshii, **J. Org. Chem.**, (1985), 50, 4673.

4. T. Durst, E.C. Kozma and J.L. Charlton, **J. Org. Chem.**, (1985), 50, 4829.

5. H.O. House, G.S. Nomura, D. Van Derveer and J.E. Wissinger, **J. Org. Chem.**, (1986), 51, 2408.

EXAMPLES:

Ph-S(→O)-$(CH_2)_3$-CO_2H —[Ac_2O, TsOH / Toluene, 1 hr]→ PhS-lactone (75%) ① [3]

1.) $BrSiMe_3$, PhH
2.) Ac_2O, Py, DMAP, CH_2Cl_2
(74%) ② [1]

1.) Ac_2O, DMAP, Et_3N
2.) HF-MeCN
(80%) ③ [1]

Ac_2O
(100%, cis:trans=40:60) ④ [1]

Ac_2O, $HClO_4$
Anhydrous CCl_4
N_2
(72%) ⑤ [3]

REAGENT: ALKYLALUMINUM HALIDES

STRUCTURE:

	Et_2AlCl	Me_2AlCl	$EtAlCl_2$
MW =	120.56	92.51	126.95
BP =			115° @ 50 mm Hg
d =	0.887	0.701	

Fieser's Reagents for Organic Synthesis: Vol., pages

1	**4**	**7** 146	**10** 177- 81
2	**5**	**8**	**11** 7- 12
3	**6** 251-52	**9**	**12** 5- 11

March's Advanced Organic Chemistry: 711

MAJOR USES:

1. Cycloadditions
2. **Ene reaction**
3. Cyclizations
4. Rearrangements
5. Lewis acid catalysts

PREPARATION:

1. All of these reagents are commercially available in a variety of solvents and concentrations.

2. For preparations, see: B. Snider, D. Rodind, M. Karras, T. Kirk, E. Deutsch, R. Cordove and R. Price, **Tetrahedron**, (1981), 37, 3927.

PRECAUTION:

Pyrophoric, flammable, moisture-sensitive.

NOTES:

REFERENCES:

1. A. Batcho, D. Berger, S. Davoust, D. Wovkulich and M. Uskokovic, **Helv.**, (1981), 64, 1682.

2. B. Snider, D. Rodini, T. Kirk and R. Cordova, **J. Am. Chem. Soc.**, (1982), 104, 555.

3. B. Snider and E. Deutsch, **J. Org. Chem.**, (1982), 47, 745.

4. Y. Matsumura, J. Fujiwara, K. Maruoka and H. Yamamoto, **J. Am. Chem, Soc.**, (1983), 105, 6312.

5. G. Majetich, R.W. Desmono, Jr., and J.J. Soria, **J. Org. Chem.**, (1986), 51, 1753.

EXAMPLES:

$CH_2=CHCO_2CH_3$ — $EtAlCl_2$, CH_2Cl_2 → (85%) ① [2]

$CH_3CH=CHCH_3$ + HCHO — Me_2AlCl →

HOCH$_2$ ... 1 eq, (20%); 1.5 eq, (73%)

(39%) (2%) ② [5]

Me_2AlCl, -20° → Me_2AlCl, 25° → ③ [2]

OSiMe$_3$ — N-OMs — Et_2AlCl → NH O (95%) ④ [5]

SiMe$_3$ — $EtAlCl_2$, Toluene → (70%) ⑤

REAGENT: ALUMINA (ALUMINUM OXIDE)

==

STRUCTURE:

MW = 101.96
MP = $\tilde{-}$ 2000°
d = 3.970

Al_2O_3

__Fieser's Reagents for Organic Synthesis__: **Vol., pages**

1	19- 20	**4**	8	**7**	5- 7	**10**	8- 9
2	17	**5**		**8**	9- 13	**11**	22- 24
3	6	**6**	16- 17	**9**	8- 11	**12**	

March's __Advanced Organic Chemistry__: 901- 904

MAJOR USES:

1. Chromatography 2. Dehydration 3. Catalyst for **Michael reactions**

4. Condensations 5. Oxidations 6. Catalyst

7. Rearrangements 8. Cyclizations

PREPARATION:

1. Occurs naturally as Bauxite, Bayerite, Boehmite, Corundum, Diaspore and Gibbsite

2. Commercially available in high purity and in a variety of forms, meshes and activations for specific applications.

3. For a preparation of alumina useful for a reaction surface, see: **Reagents**, __6__, 16; __8__, 10.

PRECAUTION:

NOTES:

REFERENCES:

1. A. Millar, K.H. Kim, D.K. Minster, T. Ohgi and S.M. Hecht, **J. Org. Chem.**, (1986), __51__, 189.

2. H. Parlar and R. Baumann, **Angew. Chem. Int. Ed.**, (1981), __20__, 1014.

3. S. Tsuroi, T. Masauda, h. Makino and A. Takeda, **J. Org. Chem.**, (1982), __47__, 4478.

4. F. Texier-Boullet and A. Foucaud, **Tetrahedron Lett.**, (1982), 4927.

5. S. Pelletier, A. Venkov, J. Filner-Moore and N. Mody, **Tetrahedron Lett.**, (1980), 809.

EXAMPLES:

Reaction 1: sugar carbamate (AcO, AcO, O–C(=O)–NHAc, orthoester with OEt) → Al_2O_3, Benzene → corresponding carbamate with NH_2 (83%) ①

Reaction 2: cyclopentadiene + CO_2Me–CH=CH_2 → Al_2O_3 → endo/exo norbornene-CO_2Me adducts (NYA) ②

Reaction 3: Et(H)C=C=C(H)CH_2CO_2Me → Al_2O_3 → H(Et)C=C(H)–C(H)=C(H)CO_2Me + (2E,4E)-Isomer (82%, 96:4) ③

Reaction 4: $Me_2C{=}O$ + CN–CH_2–CO_2Me → Al_2O_3 → $Me_2C{=}C(CN)CO_2Me$ (53%) ④ 4

Reaction 5: 2-methylene-1-tetralone (O, =CH_2) → Al_2O_3, $HN(Et)_2$ → 2-(CH_2NEt_2)-1-tetralone (Quant.) ⑤ 3

REAGENT: ALUMINUM CHLORIDE ⚠1

STRUCTURE:

MW = 133.34
MP = 190°
d = 2.440

$AlCl_3$

Fieser's Reagents for Organic Synthesis: Vol., pages

1	24- 34	**4**	10- 15	**7**	7- 9	**10**	9- 11
2	21- 23	**5**	10- 13	**8**	13- 15	**11**	25- 28
3	7- 9	**6**	17- 19	**9**	11- 13	**12**	26- 29

March's Advanced Organic Chemistry: 48- 87, 499,510, 730, 961, 990.

MAJOR USES:

1. Strong Lewis acid
2. Catalyst for aromatic electrophilic substitution reactions

PREPARATION:

The reagent can be prepared from aluminum and gaseous HCl; however is commercially available.

PRECAUTION:

1. Moisture sensitive, and reacts explosively with water.
2. Corrosive. Use with caution.

NOTES:

1. For a discussion of techniques for using this reagent, see: **Reagents**, Vol. 1, 24- 25, Vol.3, 21.

REFERENCES:

1. M. Karpf, **Tetrahedron Lett.**, (1982), 4923.

2. Z. Ismail ad H. Hoffmann, **J. Org. Chem.**, (1981), 46, 3549.

3. M. Jung and K. Halweg, **Tetrahedron Lett.**, (1981), 2735.

4. G. Hartman, W. Halczenko and B. Phillips, **J. Org. Chem.**, (1986), 51, 142.

5. R.B. Gammill, **Tetrahedron Lett.**, (1985), 1385.

EXAMPLES:

$$\text{(CH}_3)_2\text{C(COCl)C(CH}_3)\text{=CH}_2 + HC{\equiv}CCH_3 \xrightarrow[CH_2Cl_2]{AlCl_3}$$ Me$_2$ClC-substituted cyclopentenone (53%) + $H_2C{=}C(Me)$-substituted cyclopentenone (11%) ①

Cyclopentadiene + $H_2C{=}C{=}CH{-}CO_2Et$ $\xrightarrow[\text{Benzene, 25°}]{AlCl_3}$ bicyclic CO_2Et / $=CH_2$ adducts (95%, endo:exo =86:14) ② [1]

Et-substituted alkenyl propiolate ester $\xrightarrow{AlCl_3}$ bicyclic lactone (Et, H) (16%) ③ [1]

2-Bromophenyl(3-furyl)methanol (H, OH, Br) $\xrightarrow{AlCl_3,\ LiAlH_4}$ 3-(2-bromobenzyl)furan (Br) (76%) ④

Dihydrobenzofuran diacetate (OAc, AcO) $\xrightarrow[PhNO_2,\ 60°]{AlCl_3}$ acetyl (CH_3, O) dihydroxy dihydrobenzofuran (OH, HO) (85%) ⑤ [2]

REAGENT: ALUMINUM ISOPROPOXIDE

STRUCTURE:

MW = 204.23
MP = 119°C

$$Al\left[-O-\underset{CH_3}{\overset{CH_3}{CH}}\right]_3$$

Fieser's Reagents for Organic Synthesis: Vol., pages

1	35- 37	**4**	15- 16	**7**		**10**	
2		**5**	14	**8**		**11**	29
3	10	**6**	19	**9**	14- 15	**12**	

March's Advanced Organic Chemistry: 785, 811, 813

MAJOR USES:

1. **Meerwein- Ponndorf-Verley Reduction** and **Oppenauer Oxidation**

2. Rearrangement of epoxides to allylic alcohols

PREPARATION:

1. See: A. Wilds, **Organic Reactions**, (1944), 2, 178.

PRECAUTION:

1. The reagent should be kept very dry, flammable solid.

NOTES:

REFERENCES:

1. R.J. Giguere and H.M.R. Hoffmann, **Tetrahedron Lett.**, (1981), 5039.

2. F. Scheidl, **Synthesis**, (1982), 728.

3. S. Terao, M. Shiraishi and K. Kato, **Synthesis**, (1979), 467.

4. H.J. Ringold, B. Loken, G. Rosenkrantz and F. Sondheimer, **J. Am. Chem. Soc.**, (1956), 78, 816.

5. R.D. Hoffsommer, D. Taub and N.L. Wendler, **Chem and Ind.**, (1964), 482.

EXAMPLES:

Al(Oi-Pr)$_3$ 110° (90%, 70:30) ① [2]

Al(Oi-Pr)$_3$ 100° (87%) Al(Oi-Pr)$_3$ 150° (69%) ② [2]

Al(Oi-Pr)$_3$ / Toluene Δ (99%) ③ [2]

Al(Oi-Pr)$_3$ / Xylene Cyclohexanone (86%) ④

Al(OCHMe$_2$)$_3$ 250° (95%) ⑤

REAGENT: BENZENESELENINIC ANHYDRIDE, (**BSA**, DIPHENYLSELINIC ANHYDRIDE)

STRUCTURE:

MW = 360.13
MP = 170 -173°

$[PhSe(=O)]_2O$

Fieser's Reagents for Organic Synthesis: Vol., pages

1		**4**		**7**	139	**10**	22- 29
2		**5**		**8**	29- 32	**11**	37- 39
3		**6**	240- 41	**9**	32- 34	**12**	

March's Advanced Organic Chemistry: 331, 536, 785, 1054, 1061, 1077- 79, 1083.

MAJOR USES:

1. A versatile oxidizing agent 2. Dehydrogenation

PREPARATION:

1. Can be prepared by the reaction of diphenyldiselenide with ozone (G. Ayrey, D. Barnard and D.T. Woodbridge, **J. Chem. Soc.**, (1962), 2089)or from oxygen transfer from Ph-IO_2 (D.H.R. Barton, J.W. Morzycki, W.B. Motherwell and S.V. Ley, **Chem. Commun.**, (1981), 1044).

2. The reagent is commercially available.

PRECAUTION:

1. Moisture sensitive and highly toxic. Use with caution.

NOTES:

REFERENCES:

1. D.H.R. Barton, and J. Bolvin, **Tetrahedron Lett.**, (1985), 1229.

2. J.S.E. Holker, E. O'Brian and B.K. Park, **J. Chem. Soc, Perkin I**, (1982), 1915.

3. K. Yamakawa, T. Satoh, N. Ohba, R. Sakaguchi, S. Takita and N. Tamura, **Tetrahedron**, (1981), 37, 473.

4. D.H.R. Barton, X. Lusinchi and P. Milliet, **Tetrahedron Lett.**, (1982), 4949.

5. D.H.R. Barton, D.J. Lester and S.V. Ley, **J. Chem. Soc.,Perkin I**, (1980), 1212.

EXAMPLES:

$[PhSe(O)]_2O$ / CH_2Cl_2; Me_3SiCN — (95%) ①

(Estrone) — $(PhSeO)_2O$; $HN[SiMe_3]_2$ — NSePh (64%) + PhSeN (12%) ②

$(PhSeO)_2O$; PhMe, 110° — AcO, HO, (57%) + AcO, HO, (12%) ③ [1]

$(PhSeO)_2O$; THF — CH_3 (53%) ④ [2]

$(PhSeO)_2O$, THF; Dark, 18 hrs — NO_2 (81%) ⑤

REAGENT: BENZENESELENYL HALIDE (PHENYLSELENYL HALIDE)

==

STRUCTURE:

	Ph-Se-Br	XSe–C₆H₅	**Ph-Se-Cl**
MW =	235.98		191.52
MP =	60- 62°		64- 63°
BP =	107- 108 @ 15 mm.		120 @ 20 mm

Fieser's Reagents for Organic Synthesis: Vol., pages

1		**4**		**7**	286- 287	**10**	16- 21
2		**5**	518- 522	**8**		**11**	34- 37
3		**6**	459- 460	**9**	25- 32	**12**	39- 41

March's Advanced Organic Chemistry: 361, 531

MAJOR USES:

1. Addition to alkenes (phenylselenyation) 2. Eliminations

3. Dehydrogenations 4. Addition to enolates.

PREPARATION:

Reagents are commercially available. For a preparation, see: H. Reich, M. Cohen and P. Clark, **Org. Synthesis**, (1979), 59, 141.

PRECAUTIONS:

The reagents are corrosive and are highly toxic.

NOTES:

The reaction involves initial selenylation of the double bond.

2. **HMPA** = hexamethylphosphoramide

REFERENCES:

1. S. Ley and B. Whittle, **Tetrahedron Lett.**, (1981), 3301.

2. L. Tietz, G. V. Kiedrowski and B. Berger, **Tetrahedron Lett.**, (1982), 51.

3. K. Nicolaou, R. Maguloa, W. Sipio, W. Barnette, Z. Lysenko and M. Joullie, **J. Am. Chem. Soc.**, (1989), 102, 3784.

4. P. Grieco, M. Mishizawa, T. Ogura, S. Burke and N. Marinovic, **J. Am. Chem. Soc.**, (1977), 99, 5773.

5. M.E. Kuehne and P.J. Seaton, **J. Org. Chem.**, (1985), 50, 4790.

EXAMPLES:

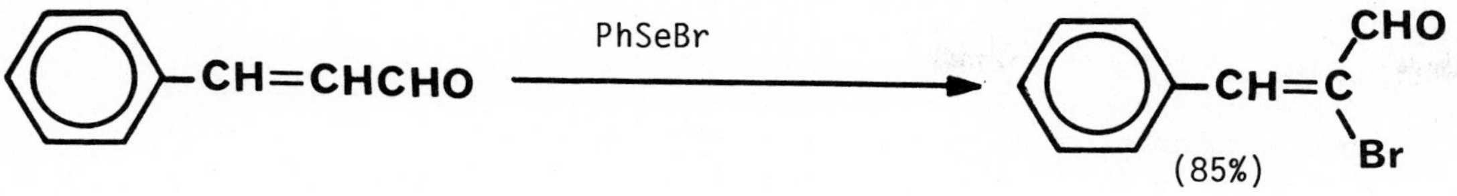

C_6H_5–CH=CHCHO $\xrightarrow{\text{PhSeBr}}$ C_6H_5–CH=C(CHO)Br (85%) ① [1] [2] △1

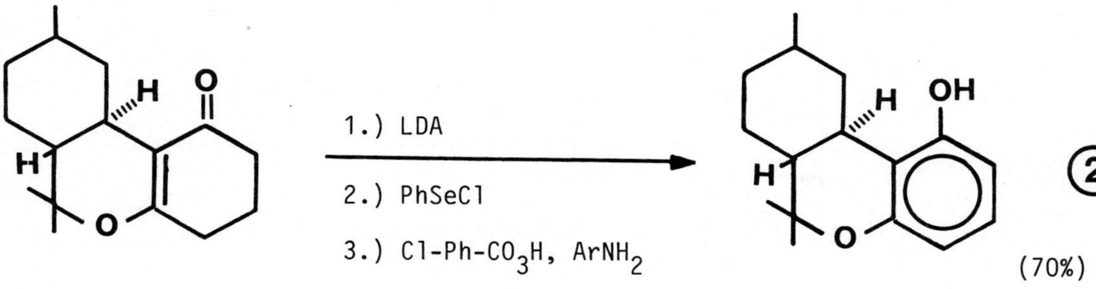

1.) LDA
2.) PhSeCl
3.) Cl-Ph-CO_3H, $ArNH_2$

(70%) ② [1] [4] △1

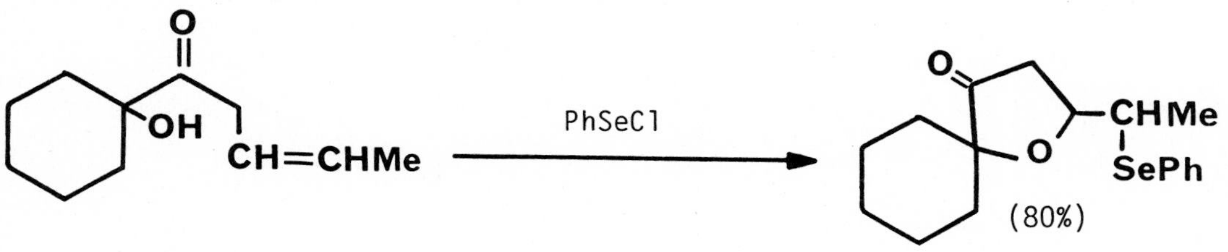

PhSeCl

CH=CHMe → CHMe(SePh) (80%) ③ [1]

1.) LDA, PhSeCl, THF
2.) LDA, THF-HMPA, $BrCH_2CH=CMe_2$

OMe, SePh (83%) ④ [4] △2

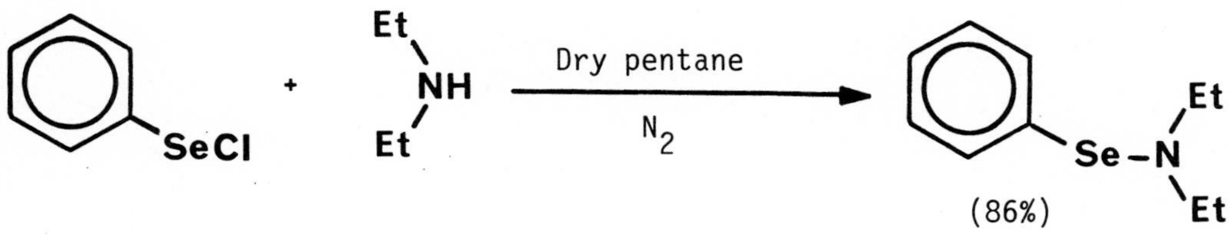

C_6H_5SeCl + Et_2NH $\xrightarrow[N_2]{\text{Dry pentane}}$ C_6H_5Se–NEt_2 (86%) ⑤

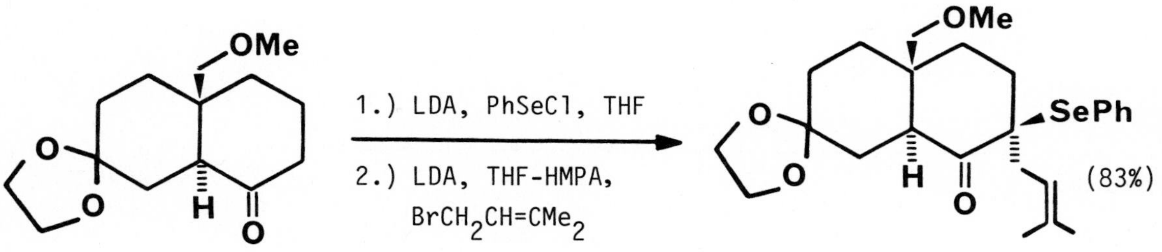

REAGENT: BENZENESULFENYL CHLORIDE (PHENYLSULFENYL CHLORIDE)

STRUCTURE:

C_6H_5–S–Cl

MW = 144.62
BP = 66° @ 4mm Hg

Fieser's Reagents for Organic Synthesis: Vol., pages

1		**4**		**7**		**10**	24
2		**5**	523- 24	**8**	32- 34	**11**	39- 40
3		**6**	30 -32	**9**	35- 38	**12**	42- 44

March's Advanced Organic Chemistry: Not indexed

MAJOR USES:

1. Chlorosulfenylation 2. Homoconjugate addition 3. Rearrangements

PREPARATION:

This reagent can be prepared from benzenethiol and sulfuryl chloride (**Reagents, Vol. 5,** 523); or from benzenethiol and chlorine (see: E. Kuhle, **Synthesis**, (1970), 561.

PRECAUTION:

1. Highly toxic

2. Corrosive

NOTES:

REFERENCES:

1. W. Reischl and W.H. Okamura, **J. Am. Chem. Soc.**, (1982), 104, 6115.

2. M. Ihara and K. Fukumoto, **Heterocycles**, (1982), 19, 1435.

3. D. Heissler and J.-J. Riehl, **Tetrahedron Lett.**, (1979), 3957.

4. D.A. Armitage, **Synthesis**, (1984), 1042.

5. W.L. Brown and A.G. Fallis, **Tetrahedron Lett.**, (1985), 607.

EXAMPLES:

OH
CCH_3
C≡CH

PhSCl, Et_3N

(80%)
SOPh

① [1]

1.) PhSCl
2.) 25°

O=
NHBzl

Cl
SPh
O=
NHBzl (85-90%)

② [1]

CH_2

PhSCl, CH_2Cl_2, 0°

PhS
(85%)
CH_2Cl

③ [2]

SCl + NaOMe

Et_2O

SOMe
(80%)

④

HO

1.) n-BuLi
2.) PhSCl

O
‖
SPh (45%)

⑤ [3]

REAGENT: BENZYLTRIMETHYLAMMONIUM HYDROXIDE (**TRITON B**)

STRUCTURE:

MW = 167.25
d = 0.920

$\overset{\ominus}{HO}\overset{\oplus}{N}(CH_3)_3$—$CH_2$—$C_6H_5$

<u>Fieser's Reagents for Organic Synthesis</u>: Vol., pages

1	1252- 54	**4**		**7**		**10**	
2		**5**	29	**8**	36- 37	**11**	
3		**6**		**9**		**12**	48- 49

March's <u>Advanced Organic Chemistry</u>: Not indexed

MAJOR USES:

1. Base

PREPARATION:

1. Commercially available as a methanol or water solution

PRECAUTION:

1. The MeOH solution is highly toxic and flammable. The water solution is corrosive and toxic.

NOTES:

REFERENCES:

1. A. Brown, D. Corbett and T. Howarth, **Chem. Commun.**, (1977), 359.

2. W.J. Horton and L.L. Pitchforth, **J. Org. Chem.**, (1960), <u>25</u>, 131.

3. E. Ghera and Y. Sprinzak, **J. Am. Chem. Soc.**, (1960), <u>82</u>, 4945.

4. M. Auramoff and Y. Sprinzak, **J. Org. Chem.**, (1961), <u>25</u>, 131.

5. H.J. Bestmann and H. Frey, **Ann.**, (1980), 2061.

EXAMPLES:

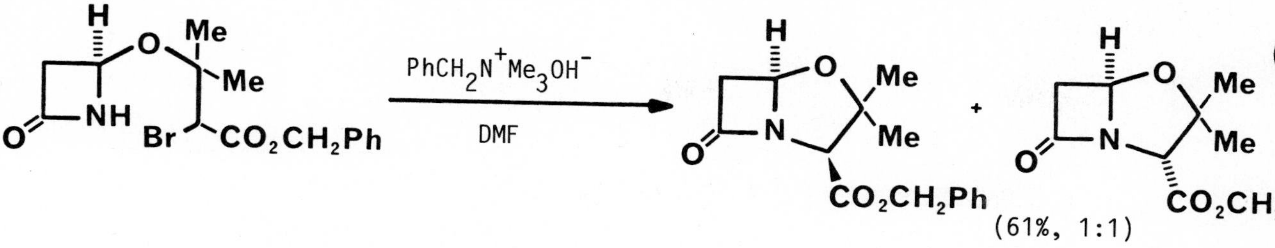

$PhCH_2N^+Me_3OH^-$
DMF
(61%, 1:1)

(1)

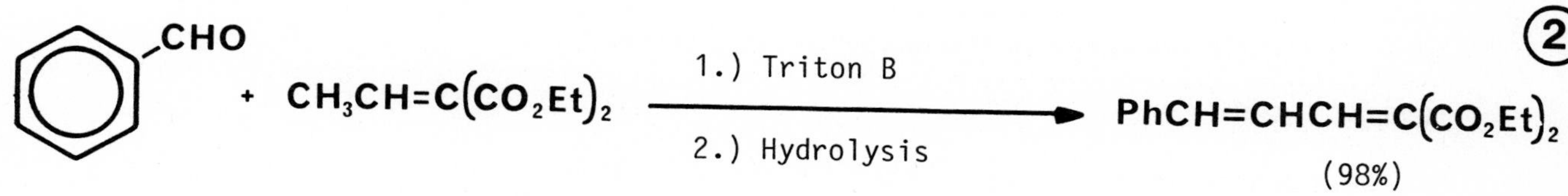

1.) Triton B
2.) Hydrolysis
$PhCH=CHCH=C(CO_2Et)_2$
(98%)

(2)

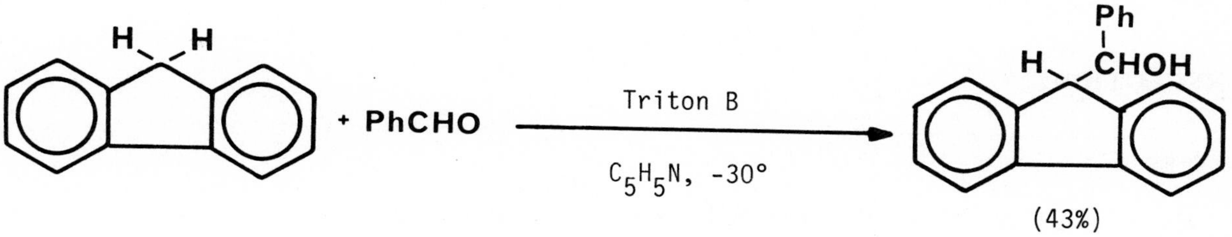

PhCHO
Triton B
C_5H_5N, -30°
(43%)

(3)

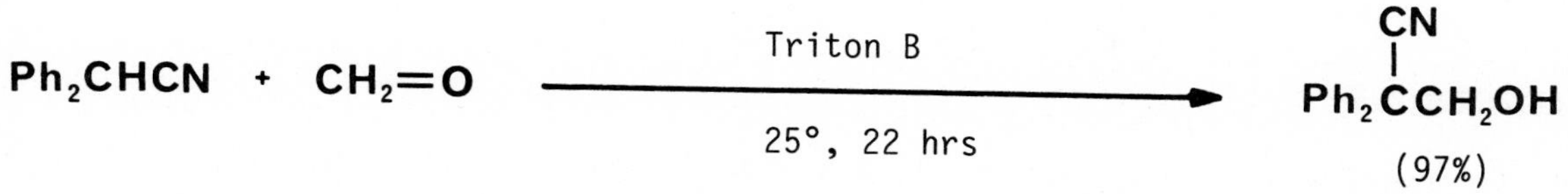

Ph_2CHCN + $CH_2=O$
Triton B
25°, 22 hrs
$Ph_2C(CN)CH_2OH$
(97%)

(4)

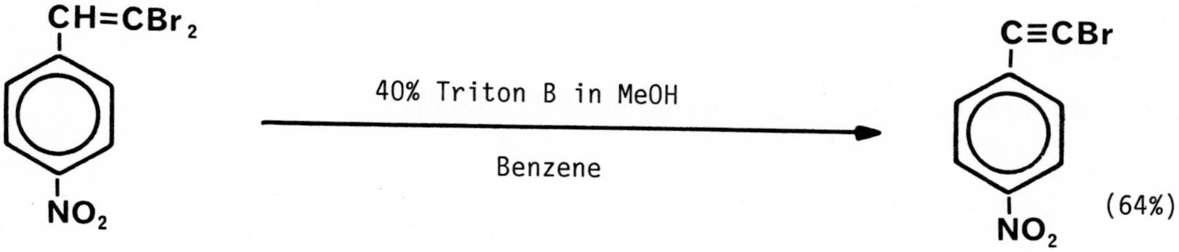

40% Triton B in MeOH
Benzene
(64%)

(5)

REAGENT: BIS(ACETONITRILE)DICHLOROPALLADIUM 1

==

STRUCTURE:

MW = 259.41

MP = >300°

$(CH_3CN)_2PdCl_2$

Fieser's Reagents for Organic Synthesis: Vol., pages

1		**4**		**7**	21- 22	**10**	30-31
2		**5**		**8**	39	**11**	46-47
3		**6**		**9**	44	**12**	50- 51

March's Advanced Organic Chemistry: Not indexed

MAJOR USES:

1. **Cope rearrangement**
2. Alkene alkylation

PREPARATION:

Commercially available

PRECAUTION:

NOTES:

1. Also known as Bis(acetonitrile)palladium(II) chloride

2. This is a **Claisen-type rearrangement**, where the carboxyl carbonyl group acts as one of the unsaturated moieties.

REFERENCES:

1. L.E. Overman and E.J. Jacobsen, **J. Am. Chem. Soc.**, (1982), 104, 7225.

2. F.E. Ziegler, U.R. Chakraborty and R.B. Weisenfeld, **Tetrahedron**, (1981), 37, 4035.

3. L.V. Dunkerton and A.J. Serino, **J. Org. Chem.**, (1982), 47, 2512.

4. L.E. Overman and A.F. Renaldo, **Tetrahedron Lett.**, (1983), 3757.

5. N. Kurokawa and Y. Ohfune, **Tetrahedron Lett.**, (1985), 83.

EXAMPLES:

(MeCN)$_2$PdCl$_2$, THF

20°

Ph

H$_2$C

[2Z,5R] (86%) [2E,5S]

[97% ee] (7:3) [96% ee]

① [1]

$CH_3(CH_2)_5$ C=C (H, I, H) + $CH_2{=}CHCCH_3$ (O) —Pd(II), CH_3CN→ $CH_3(CH_2)_5CH{=}CHCH{=}CHCCH_3$ (O)

(81%, E,E:E,Z=20:1)

② [2]

1.) Pd(CH$_3$CN)$_2$Cl$_2$, Et$_3$N

2.) Na$^+$C$^-$(COOEt)$_2$ (CH$_3$)

CH$_3$ C(CO$_2$Et)$_2$

(50%)

③ [2]

MeC(O) Me

(CH$_3$CN)$_2$PdCl$_2$

CH$_2$Cl$_2$

MeC(O) Me

(94%)

④ [1]

H NHBoc CO$_2$Me AcO H

(MeCN)$_2$PdCl$_2$

C$_6$H$_6$

AcO H NHBoc CO$_2$Me

(60%)

⑤ [2] △2

REAGENT: 9-BORABICYCLO[3.3.1]NONANE **(9-BBN)**

STRUCTURE:

MW = 122.02
FP = -22°C

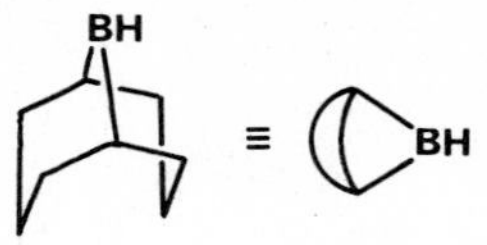

Fieser's Reagents for Organic Synthesis: Vol., pages

1		**4**	41	**7**	29-31	**10**	48-49
2	31	**5**	46-47	**8**	47-49	**11**	68
3	24-29	**6**	62-64	**9**	57-58	**12**	

March's Advanced Organic Chemistry: 426-7, 694, 704, 707, 810, 812, 1095

MAJOR USES:

1. Hydroboration 2. Aldehyde synthesis 3. Carboethoxymethylation

4. Carbonylation 5. Alkylation and arylation 6. Reductions

PREPARATION:

For the preparation from diborane and 1,5-cyclooctadiene, see: H.C. Brown, E. Knights and C. Scouten, **J. Am. Chem. Soc.**, (1974), 96, 7765. For a simplified preparation from $BH_3.SMe_2$, see: J. Sonderquist and H.C. Brown, **J. Org. Chem.**, (1981), 46, 4599. The reagent is commercially available.

PRECAUTION:

Although much more stable than most borane derivatives, the reagent is moisture sensitive and is potentially pyrophoric.

NOTES:

1. This reaction uses a 100% excess of the reagent and no solvent.

2. This derivative is known as "**Midland's Reagent**"- M.M. Midland, S. Greer, A. Tramontano and S.A. Zderil, **J. Am. Chem. Soc.**, (1979), 101, 2352.

REFERENCES:

1. Z. Benmaarouf- Khallaayoun, M. Baboulene, V. Speziale and A. Lattes, **Synth. Commun.**, (1985), 15, 233.

2. F.M. Hauser and R.P. Rhee, **J. Org. Chem.**, (1981), 46, 229.

3. H.C. Brown and G.G. Pai, **J. Org. Chem.**, (1983), 48, 1784.

4. H.C. Brown ad J.V.N. Vera Prasad, **J. Org. Chem.**, (1985), 50, 3002.

EXAMPLES:

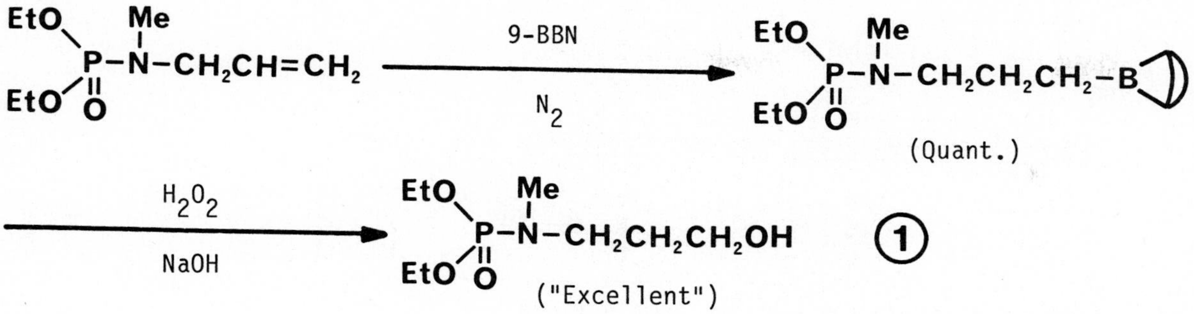

EtO(EtO)P(O)–N(Me)–$CH_2CH=CH_2$ —(9-BBN, N_2)→ (EtO)(EtO)P(O)–N(Me)–$CH_2CH_2CH_2$–B (Quant.) —(H_2O_2, NaOH)→ (EtO)(EtO)P(O)–N(Me)–$CH_2CH_2CH_2OH$ ("Excellent") ①

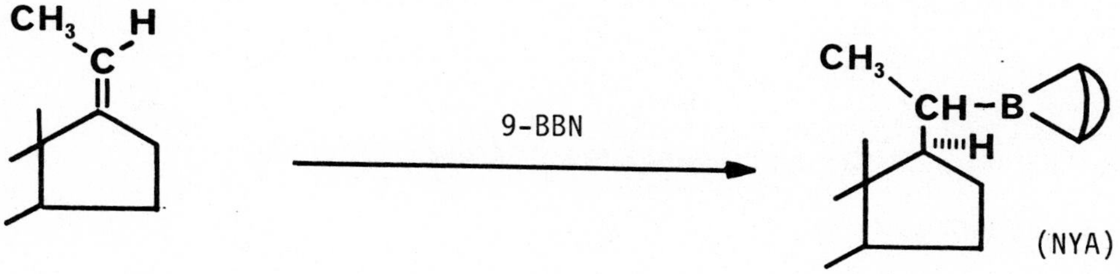

CH_3, H, C —(9-BBN)→ CH_3, CH–B, H (NYA) ② [1]

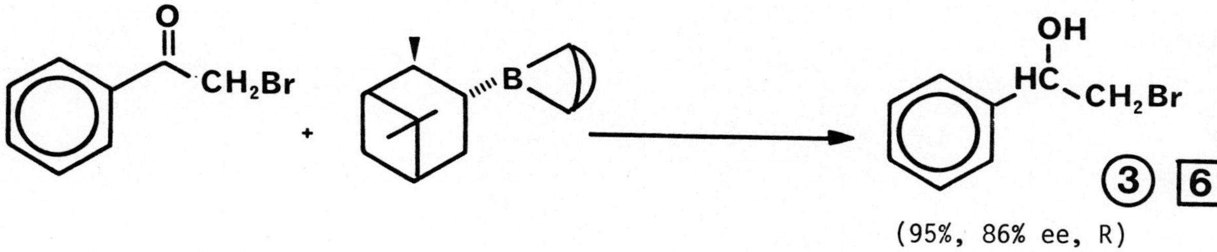

O, CH_2Br + B → OH, HC, CH_2Br (95%, 86% ee, R) △1 △2 ③ [6]

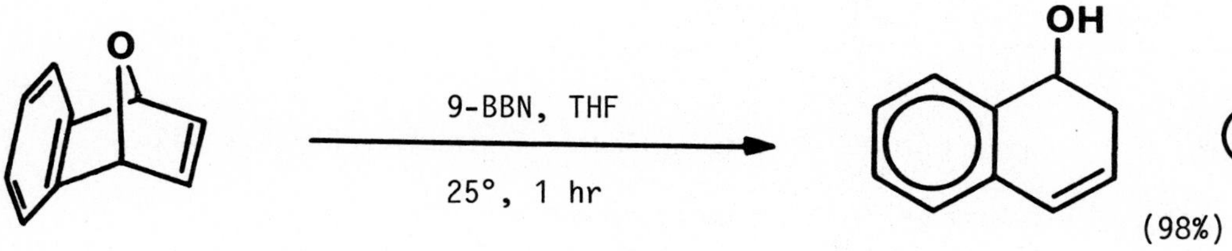

O —(9-BBN, THF, 25°, 1 hr)→ OH (98%) ④ [1]

REAGENT: BORANE-DIMETHYL SULFIDE (BMS)

STRUCTURE:

MW = 75.97
FP = 15°C
d = 0.790

$(CH_3)_2S \cdot BH_3$

Fieser's Reagents for Organic Synthesis: Vol., pages

1	**4** 124, 191	**7**	**10** 49- 50
2	**5** 47	**8** 49- 50	**11** 69
3	**6**	**9**	**12** 64

March's Advanced Organic Chemistry: Not indexed

MAJOR USES:

1. Stable substitute to diborane
2. Reducing agent

PREPARATION:

1. See A. Burg and R. Wagner, **J. Am. Chem. Soc.**, (1954), 76, 3307.

2. Also commercially available.

PRECAUTION:

1. All solutions are flammable and moisture sensitive.

NOTES:

REFERENCES:

1. H.P. Flaumann, J.G. Smith and R. Rodrigo, **Chem. Commun.**, (1980), 354.

2. S. Krishnamurthy, **Tetrahedron Lett.**, (1982), 3315.

3. H.C. Brown and Y.M. Choi, **Synthesis**, (1981), 439.

4. B.L. Jensen, J. jewett-Bronson, S.B. Hadley and L.G. French, **Synthesis**, (1982), 732.

5. H.M. Hugel, **Synthesis**, (1983), 935.

EXAMPLES:

$BH_3 \cdot SMe_2$

THF, Δ

(NYA)

① [2]

NHCHO, CO_2Et → $NHCH_3$, CO_2Et

BMS, THF

65°, 2 hrs

(96%)

② [2]

CO_2Et → CH_2OH

BMS, THF

4 hrs

(90%)

③ [2]

O, Br, Br → OH, Br, Br

BMS

Anhydrous ether, Reflux

(88%)

④ [2]

MeO, Cl, NH_2 → MeO, Cl

1.) BMS THF, Reflux

2.) Ac_2O, Pyridine

(76%)

⑤ [2]

REAGENT: BORON TRIFLUORIDE (BORON TRIFLUORIDE ETHERATE)

STRUCTURE:

	BF_3		$BF_3 \cdot (C_2H_5)_2O$
MW = 67.82		141.94	
BP = -127°		126°	

Fieser's Reagents for Organic Synthesis: Vol., pages

1	68-72	**4**	44-45	**7**	31-32	**10**	50-56
2	35-36	**5**	51-55	**8**	51-52	**11**	71-75
3	32-33	**6**	65-67	**9**	64-65	**12**	66- 70

March's Advanced Organic Chemistry: 7, 437, 785, 990, 1100

MAJOR USES: 1

1. Lewis acid catalyst (less reactive than $AlCl_3$) 2. Used in generating diborane 3. Used for acid-catalyzed cyclizations 4. Epoxide rearrangement 5. Dehydration

PREPARATION:

This reagent is most often used as the etherate complex. In this form it is often dark-colored, but is purified by distillation. See: **Org. Reactions**, 13, 28, (1963).

PRECAUTION:

Avoid inhaling vapors. Toxic. Irritating to the eyes and mucus membranes. Corrosive to skin.

NOTES:

1. The order of reactivity of other boron trihalides is:

$$BI_3 > BBr_3 > BCl_3 > BF_3$$

REFERENCES:

1. G. Schiemen and U. Schmidt, **Ann. Chem.**, (1982), 1509.

2. J. Tou and W. Reusch, **J. Org. Chem.**, (1980), 45, 5012.

3. A.B. Smith, III and R. Dieter, **J. Am. Chem. Soc.**, (1981), 103, 2009, 2017.

4. K. Fuji, T. Kawabata, M. Node, and E. Fujita, **Tetrahedron Lett.**, (1981), 875.

5. Y. Ito, H. Imai, T. Matsuura and T. Saegusa, **Tetrahedron Lett.**,(1984), 3091.

EXAMPLES:

MeO, OMe, OMe; $PhCH{=}CHCO_2H + BF_3$, 80° → MeO, O, BF_2 (⊖), O⊕, OMe, CH=CHPh (94%) ① [1]

O, O, MeO + (diene); BF_3, 0° → O, MeO, O, H; O, MeO, O, H (70-80%, 2.4:1) ② [1]

$COCH_2NH_2$, $C{=}CH_2$, CH_3; $BF_3\cdot(C_2H_5)_2O$ → O (40%) ③ [1]

CN, CH=C, CO_2Et; $BF_3\cdot Et_2O$, EtSH → $CH(SEt)_2$ (88%) ④

Li, *t*-Bu–N=C–*n*-Bu + CuI; (cyclohexenone, O), $BF_3\cdot OEt_2$ → *n*-Bu, C=N–*t*-Bu, O (71%) ⑤ [1]

REAGENT: BROMINE

STRUCTURE:

MW = 159.82
BP = 58.8° **MP** = -7.2° Br_2
d = 3.119

Fieser's Reagents for Organic Synthesis: Vol., pages

1		**4**	46.7	**7**	33-35	**10**	56
2		**5**	55-57	**8**	52-53	**11**	75-76
3	34	**6**	70-73	**9**	65-66	**12**	70- 71

MAJOR USES:

1. Bromination
2. Oxidative cleavage
3. Reduction
4. Ring contraction
5. Selective oxidation

PREPARATION:

The reagent is readily available from commercial sources.

PRECAUTION:

Bromine is a strong oxidizing agent, and is highly toxic.

NOTES:

1. The bromine initiates the 1,4-addition of Br and -OMe to the "diene system" of the furan derivative. Then, the reactive allylic bromine is displaced by MeOH.

REFERENCES:

1. I. Fleming and J. Goldhill, Jr., **J.C.S. Perkin I,** (1980), 1493.

2. A.J. Biloski, R.D. Wood, and B. Ganem, **J. Am. Chem. Soc.**, (1982), 104, 3233.

3. B.P. Gunn, **Tetrahedron Lett.**, (1985), 2869.

4. A.M. Morella and A.D. Ward, **Tetrahedron Lett.**, (1985), 2899.

5. P. Knochel and J.F. Normant, **Tetrahedron Lett.**, (1985), 425.

EXAMPLES:

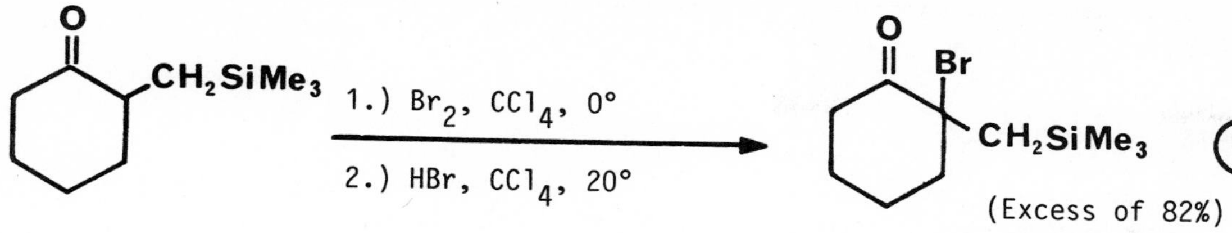

1.) Br_2, CCl_4, 0°

2.) HBr, CCl_4, 20°

CH_2SiMe_3 → Br, CH_2SiMe_3 ① [1]

(Excess of 82%)

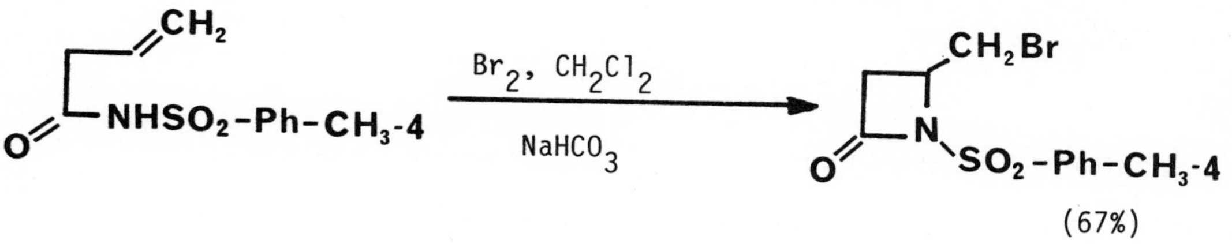

CH_2, $NHSO_2$-Ph-CH_3-4

Br_2, CH_2Cl_2

$NaHCO_3$

CH_2Br, N, SO_2-Ph-CH_3-4 ②

(67%)

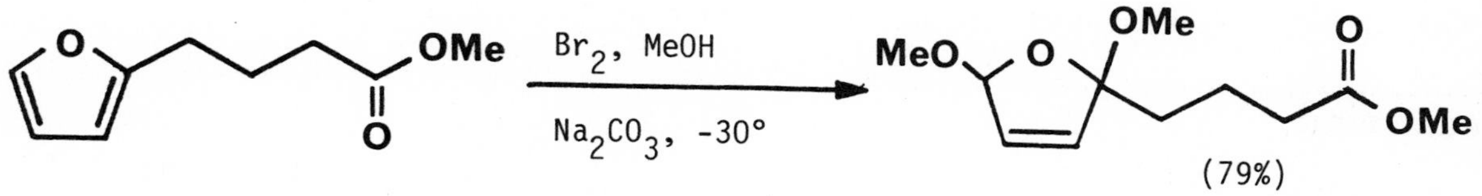

OMe

Br_2, MeOH

Na_2CO_3, -30°

MeO, O, OMe, OMe ③ △1

(79%)

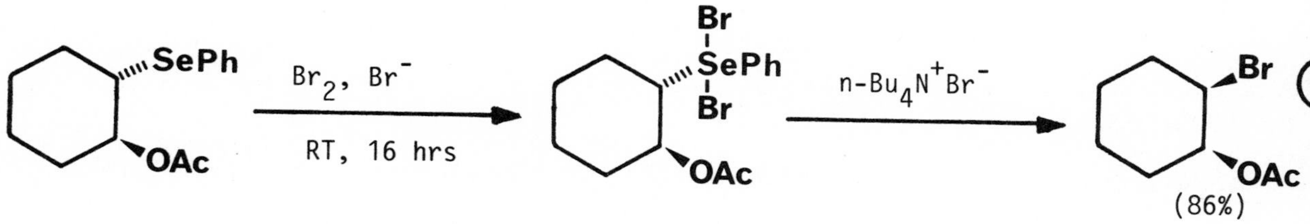

SePh, OAc

Br_2, Br^-

RT, 16 hrs

Br, SePh, Br, OAc

n-$Bu_4N^+Br^-$

Br, OAc ④ [5]

(86%)

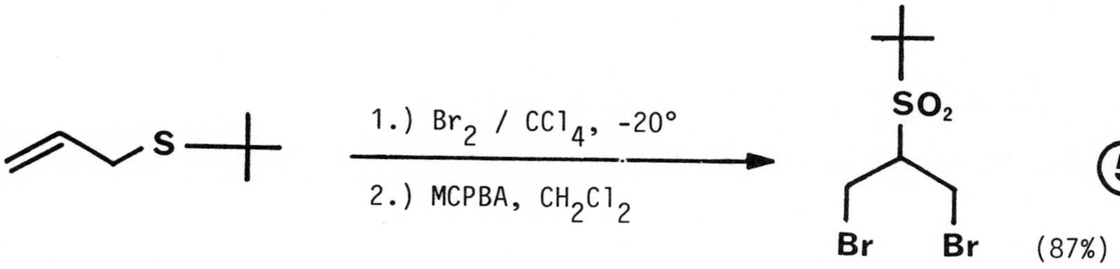

S

1.) Br_2 / CCl_4, -20°

2.) MCPBA, CH_2Cl_2

SO_2, Br, Br ⑤ [1]

(87%)

REAGENT: N-BROMOSUCCINIMIDE (**NBS**)

==

STRUCTURE:

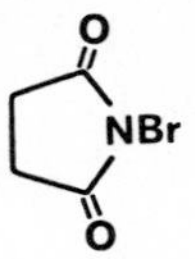

MW = 178.0
MP = 173-75°

Fieser's Reagents for Organic Synthesis: Vol., pages

1	78- 80	**4**	49- 53	**7**	37- 40	**10**	57- 59
2	40- 42	**5**	65- 66	**8**	54- 56	**11**	79
3	34- 36	**6**	74- 76	**9**	70- 72	**12**	

March's Advanced Organic Chemistry: 384, 477, 522, 529, 531, 574, 621, 624-26, 997, 1059, 1061, 1063, 1083.

MAJOR USES:

1. Allylic bromination   2. Oxidizing agent   3. Brominating agent

---

PREPARATION:

1. Commercially available

2. The product is often colored and can be recrystallized from water.

PRECAUTION:

1. See: **Nature**, (1951), 168, 32 for a note regarding an explosion using **NBS**.

2. Highly irritating to eyes, skin and mucous membranes.

NOTES:

REFERENCES:

1. M. Kodama, S. Yokoo, H. Yamada and S. Ito, **Tetrahedron Lett.**, (1978), 3121.

2. N. Stojanac, Z. Stojanan, P.S. White and S. Valenta, **Can. J. Chem.**, (1979), 57, 3346.

3. S. Danishefsky and K. Tsuzuki, **J. Am. Chem. Soc.**, (1980), 102, 6891.

4. R. Inhof, E. Grossinger, W. Graf, L. Benes-Fenz, H. Berner, R. Schaufelberger, and H. Wehrle, **Helv.**, (1973), 56, 139.

5. G. Hartman, W. Halczenko and B.T. Phillips, **J. Org. Chem.**, (1986), 51, 142.

EXAMPLES:

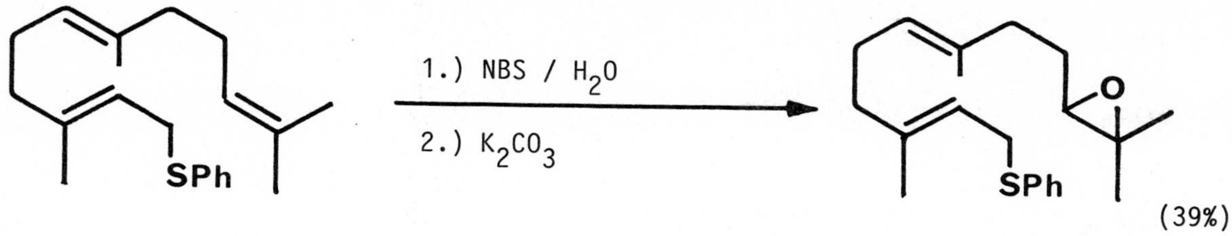

1.) NBS / H_2O

2.) K_2CO_3

SPh → SPh (39%) ① [3]

NBS / CCl_4

t-BuOOH

CO_2Et, O, O → CO_2Et, Br, O, O (NYA) ② [1]

NBS

PhC(O)OOC(O)Ph

CO_2Me → CO_2Me, CH_2Br (NYA) ③ [1]

1.) NBS

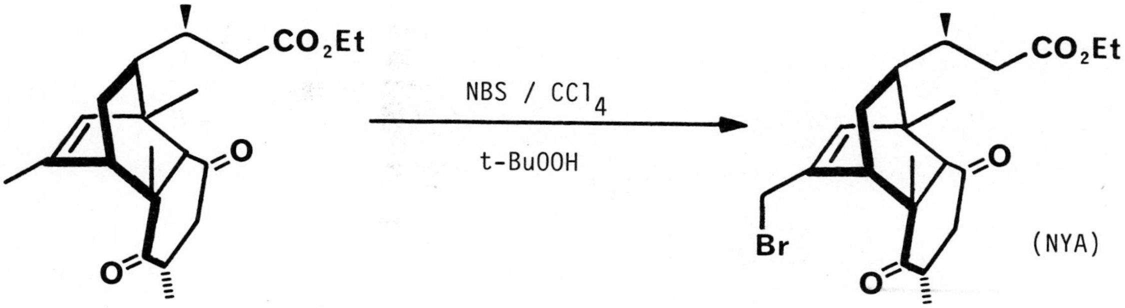

2.) Base

3.) NBS

4.) Base

AcO, AcO, O, Me, MeO, O, OAc → AcO, O, Me (39%) ④ [3]

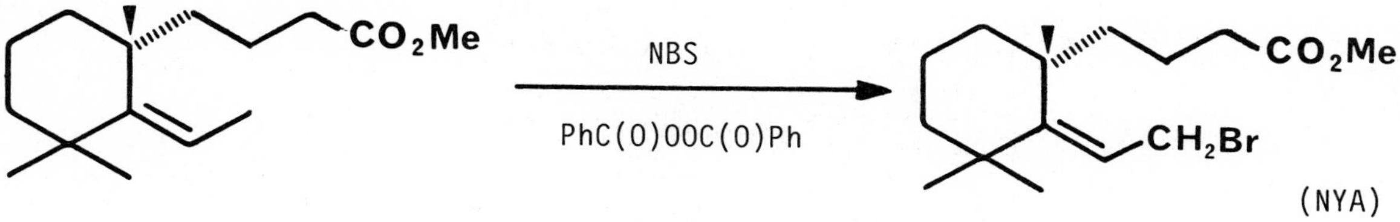

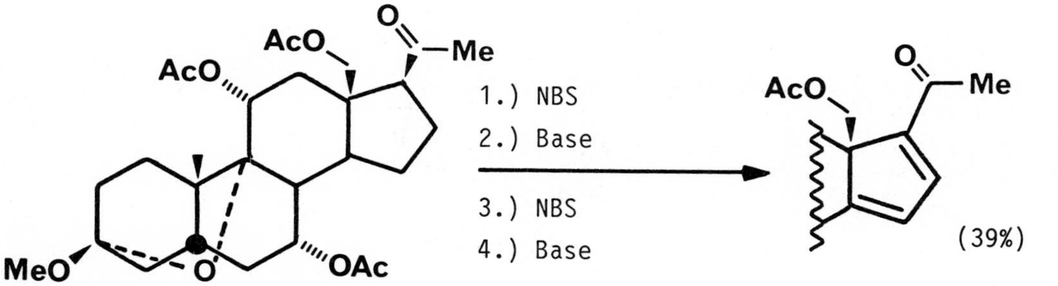

NBS/ CCl_4

Benzoyl peroxide, h

Me, S → CH_2Br, S (76%) ⑤ [1]

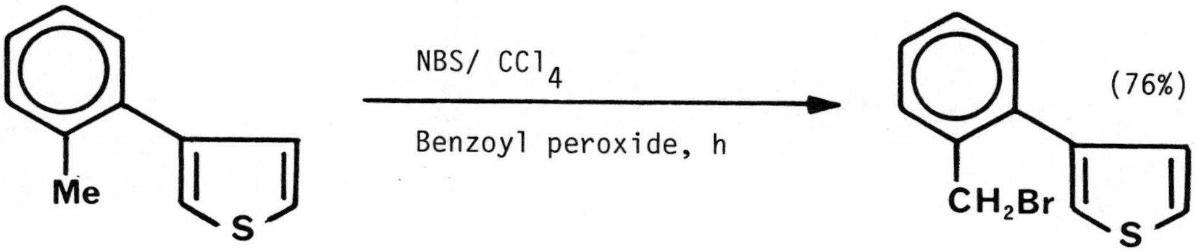

REAGENT: t-BUTYLHYPOCHLORITE

STRUCTURE:

MW = 108.57
BP = 77- 78°C

$(CH_3)_3C-O-Cl$

Fieser's Reagents for Organic Synthesis: Vol., pages

1	90- 94	**4**	58- 60	**7**		**10**	66- 67
2	50	**5**	77- 78	**8**		**11**	
3	38	**6**	82	**9**		**12**	

March's Advanced Organic Chemistry: 532, 574, 621-22, 624, 626, 1089

MAJOR USES:

1. Halogenation 2. Oxidizing agent 3. N-Chlorination

PREPARATION:

1. Chlorine in an alkaline solution of t-BuOH (**Org. Synth.**, (1963), Coll. Vol. 4, 125).

2. A simpler procedure uses commercial bleach: **Org. Synth.**, (1969), 49, 9.

PRECAUTION:

1. Explosions have been reported during the preparation using Cl_2.

2. Do not let the temperature exceed 20° during the preparation.

3. Harmful to mucous membranes and eyes.

4. Violent reactions reported when the reagent is exposed to rubber, strong light or over-heating.

NOTES:

REFERENCES:

1. R.S. Guass, M. Hojjatie, W.N. Setzer and G.S. Wilson, **J. Org. Chem.**, (1986), 51, 1815.

2. A.J. Mancuso and D. Swern, **Synthesis**, (1981), 165.

3. Z. Lidert and S. Gronowitz, **Synthesis**, (1980), 322.

4. M. Haake and H. Gebbing, **Synthesis**, (1979), 98.

5. M.R. Detty, **J. Org. Chem.**, (1980), 45, 274.

EXAMPLES:

H_3C-S ... $C(CH_3)_2OH$ — 1.) Me_3COCl, CH_2Cl_2, Argon; 2.) $HgCl_2$ → $H_3C-S^{\oplus}$ ... O ... CH_3, CH_3 (Quant.) ① [1]

Cl–C_6H_4–NH_2 + DMSO — Me_3COCl / CH_2Cl_2 → Cl–C_6H_4–$N{=}S(O)Me_2$ (68%) ② [3]

$PhCH_2NHCH_2C(O)NHPh$ — 1.) t-BuOCl; 2.) NaOMe → $PhCH_2N{=}CHC(O)NHPh$ (83%) ③ [3]

N-S-Ph + MeOH — 1.) t-BuOCl / CH_2Cl_2; 2.) $AgBF_4$ → N-$S^{\oplus}$(OMe)-Ph + $\overset{\ominus}{B}F_4$ (80%) ④ [1]

PhSe ... O ... SePh — t-BuOCl, MeOH / CH_2Cl_2 → O (55%) ⑤ [1]

REAGENT: n-BUTYLLITHIUM

STRUCTURE:

$CH_3(CH_2)_3Li$

MW = 64.06
FP = -12°C
d = .680 for a 1.6M hexane solution

Fieser's Reagents for Organic Synthesis: Vol., pages

1	95- 96	**4**	60- 63	**7**	45- 47	**10**	68- 71
2	51- 53	**5**	78	**8**	65- 66	**11**	101- 103
3		**6**	85- 91	**9**	83- 87	**12**	96

March's Advanced Organic Chemistry: 400, 545

MAJOR USES:

1. Strong base
2. Lithiations

PREPARATION:

1. The reagent can be prepared from n-butyl halide and lithium wire (R.G. Jones and H. Gilman, **Org. Reactions**, (1951), 6, 352). Also commercially available.

PRECAUTION:

1. Solutions are pyrophoric, corrosive and very moisture sensitive. Best to keep the reagents under an argon atmosphere.

NOTES:

1. **Wittig rearrangement**.

REFERENCES:

1. T. Nakai, K. Mikami, S. Taya and Y. Fujita, **J. Am. Chem. Soc.**, (1981), 103, 6492.

2. P. Fischer and G. Schaefer, **Angew. Chem. Int. Ed.**, (1981), 20, 863.

3. A. Hosomi, Y. Araki and H. Sakurai, **J. Am. Chem. Soc.**, (1982), 104, 2081.

4. R.L. Shone, J.R. Deason and M. Miyano, **J. Org. Chem.**, (1986), 51, 268.

5. Y. Hanzawa, K. Kawagoe, N.Kobayashi, T. Oshima and Y. Kobayashi, **Tetrahedron Lett.**, (1985), 26, 2877.

EXAMPLES:

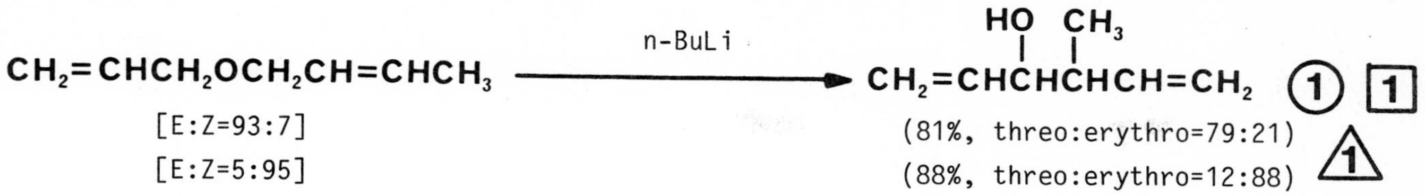

$CH_2{=}CHCH_2OCH_2CH{=}CHCH_3$ [E:Z=93:7] [E:Z=5:95] —n-BuLi→ $CH_2{=}CHCH(OH)CH(CH_3)CH{=}CH_2$ (81%, threo:erythro=79:21) (88%, threo:erythro=12:88) ① [1] △1

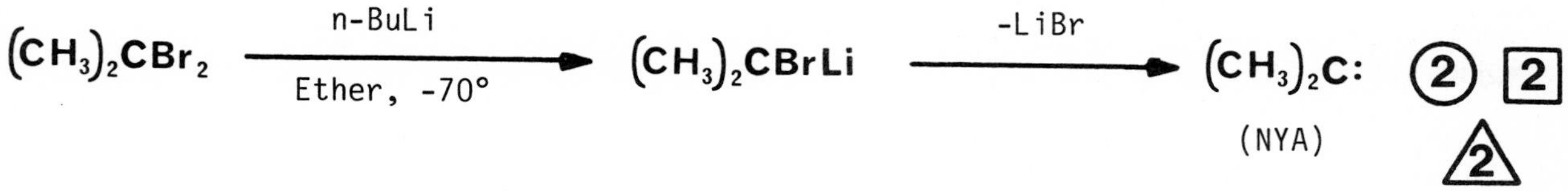

$(CH_3)_2CBr_2$ —n-BuLi, Ether, -70°→ $(CH_3)_2CBrLi$ —-LiBr→ $(CH_3)_2C:$ (NYA) ② [2] △2

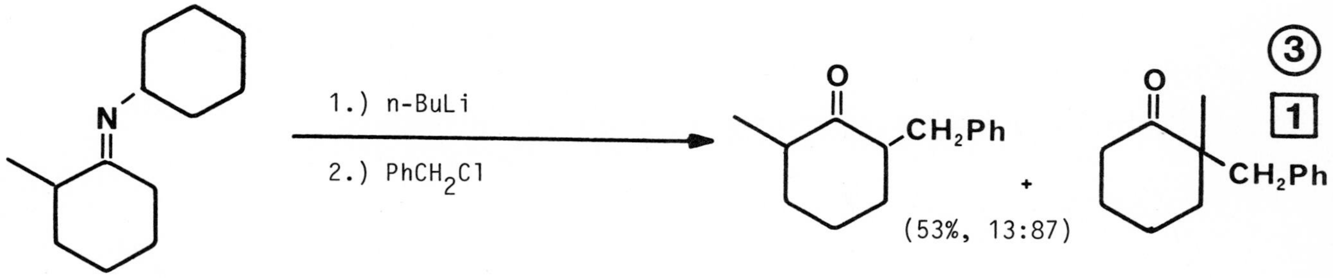

N-cyclohexyl imine of 2-methylcyclohexanone —1.) n-BuLi; 2.) $PhCH_2Cl$→ 2-methyl-6-benzylcyclohexanone + 2-methyl-2-benzylcyclohexanone (CH_2Ph) (53%, 13:87) ③ [1]

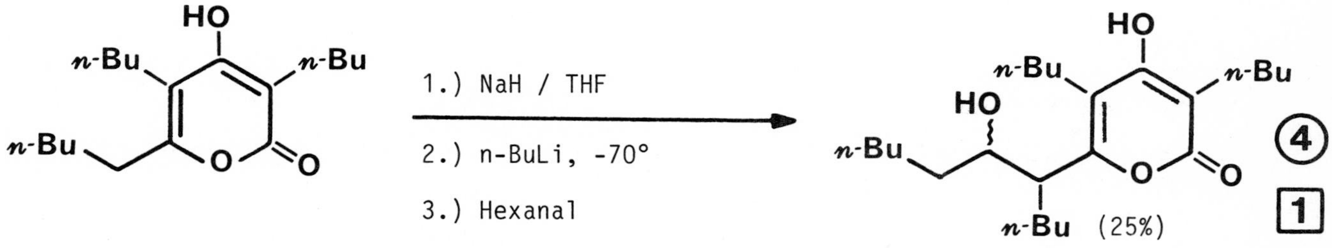

4-hydroxy-3,5-di-n-Bu-6-(n-Bu-CH₂)-2-pyrone —1.) NaH / THF; 2.) n-BuLi, -70°; 3.) Hexanal→ HO, n-Bu, n-Bu, HO, n-Bu, n-Bu product (25%) ④ [1]

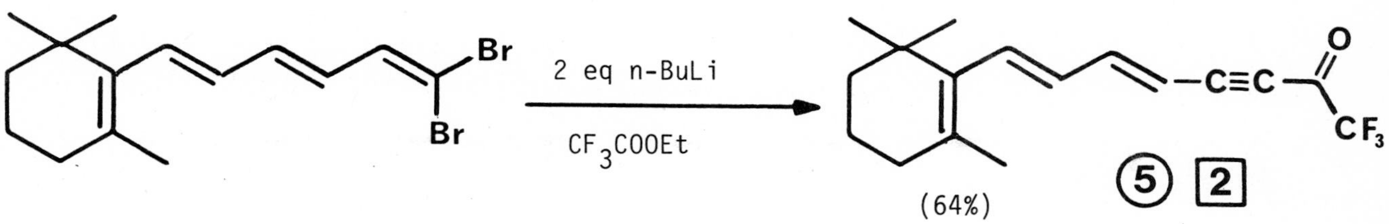

Dibromo-triene (Br, Br) —2 eq n-BuLi, CF_3COOEt→ —C≡C—C(=O)CF_3 (64%) ⑤ [2]

REAGENT: t-BUTYLLITHIUM

STRUCTURE:

$$CH_3-C(CH_3)_2Li$$

MW = 64.06
FP = -6° (a 1.7M solution in pentane)
d = 0.669 (a 1.7M solution in pentane)

<u>Fieser's Reagents for Organic Synthesis</u>: Vol., pages

Vol.	pages	Vol.	pages	Vol.	pages	Vol.	pages
1	96-97	**4**		**7**	47	**10**	76- 77
2		**5**	79- 80	**8**	70- 72	**11**	103- 105
3		**6**		**9**	89	**12**	100

March's <u>Advanced Organic Chemistry</u>: 400

MAJOR USES:

1. Lithiations
2. Br-Li exchange
3. Cyclizations
4. Strong base

PREPARATION:

1. This reagent can be prepared from t-butylchloride and lithium, see: **Reagents**, Vol. 1, page 96.

2. Commercially available as a pentane solution.

PRECAUTION:

Pyrophoric. This reagent is very reactive to protic solvents. Reactions should ensure anhydrous conditions and be maintained under an inert atmosphere such as argon.

NOTES:

1. **Wurtz reaction.**

REFERENCES:

1. W.F. Bailey and R.P. Gagnier, **Tetrahedron Lett.**, (1982), 5123.

2. C.J. Kowalski and K.W. Fields, **J. Am. Chem. Soc.**, (1982), <u>104</u>, 321.

3. C.J. Kowalski, M.L. O'Dowd, M.L. Burke and K.W. Fields, **J. Am. Chem. Soc.**, (1980), <u>102</u>, 5411.

4. P.E. Peterson, D.J. Nelson and R. Risener, **J. Org. Chem.**, (1986), <u>51</u>, 2381.

5. W.D. Wulff, G.A. Peterson, W.E. Bauta, K.-S. Chan, K.L. Faron, S.R. Gilbertson, R.W. Kaesler, D.C. Yang and C.K. Murray, **J. Org. Chem.**, (1986), <u>51</u>, 277.

EXAMPLES:

CH_2I / CH_2I — t-BuLi → H / H (97%) ① [1] △1

OLi
t-Bu–C=CBr_2 — t-BuLi, THF, -78°, then RT → PhCHO, -78°, then RT → Ph, t-Bu, COOH (75%) ②

=O, Br — 1.) $LiN[Si(CH_3)_3]_2$ 2.) t-BuLi → OLi, Li (NYA) ③ [1]

O, Br, O — t-BuLi, Me_3SiCl → O, $SiMe_3$, O (76%) ④ [1]

Br — 1.) t-BuLi 2.) $Cr(CO)_6$ 3.) $MeSO_3F$ → $(CO)_5Cr=C$, OMe (71%) ⑤ [1]

REAGENT: CERIC AMMONIUM NITRATE (**CAN**, AMMONIUM CERIUM(IV) NITRATE)

STRUCTURE:

$$(NH_4)_2Ce(NO_3)_6$$

MW = 548.23

Fieser's Reagents for Organic Synthesis: Vol., pages

1	120-1	**4**	71-4	**7**	55-6	**10**	79-81
2	63-5	**5**	101-2	**8**		**11**	114
3	44-5	**6**	99	**9**	99	**12**	

March's Advanced Organic Chemistry: 632, 787, 1058, 1077, 1079

MAJOR USES:

1. Oxidations   2. Oxidative cleavage   3. Aromatic substitution

PREPARATION:

This reagent is commercially available in 99.99% purity. The reagent can be prepared by dissolving $CeO\cdot H_2O$ in hot, concentrated nitric acid.

PRECAUTION:

This is an oxidizing agent and an irritant.

NOTES:

REFERENCES:

1. H. Cristau, B. Chabaud, R. Lababdiniere and H. Christol, **Synth. Commun.**, (1981), 11, 423.

2. H. Uno, **J. Org. Chem.**, (1986), 51,, 350.

3. M.P. Sibi, J.W. Dankwardt, and V. Snieckus, **J. Org. Chem.**, (1986), 51, 271.

4. F. Chioccara and E. Novellino, **Synth. Commun.**, (1986), 16, 967.

5. T. Chorn, R. Giles, P. Mitchell and I. Green, **J.C.S. Chem. Commun.**, (1981), 534.

EXAMPLES:

2,2-Dimethyl-1,3-dithiane —(2 CAN)→ —(2 CAN, −2 H^+)→ $(CH_3)_2C{=}O$ (75%) + 1,3-dithiane 1-oxide (60%) ① [2]

8-Methoxy-1-hydroxy-4-methoxy-2-acetylnaphthalene —(CAN, Acetonitrile, H_2O)→ 5-methoxy-2-acetyl-1,4-naphthoquinone (92%) ② [2]

1,4-Dimethoxy-5-acetoxynaphthalene (OMe, OAc, OMe) —(CAN)→ 5-acetoxy-1,4-naphthoquinone (OAc) (90%) ③ [2]

NH_2, CH_2CHCO_2H, HO, OH (DOPA) —(1.) CAN, H_2SO_4; 2.) L-cysteine, H_2SO_4)→ NH_2, CH_2CHCO_2H, HO, OH, SCH_2CHCO_2H (NH_2) (62%) ④ [3]

MeO, CH_2OH, MeO, $CH{=}CHCH_2CH_2CH_3$ naphthalene —(CAN)→ pyranonaphthoquinone with OH and $CH_2CH_2CH_3$ (54%) + 2-CH_2OH-3-($CH{=}CHCH_2CH_2CH_3$)-1,4-naphthoquinone (29%) ⑤ [2]

REAGENT: CESIUM FLUORIDE

==

STRUCTURE:

MW = 151.90
MP = 682°C
d = 4.115

CsF

Fieser's Reagents for Organic Synthesis: Vol., pages

1	121	**4**		**7**	57- 58	**10**	81- 84
2		**5**		**8**	81- 82	**11**	115- 17
3		**6**	100	**9**	100	**12**	108

March's Advanced Organic Chemistry: Not Indexed

MAJOR USES:

1. Desilylations 2. Eliminations

PREPARATION:

1. The reagent is commercially available in 99% purity

PRECAUTION:

1. CsF is hygroscopic and is an irritant.

NOTES:

REFERENCES:

1. Y. Ito, S. Miyata, M. Nakatsuka and T. Saegusa, **J. Am. Chem. Soc.**, (1981), 103, 5250.

2. N.V. Bac and Y. Langlois, **J. Am. Chem. Soc.**, (1982), 104, 7666.

3. A. Ricci, M. Fiorenza, M.A. Grifagni and G. Bartolini, **Tetrahedron Lett.**, (1982), 5079.

4. M. Fiorenza, A. Mordini, S. Papaleo, S. Pastorelli and A. Ricci, **Tetrahedron Lett.**, (1985), 787.

5. Y. Ito, E. Nakajo, K. Sho and T. Saegusa, **Synthesis**, (1985), 698.

EXAMPLES:

$o\text{-}(Me_3\overset{\oplus}{N}CH_2)C_6H_4N(Me)SiMe_3\ I^{\ominus}$ —CsF, 25°→ $o\text{-}CH_2{=}C_6H_4{=}NCH_3$ (Excess of 77%) ① [2]

1,1-Me_2-2-($CH(CH_3)SiMe_2t\text{-}Bu$)-tetrahydropyridinium $I^{\ominus}$ —CsF, CH_3CN, Δ→ Me_2N-dienyl chain (77%) ② [2]

2-Cyclohexenone + $CH_2{=}CHCH_2SiMe_3$ —CsF, THF→ 3-($CH_2CH{=}CH_2$)cyclohexanone (65%) ③ [1]

$PhCOCH_2SiMe_3$ + PhCHO —CsF, THF, 4 hrs→ $PhCOCH{=}CHPh$ (90%) ④ [1]

$o\text{-}(Me_3\overset{\oplus}{N}CH_2)C_6H_4$-C(S,S-dithiane)$SiMe_3$ $F_3CSO_3^{\ominus}$ —CsF→ [o-quinodimethane dithiane intermediate] → dimer (bis-spirodithiane dibenzocyclooctadiene) (96%) ⑤ [1] [2]

REAGENT: CHLORAMINE- T (SODIUM N-CHLORO-p-TOLUENESULFONATE)

==

STRUCTURE:

MW = 227.67

MP = 167- 170°C (dec)

$CH_3-C_6H_4-SO_2-N^{\ominus}(Cl)\ Na^{\oplus} \equiv TsN^{\ominus}ClNa^{\oplus}$

Fieser's Reagents for Organic Synthesis: Vol., pages

1		**4**	75, 445	**7**	58	**10**	85
2		**5**	101	**8**	83	**11**	118
3		**6**		**9**	101- 02	**12**	

March's Advanced Organic Chemistry: 726, 738

MAJOR USES:

1. Chlorolactonization
2. Removal of protecting groups
3. Tosylamination of alkenes

PREPARATION:

1. Commercially available as the hydrate

PRECAUTION:

1. An explosion has been reported from use of this reagent, and should be considered able to explode if heated over 130°C. See: I. Klundt, **Chem. and Eng. News**, (1977), 56.

NOTES:

REFERENCES:

1. B. Damin, A. Forestiere, J. Garapon and B. Sillion, **J. Org. Chem.**, (1981), 46, 3552.

2. T. Otsubo, F. Ogura, H. Yamaguchi, H. Higuchi, Y. Sakata and S. Misumi, **Chem. Letters**, (1981), 447.

3. Y. Ito, E. Nakajo, K. Sho and T. Saegusa, **Synthesis**, (1985), 698.

4. J.E. Fankhauser, R.M. Peevey and P.B. Hopkins, **Tetrahedron Lett.**, (1984), 15.

5. G.W. Kabalka and E.E. Gooch, **J. Org. Chem.**, (1981), 46, 2582.

EXAMPLES:

$CH_2=CH(CH_2)_2CO_2H$ + $TsN^{\ominus}ClNa^{\oplus}$ —(CH_3SO_3H)→ $ClCH_2$-γ-butyrolactone (63%) ① [1]

② n-$C_{10}H_{21}TePh$ + 4-Me-PhSO_2NClNa —(THF)→ n-$C_{10}H_{21}TePh$(=NSO_2Ph-Me-4) —(Δ)→ n-$C_8H_{17}CH=CH_2$ + PhTeNHSO_2Ph-Me-4 (66%)

—(Chloramine T)→ CO_2Me (60%) ③ [2]

—(Anhydrous Chloramine T; MeOH / Argon)→ NHTs (76%) ④ [2] [3]

SePh

—(BH_3-THF)→ $[(C_6H_{11})_3B]$ —(Chloramine T; NaI)→ I (99%) ⑤

REAGENT: CHLORANIL AND o-CHLORANIL (TETRACHLORO-1,4- AND 1,2-BENZOQUINONE)

==

STRUCTURE:

MW = 245.89

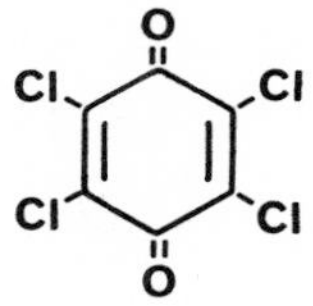

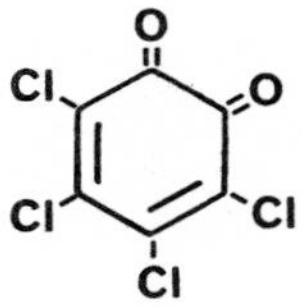

MP = 290° 127-29°

<u>**Fieser's Reagents for Organic Synthesis**</u>**: Vol., pages**

1	127- 29	**4**	75- 76	**7**	355- 56	**10**	85- 86
2	66- 67	**5**		**8**		**11**	
3	46	**6**		**9**	102	**12**	

March's <u>**Advanced Organic Chemistry**</u>: 1053

MAJOR USES:

1. Oxidative coupling of furans with quinones 2. Oxidizing agent

3. Dehydrogenation

PREPARATION:

1. Commerciallly available.

PRECAUTION:

1. Irritating to skin and mucous membranes.

NOTES:

REFERENCES:

1. J. Bridson, S. Bennett and G. Butler, **Chem. Commun.**, (1980), 413.

2. P. Fu, C. Cortez, K. Sukumaran and R. Harvey, **J. Org. Chem.**, (1979), <u>44</u>, 4265.

3. R.I. Fryer, J.V. Earley, E. Evans, J. Schneider and L.H. Sternbach, **J. Org. Chem.**, (1970), <u>35</u>, 2455.

4. O.C. Musgrave and C.J. Webster, **J. Chem. Soc.**, (1971), <u>Part C</u>, 1393.

5. H. Ishikawa and T. Mukaiyama, **Bull. Chem. Soc. Japan**, (1972), <u>45</u>, 967.

EXAMPLES:

Chloranil

(70%) ① [1]

o-Chloranil

(97%) ② [3]

Chloranil

(44%) ③ [3]

2 → Chloranil, H_2SO_4

(76%) ④

2 $Bu_3SnNHNHPh$ → Chloranil, Benzene, RT → Ph-Ph (93%) + Bu_3SnO-(tetrachlorophenyl)-OH ⑤

REAGENT: m-CHLOROPEROXYBENZOIC ACID (**MCPBA**)

===

STRUCTURE:

O=C-O-O-H (benzene ring, Cl)

MW = 172.57
MP = 92.4° (Decomposition)

Fieser's Reagents for Organic Synthesis: Vol., pages

1	135- 139	**4**		**7**	62- 64	**10**	92- 93
2	68- 69	**5**		**8**	97- 102	**11**	122- 123
3	49- 50	**6**	110- 114	**9**	108- 110	**12**	118- 127

March's Advanced Organic Chemistry: 628, 923, 991, 1087

MAJOR USES:

1. Epoxidation
2. General oxidation
3. **Baeyer-Villiger reaction**
4. Deamination of amines
5. Preparation of N-oxides
6. Preparation of sulfoxides

PREPARATION:

1. This peroxyacid can be prepared by reacting m-chlorobenzoyl chloride with hydrogen peroxide in the presence of $MgSO_4 \cdot 7H_2O$ n NaOH and dioxane in a polyethylene flask. (R. McDonald, R. Stepple and J. Dursay, **Org. Synth.**, (1970), 50, 15.

2. This is commercially available (80- 85% purity). The reagent is relatively stable; and stability can be improved by removing the chlorobenzoic acid by-product.

PRECAUTION:

Compared to many peroxy compounds, this is relatively stable. At room temperature there is less than 1% decomposition per year (see, however, **Reagents**, 8, 97 for a report of an explosion). As with all peroxy- compounds, this should be treated with caution.

NOTES:

1. See the **Baeyer-Villiger reaction** in this volume.
2. A sulfoxide elimination

REFERENCES:

1. G. Rubottom, J. Gruber, H. Juve and D. Charleson, **Org. Synth.**,(1986), 64, 118.

2. J.A. Hirsch and V.C. Truc, **J. Org. Chem.**, (1986), 51, 2218.

3. S. Danishefsky, T. Kitihara, P.F. Schuda and S.J. Etheridge, **J. Am. Chem. Soc.**, (1977), 99, 6066.

4. K. Mori and H. Ueda, **Tetrahedron Lett.**, (1981), 461.

5. M.T. Reetz and H. Muller-Starke, **Tetrahedron Lett.**, (1984), 3301.

EXAMPLES:

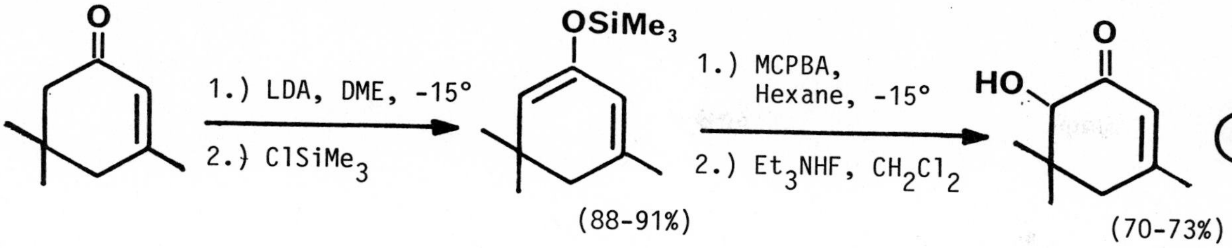

1.) LDA, DME, -15°
2.) $ClSiMe_3$

$OSiMe_3$ (88-91%)

1.) MCPBA, Hexane, -15°
2.) Et_3NHF, CH_2Cl_2

HO (70-73%) ① [1]

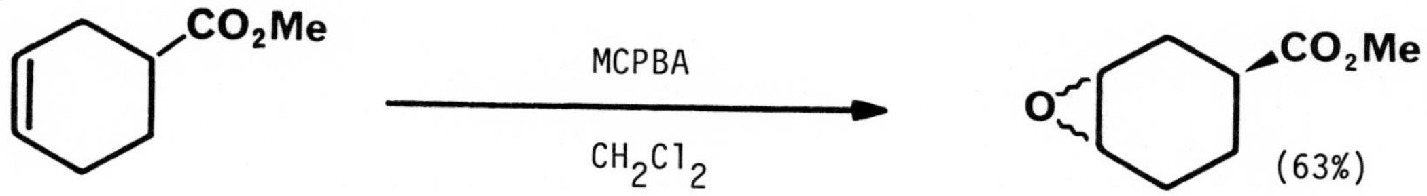

CO_2Me — MCPBA, CH_2Cl_2 → O, CO_2Me (63%) ② [1]

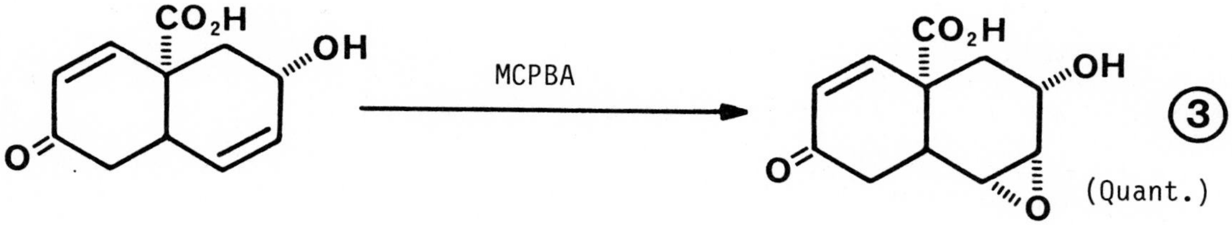

CO_2H, OH, O — MCPBA → CO_2H, OH, O, O (Quant.) ③ [1]

O — MCPBA → O, O (90%) ④ [3] △1

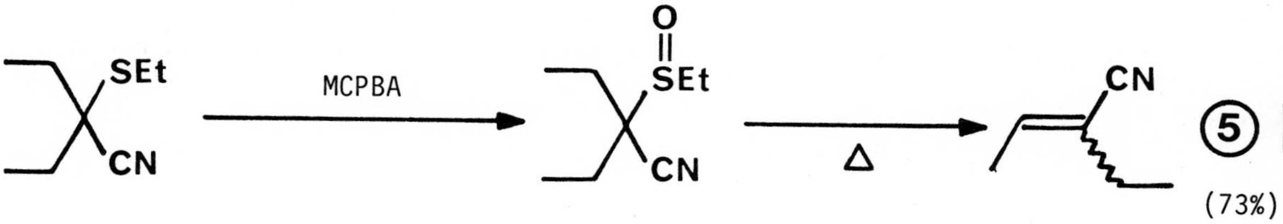

SEt, CN — MCPBA → O, SEt, CN — △ → CN (73%) ⑤ [6] △2

REAGENT: CHLOROSULFONYL ISOCYANATE ⚠1

STRUCTURE:

MW = 141.53
MP = -44° C
BP = 107° C
d = 1.626

$Cl-S(=O)_2-N=C=O$

Fieser's Reagents for Organic Synthesis: Vol., pages

1	117- 8	**4**	90- 94	**7**	65- 66	**10**	94- 95
2	70	**5**	132- 36	**8**	105- 06	**11**	125
3	51- 53	**6**	122	**9**		**12**	

March's Advanced Organic Chemistry: 350, 372, 869, 933

MAJOR USES:

1. -Lactam synthesis 2. Preparation of anhydrides 3. Oxidations

---

PREPARATION:

1. Can be prepared from cyanogen chloride and sulfur trioxide (**Org. Synth.**, (1966), 46, 23).

2. Commercially available.

PRECAUTION:

1. This reagent is corrosive and a lachrymator.

NOTES:

1. For a review, see: J.K. Rasmussen and A. Hassner, **Chem. Rev.**, (1976), 76, 389.

2. Also known as N-caronylsulfamyl chloride.

3. Using $Me_2S=N-SO_2-Cl$ as the oxidizing agent.

REFERENCES:

1. T. Tanaka and T. Mitadera, **Synthesis**, (1982), 1497.

2. F.H. Hauser and R.P. Rhee, **J. Org. Chem.**, (1981), 46, 227.

3. K.S. Keshavamurthy, Y.D. Vankar and D.N. Dhar, **Synthesis**,(1982), 506.

4. C. Bottechi, C. Chelucci and M. Marchetti, **Synth. Commun**, (1982), 12, 25.

5. G.A. Olah, Y.D. Vankar and M. Arvanaghi, **Synthesis**, (1980), 141.

EXAMPLES:

$CH_2=CHCH_2I + O=C=NSO_2Cl \longrightarrow$ 4-(iodomethyl)azetidin-2-one (37%) ① [1]

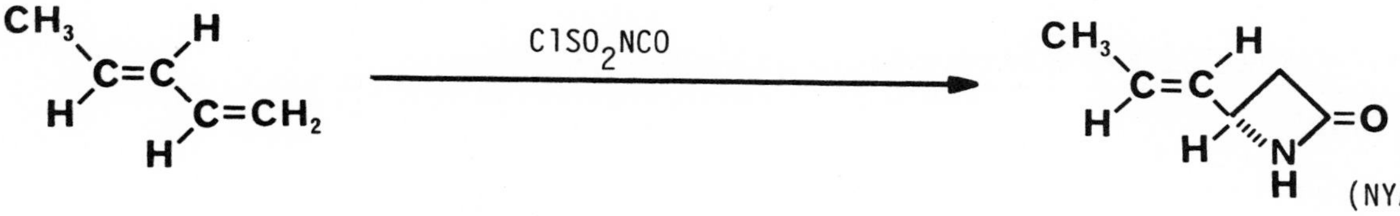

$CH_3CH=CH-CH=CH_2 \xrightarrow{ClSO_2NCO}$ 4-(prop-1-enyl)azetidin-2-one (NYA) ② [1]

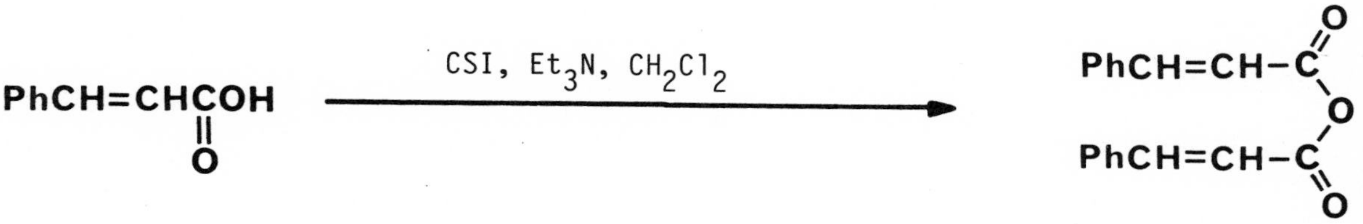

$PhCH=CHCOOH \xrightarrow{CSI,\ Et_3N,\ CH_2Cl_2} (PhCH=CH-CO)_2O$ (92%) ③ [2]

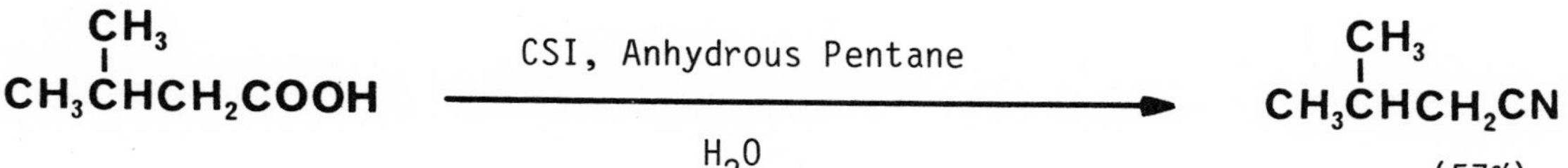

$CH_3CH(CH_3)CH_2COOH \xrightarrow[H_2O]{CSI,\ Anhydrous\ Pentane} CH_3CH(CH_3)CH_2CN$ (57%) ④

$C_6H_5CH_2OH \xrightarrow[CH_2Cl_2,\ Et_3N]{CSI\ /\ DMSO} C_6H_5CHO$ (90%) ⑤ [3] △3

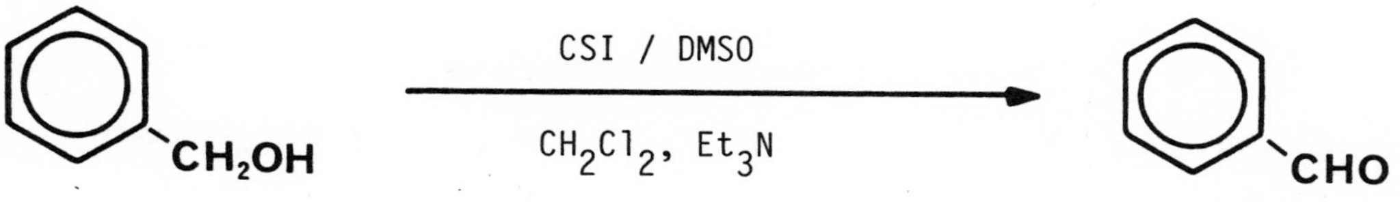

REAGENT: CHLOROTRIMETHYLSILANE (**TMSCl**)
TRIMETHYLCHLOROSILANE

===

STRUCTURE:

MW = 108.64 $ClSiMe_3$
BP = 57°C
d = 0.856

Fieser's Reagents for Organic Synthesis: Vol., pages

1	1232	**4**	537- 39	**7**	66- 67	**10**	96
2	435- 38	**5**	709- 713	**8**	107- 09	**11**	125- 27
3	310- 12	**6**	626- 28	**9**	112- 13	**12**	126

March's Advanced Organic Chemistry:

MAJOR USES:

1. Silylating reagent 2. Deoxygenation

PREPARATION:

1. Commercially available

PRECAUTION:

1. Flammable and corrosive.

NOTES:

REFERENCES:

1. S. Lane, S.J. Quick and R.J.K. Taylor, **Tetrahedron Lett.**, (1984), 1039.

2. G.C. Andrews, T.C. Crawford and L.G. Contillo, Jr., **Tetrahedron Lett.**, 3803.

3. J.M. Aizpurua and C. Palomo, **Tetrahedron Lett.**, (1984), 1103.

4. K. Sasaki, Y. Aso, T. Otsuzo and F. Ogura, **Tetrahedron Lett.**, (1985), 453.

5. A. Millar, K.H. Kim, D.K. Minster, T. Ohgi and S.M. Hecht, **J. Org. Chem.**, (1986), 51, 189.

EXAMPLES:

Me_3SiCl

RT

(86%) ① [2]

$ClSiMe_3$

$OSiMe_3$

Cl (99%) ②

Cl

NaI / $ClSiMe_3$ / $[HSi(Me)_2]_2O$

CHO

Cl

CH_2I (95%) ③

TeLi

Me_3SiCl

$TeSiMe_3$

(73%) ④ [1]

AcO

AcO

OEt

NH_2

Me_3SiCl, CH_2Cl_2

AcO

AcO

OAc

NH_2 (93%) Cl ⑤

REAGENT: CHROMIC ANHYDRIDE (CHROMIUM TRIOXIDE)
(CHROMIUM(III) OXIDE; CHROMIC ACID)

==

STRUCTURE:

CrO_3

MW = 99.99
MP = 195° (dec. 250° C)
d = 2.700

Fieser's Reagents for Organic Synthesis: Vol., pages

1	144- 147	**4**	96- 97	**7**	70	**10**	99
2	72- 75	**5**	140- 141	**8**		**11**	
3	54- 57	**6**		**9**	115	**12**	133

March's Advanced Organic Chemistry: See March index

MAJOR USES: ⚠1

A versatile oxidizing agent

PREPARATION:

Commercially available in 99.9% purity. For preparation of oxidizing solutions such as **Fieser's Reagent** (CrO_3 - HOAc), **Sarett Reagent** (CrO_3 - pyridine), **Cornforth Reagent** (Cro$_3$- pyridine- H_2O), **Thiele Reagent** (CrO_3- Acetic anhydride- H_2SO_4), and **Jones Reagent** (Cro$_3$- H_2SO_4), see: **Reagents.**

PRECAUTION:

This is a powerful oxidizing agent, and is a suspected carcinogen.

NOTES:

1. As the content of water in solvent increases, the oxidizing strength decreases.

REFERENCES:

1. M. Allen, N. Darby, P. Salisbury, E. Sigurdson, and T. Money, **Can. J. Chem.**, (1979), 57, 733.

2. H. Molin and B.G. Pring, **Tetrahedron Lett.**, (1985), 677.

3. L.A. Paquette, D.T. Belmont and Y.-L. Hsu, **J. Org. Chem.**, (1985), 50, 4667.

4. R.J. Isreal and R.K. Murray, Jr., **J. Org. Chem.**, (1985), 50, 4703.

5. P.E. Peterson, R.L. Breedlove Leffew and B.L. Jensen, **J. Org. Chem.**, (1986), 51, 1948.

EXAMPLES:

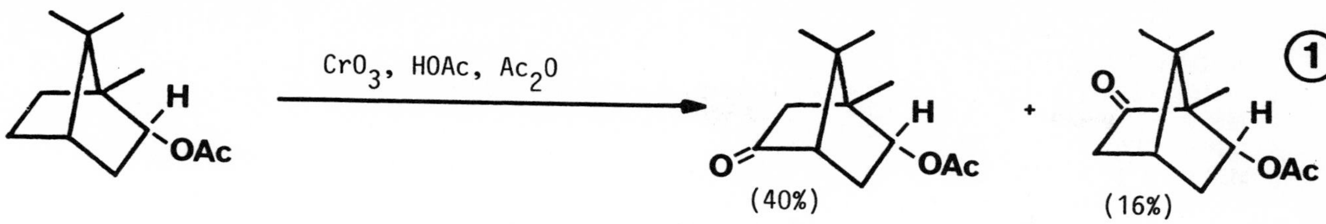

①

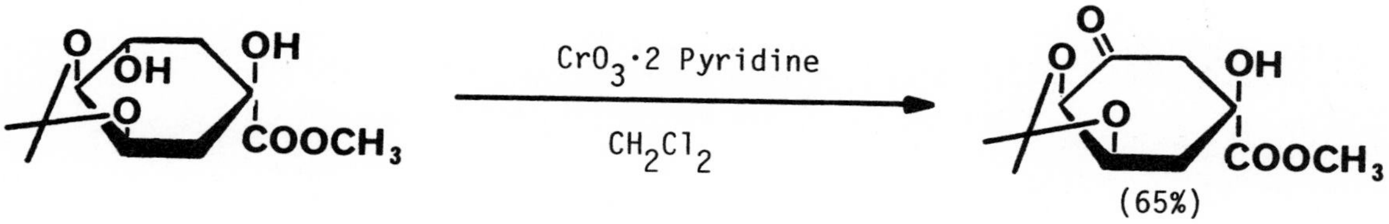

②

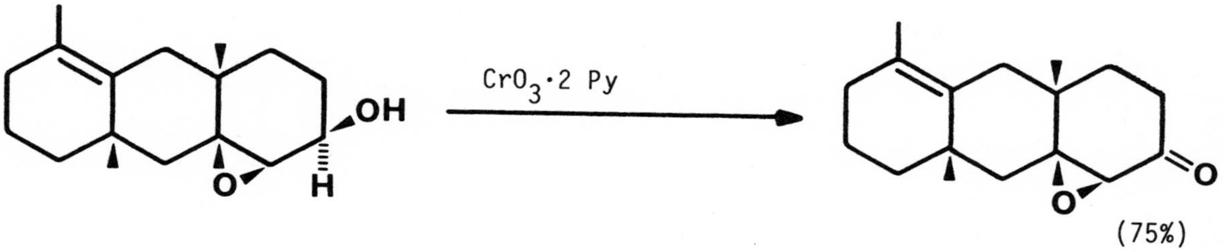

③

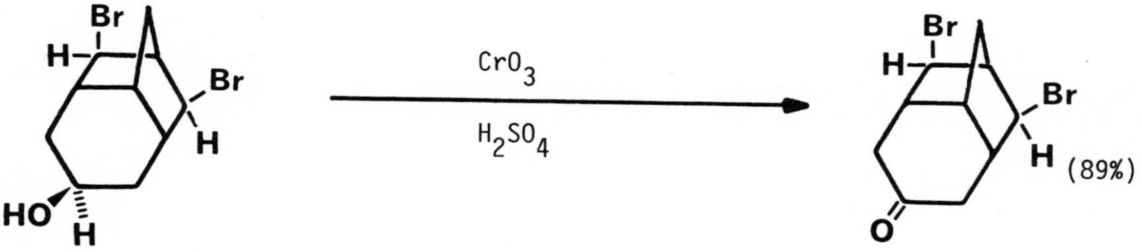

④

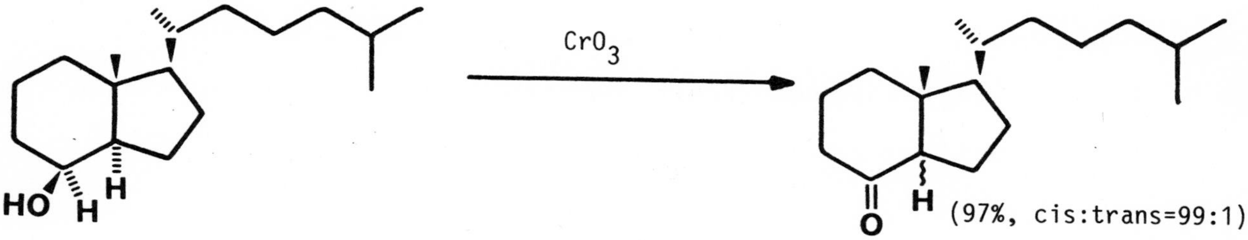

⑤

REAGENT: CHROMIUM (II) CHLORIDE (CHROMOUS CHLORIDE)

==

STRUCTURE:

MW = 122.90 $CrCl_2$
MP = 820°C
d = 2.900

Fieser's Reagents for Organic Synthesis: Vol., pages

1	149- 50	**4**		**7**	73	**10**	
2	76- 77	**5**	144	**8**		**11**	132- 34
3	60- 61	**6**		**9**		**12**	

March's Advanced Organic Chemistry: 406, 924, 1109

MAJOR USES:

1. Reducing agent 2. Dehalogenation 3. Organohalogen coupling

4. Homoallylic alcohol synthesis

PREPARATION:

1. This reagent can be prepared from chromic chloride hexahydrate (see: G. Rosenkrantz, D. Mancera, J. Gatica and C. Djerassi, **J. Am. Chem. Soc.**, (1950), 73, 4077).

2. The reagent is also commercially available in anhydrous form.

PRECAUTION:

1. The reagent is very sensitive to moisture and is an irritant.

NOTES:

1. In this procedure, the reagent is continuously reduced electrochemically from Cr (III) to Cr (II).

REFERENCES:

1. T. Hiyama, K. Kimura and N. Nozaki, **Tetrahedron Lett.**, (1981), 1037.

2. Y. Okude, S. Hirano, T. Hiyama and H. Nozaki, **J. Am. Chem. Soc.**, (1977), 99, 3179.

3. T. Cohen and S. Nolan, **Tetrahedron Lett.**, (1978), 3533.

4. R. Sustmann and R. Altevogt, **Tetrahedron Lett.**, (1981), 5167.

5. J. Wellmann and E. Steckham, **Synthesis**, (1978), 901.

EXAMPLES:

PhCHO + (E)-crotyl bromide $\xrightarrow[\text{THF}]{CrCl_2}$ Ph–CH(OH)–CH(Me)–CH=CH$_2$ (96%, 100% threo) ① [4]

Cyclohexanone + allyl bromide $\xrightarrow[\text{THF}]{CrCl_2}$ 1-allylcyclohexanol (78%) ② [4]

$(PhS)_3CCH(Ph)CH_2C(O)Ph$ $\xrightarrow[\text{DMF, Argon 100°}]{CrCl_2}$ $(PhS)_2CHCH(Ph)CH_2C(O)Ph$ (94%) ③ [1]

Ph_2CHBr + Ph_3CCl $\xrightarrow[\text{THF}]{CrCl_2}$ Ph_2CHCPh_3 (89%) ④ [3]

$C_6H_5CH_2Br$ $\xrightarrow[\text{-0.3 V, Glassy carbon cathode}]{CrCl_2\text{, DMF}}$ C_6H_5-CH_2CH_2-C_6H_5 (60%) ⑤ [3] △1

REAGENT: CHROMYL CHLORIDE (DICHLORODIOXO CHROMIUM)

==

STRUCTURE:

MW = 154.92 Cl_2CrO_2
BP = 117°C

Fieser's Reagents for Organic Synthesis: **Vol., pages**

1	151	**4**	98- 99	**7**		**10**	
2	79	**5**	144- 45	**8**	112	**11**	134
3	62	**6**	126- 27	**9**		**12**	

March's Advanced Organic Chemistry: 727, 1077, 1079, 1085, 1086

MAJOR USES:

1. Oxidizing agent

PREPARATION:

PRECAUTION:

1. Will burn and blister skin. Use only in a well-ventilated hood.

NOTES:

REFERENCES:

1. K.B. Sharpless and A.Y. Teranishi, **J. Org. Chem.**, (1973), 38, 185.

2. J.E. Backvall, M.W. Young and K.B. Sharpless, **Tetrahedron Lett.**, (1977), 3523.

3. J. San Filippo, Jr., and C.-I. Chern, **J. Org. Chem.**, (1977), 42, 2182.

4. T.V. Lee and J. Toczek, **Tetrahedron Lett.**, (1982), 2917.

5. F. Freeman, P.J. Cameron ad R.H.DuBois, **J. Org. Chem.**, (1968), 33, 3970.

EXAMPLES:

$[C_{12}H_{22}]$ —— Cl_2CrO_2, Acetone / Dry Ice ——→ $[C_{12}H_{21}OCl]$ (79%) ①

n-Bu(H)C=C(H)H —— Cl_2CrO_2, CH_2Cl_2 / AcCl / -78° ——→ *n*-Bu(Cl)(H)C–C(H)(OAc)H (77%) ②

$C_6H_{11}CH(OH)CH_3$ —— CrO_2Cl_2 / CH_2Cl_2 / SiO_2-Al_2O_3 ——→ $C_6H_{11}C(=O)CH_3$ (100%) ③

OSiMe$_3$, Me —— CrO_2Cl_2 / CH_2Cl_2 / N_2, -78° ——→ O, OH, Me (76%) ④

$CH_2=C(CH_3)CH_2C(CH_3)_2CH_3$ —— CrO_2Cl_2, CH_2Cl_2 / 0-5° ——→ $OHC\text{-}CH(CH_3)CH_2C(CH_3)_2CH_3$ (76%) ⑤

REAGENT: COPPER [II] CHLORIDE

STRUCTURE:

$CuCl_2$

MW = 134.45
MP = 498°C
Decomposes to CuCl + Cl_2 at > 300°

Fieser's Reagents for Organic Synthesis: Vol., pages

1	163	**4**	105- 07	**7**	79	**10**	106
2	84- 85	**5**	158- 60	**8**	119- 20	**11**	
3	66	**6**	139- 41	**9**	123	**12**	

March's Advanced Organic Chemistry: 478, 529, 532, 642, 647, 650, 654, 725, 924, 1085, 1092.

MAJOR USES:

1. Halogenation of aromatics and carbonyl compounds 2. Ketone coupling

3. Oxidative ring closures

PREPARATION:

1. Available by heating the dihydrate at 100° [**J. Org. Chem.**, (1961), 26, 2263].

2. Also commercially available in very high purity.

PRECAUTION:

1. Toxic 2. Will cause skin irritation.

NOTES:

REFERENCES:

1. Y. Ito, M. Nakatsuka and T. Saegusa, **J. Org. Chem.**, (1980), 45, 2022.

2. L.A. Paquette, R.A. Snow, J.L. Muthard and T. Cynkowski, **J. Am. Chem. Soc.**, (1979), 101, 6991.

3. G.I. Nikishin, E.I. Troyansky and M.I. Lazareve, **Tetrahedron Lett.**, (1985), 1877.

4. F.M. Hauser and S.R. Ellenberger, **J. Org. Chem.**, (1986), 51, 50.

5. O. Attanasi, P. Filippone, A. Mei, S. Santeusanio and F. Serra-Zanetti, **Synthesis**, (1985), 157.

EXAMPLES:

$Me_3C(OSiMe_3)C{=}CH_2$ —[$CuCl_2$, DMF]→ $Me_3CC(=O)CH_2Cl$ (65%) ① [1]

—[1.) 2 eq. LDA, THF; 2.) $CuCl_2$, THF, DMF]→ (58%) ② [2]

COMe

COMe

O= =O

$CH_3(CH_2)_4NHSO_2CH_3$ —[$Na_2S_2O_8$-$CuCl_2$]→ 2-CH_3-pyrrolidine NSO_2CH_3 (37%) ③ [2]

Cl_3COCNH, OAc, SPh —[$CuCl_2$, H_2O, CH_3CN]→ Cl_3COCNH, CHO (86%) ④

$H_2NC(=O)N{=}NC(CH_3){=}CHCO_2CH_3$ + $H_3CC(=O)CH_2C(=O)CH_3$ —[$CuCl_2 \cdot 2\ H_2O$]→ $HNC(=O)NH_2$, N, H_3C, CH_3, $CH_3OC(=O)$, $C(=O)CH_3$ (77%) ⑤ [3]

REAGENT: COPPER (I) BROMIDE AND CHLORIDE

STRUCTURE:

MW = **CuBr** 143.46 **CuCl** 99.00
MP = 504° 430°

Fieser's Reagents for Organic Synthesis: **Vol., pages**

1	165- 169	**4**	108- 110	**7**	79- 81	**10**	
2	90- 92	**5**	163- 165	**8**	116- 119	**11**	140- 141
3	67- 69	**6**	143- 146	**9**	123	**12**	141

March's Advanced Organic Chemistry: 339, 400, 710, 1061, 1085

MAJOR USES:

1. **Grignard** addition catalyst
2. **Sandmeyer reaction**
3. **Gatterman- Koch reaction**
4. **Ullmann synthesis**
5. Oxidative catalyst
6. Coupling reactions
7. Dehalogenation
8. Conjugate additions

PREPARATION:

CuBr: J. Buck and W. Ide, **Org. Synth., Coll. Vol. 2**, 132, (1943)
CuCl: C. Marvel and S. McElvain, **Org. Synth., Coll. Vol. 1**, 170, (1941).

PRECAUTION:

Keep reagents out of air and sunlight.

NOTES:

1. **THP** = Tetrahydropyranyloxy

REFERENCES:

1. J. Setsune, K. Mabukawa and T. Kitao, **Tetrahedron Lett.**, (1982), 663.

2. E. Balogh-Hergovich, G. Speier and Z. Tyeklar, **Synthesis**, (1982), 731.

3. H.R. Buser, P.M. Guerin, M. Toth, G. Szocs, A. Schmid, W. Franke and H. Arn, **Tetrahedron Lett.**, (1985), 403.

4. A.A. Smaardijk, S. Noorda, F. vanBolhuis and H. Wynberg, **Tetrahedron Lett.**, (1985), 493.

5. W. Ando, Y. Kumamoto and T. Takata, **Tetrahedron Lett.**, (1985), 5187.

EXAMPLES:

2-Bromophenol + $CH_2(CO_2Et)_2$ → 1.) NaH, Dioxane; 2.) CuBr, 70° → 3-(CO_2Et)-benzofuran-2-one (93%) ① [6]

Phenanthrenequinone → 1.) CuCl, Py, O_2; 2.) HCl → biphenyl-2,2'-dicarboxylic acid (HO_2C, CO_2H) (95%) ② [5]

$CH_3(CH_2)_8C{\equiv}CCH_2Br$ + $BrMgC{\equiv}CCH_2CH_2OTHP$ → CuCl → $CH_3(CH_2)_8C{\equiv}CCH_2C{\equiv}CCH_2CH_2OTHP$ (75%) ③ △1 [6]

2-($N_2^{\oplus}Cl^{\ominus}$)-C$_6$H$_4$-CH(HO)-P(O)$(OCH_3)_2$ → CuCl → 2-Cl-C$_6$H$_4$-CH(HO)-P(O)$(OCH_3)_2$ (NYA) ④ [2]

1,2-dithiolane-$(CH_2)_4CO_2Me$ + $PhC(=N_2)Ph$ → CuCl, CH_2Cl_2, RT, 2-3 hrs → 2,2-diphenyl-1,3-dithiane-$(CH_2)_4CO_2Me$ (100%) ⑤ [6]

REAGENT: COPPER [I] IODIDE

STRUCTURE:

MW = 190.46 CuI
MP = 588- 606°

Fieser's Reagents for Organic Synthesis: Vol., pages

1	169	**4**		**7**		**10**	107
2	92	**5**	167- 68	**8**	121- 22	**11**	141
3	69- 71	**6**	147	**9**	124- 25	**12**	141- 43

March's Advanced Organic Chemistry: 406

MAJOR USES: ⚠1

1. Preparation of organocopper compounds 2. Reaction of diazo compounds

---

PREPARATION:

PRECAUTION:
Light sensitive.

NOTES:

1. For use with organocopper reagents, see the Review: E. Erdik, **Tetrahedron**, (1984), 40, 641.

REFERENCES:

1. Y. Ito, H. Imai, T. Matsuura and T. Saegusa, **Tetrahedron Lett.**, (1984), 3091.

2. T. Fujisawa, T. Itoh, M. Nakai and T. Sato, **Tetrahedron Lett.**, (1985), 771.

3. H. Maruyama and T. Hiraoka, **J. Org. Chem.**, (1986), 51, 398.

4. K.H. Duchene and F. Vogtle, **Synthesis**, (1986), 659.

5. D.S. Ross, K.D. Moran and R. Malhotra, **J. Org. Chem.**, (1983), 48, 2120.

EXAMPLES:

t-BuN=C(Li)n-Bu + CuI —(cyclohexenone, $BF_3 \cdot OEt_2$)→ 3-[C(n-Bu)=Nt-Bu]cyclohexanone (71%) ① [1]

(epoxide)CH$_2$SPh + Ph(CH$_2$)$_2$MgBr —(CuI, THF-Me_2S)→ Ph(CH$_2$)$_3$CH(OH)CH$_2$SPh (90%) ② [1]

Br-azetidinone (SO$_2$-C$_6$H$_3$(OMe)$_2$, N-C(CO$_2$Me)=CMe$_2$) —(CuI, n-BuLi; MeCHO)→ MeCH(OH)-azetidinone (47%) ③ [1]

3,5-(MeOCH$_2$)$_2$C$_6$H$_3$-C≡CH —(CuI, NH_3, EtOH)→ 3,5-(MeOCH$_2$)$_2$C$_6$H$_3$-C≡C-Cu (41%) ④ [1]

n-C_8H_{17}Br + (i-PrO)$_2$MeSiCH$_2$MgCl —(CuI)→ n-C_8H_{17}CH$_2$SiMe(i-PrO)$_2$ (92%) —(30% H_2O_2)→ n-$C_8H_{17}CH_2OH$ (88%) ⑤ [1]

REAGENT: CROWN ETHERS 

STRUCTURE:

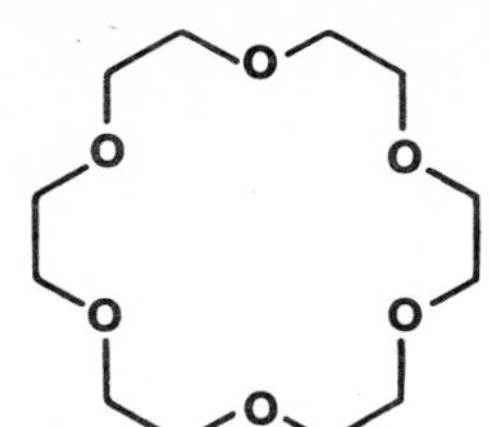

MW = 264.32
BP = 39- 40°C

18-Crown-6,(1,4,7,10,13,16-hexaoxacyclooctadecane)

Fieser's Reagents for Organic Synthesis: Vol., pages

1	**4** 142- 45	**7** 76- 79	**10** 110- 12
2	**5** 152- 55	**8** 128- 30	**11** 143- 45
3	**6** 133- 37	**9** 126- 27	**12** 147

March's Advanced Organic Chemistry: 77- 79, 105, 321- 32, 371, 395.

MAJOR USES:

1. Complexing agent
2. Solvent for increasing reactivity

PREPARATION:

1. See the references to **Reagents**.
2. Commercially available.

PRECAUTION:

1. Can be toxic

NOTES:

1. As one structural example we have shown 18-crown-6 because it is a common crown ether. Many other structural types are known.

2. For reviews, see: G. Gokel and H. Durst, **Synthesis**, (1976), 168; J. Bradshaw and P. Stott, **Tetrahedron**, (1980), 36, 461.

REFERENCES:

1. M. DeCamp and L. Viscogliosi, **J. Org. Chem.**, (1982), 46, 3918.

2. R. Baker and R. Sims, **J. Chem. Soc.**, Perkin I, (1981), 3087.

3. M. Lissel, **Tetrahedron Lett.**, (1985), 1843.

4. G. Maier and B. Wolf, **Synthesis**, (1985), 871.

5. Y. Torisawa, H. Okabe and S. Ikegami, **Chem. Letters**, (1984), 1555.

EXAMPLES:

+ $CHBr_3$ → (K_2CO_3, 18-Crown-6; 140°) Br (18-45%) ①

CH_3, CH_3, CHO; O, CO_2CH_3, $CH_2PO(OCH_3)_2$ → (NaH, Glyme; 15-Crown-5) O, CO_2CH_3, H, C=C, H, CH_3, $-CH_3$ (45%) ②

C≡CH → (CH_3I, KOH; 18-Crown-6) $C{\equiv}CCH_3$ (74%) ③

Me_3Si, CO_2Me, CO_2Me, Me_3Si → (KF / Dibenzo-18-Crown-6; Acetonitrile) CO_2Me, CO_2Me (89%) ④

OH, OH → (1.) $MeSO_2Cl$; 2.) CsOAc / 18-Crown-6) OAc, OAc (70%) ⑤

REAGENT: CYANOGEN BROMIDE

STRUCTURE:

MW = 105.93°C   Br-CN
BP = 61- 62°
MP = 52°

<u>Fieser's Reagents for Organic Synthesis</u>: Vol., pages

1	174- 76	**4**	110	**7**		**10**	
2	93	**5**	169- 70	**8**	131	**11**	
3		**6**	148- 49	**9**		**12**	

March's <u>Advanced Organic Chemistry</u>: 497, 863, 1001

MAJOR USES:

1. **von Braun Reaction**   2. Reaction with heteroatoms

PREPARATION:

1. Commercially available

PRECAUTION:

1. Extremely toxic reagent. Avoid vapors.

NOTES:

REFERENCES:

1. D. Martin and M. Bauer, **Org. Synth.**, Submitted (1979)

2. J.J. Fitt and H.W. Gschwemd, **J. Org. Chem.**, (1981), <u>46</u>, 3349.

3. O. Subba Rao and W. Lwowski, **J. Het. Chem.**, (1980), <u>17</u>, 187.

4. G. Zweifel, R.P. Fisher and J.T. Snow, **J. Am. Chem. Soc.**, (1972), <u>94</u>, 6560.

5. J.B. Bremner, E.J. Brown, V. Cmohan and B.F. Yates, **Aust. J. Chem.**, (1984), <u>37</u>, 1043.

EXAMPLES:

Br_2 + NaCN —(H_2O, -NaBr)→ BrCN —(PhOH, Et_3N, CCl_4, -5° to 10°)→ PhO–CN (75-85%) ① [2]

BrCN (86%) ② [1]

CNBr, Et_3N, THF; EtO–C(=N–CN)N_3 (93%) ③ [2]

n-Bu; H; B; CNBr, CH_2Cl_2, 0°; *n*-Bu; H (69%) ④

CNBr, Dry $CHCl_3$, K_2CO_3, 18 hrs; CN; Ph (67%) ⑤ [1]

REAGENT: CYANOTRIMETHYLSILANE (TRIMETHYLSILYLCYANIDE) 1

STRUCTURE:

MW = 99.21
BP = 114-117°
MP = 2°
d = 0.744

Me_3SiCN

Fieser's Reagents for Organic Synthesis: Vol., pages

1	**4** 542- 43	**7** 397- 99	**10** 112- 14
2	**5** 720 22	**8** 133	**11** 147 50
3	**6** 632- 33	**9** 127- 29	**12** 148- 51

March's Advanced Organic Chemistry: 430

MAJOR USES:

1. Cyanosilylation 2. Quinone protection 3. Carbonyl additions
4. Nitrile preparation

PREPARATION:

1. See: **Reagents** under appropriate pages.

2. See: M.T. Reetz and I. Chatzhosifidis, **Synthesis**, (1982), 330 and references cited.

3. The reagent is commercially available.

PRECAUTION:

1. Highly toxic liquid. Handle with care. The compound is also highly flammable.

NOTES:

1. For a review, see: W.C. Groutas and D. Felker, **Synthesis**, (1980), 561.

REFERENCES:

1. M.T. Reetz and I. Chatzhosifidis, **Angew. Chem. Int. Ed.**, (1981), 20, 1017.

2. J.C. Mullis and W.P. Weber, **J. Org. Chem.J. Org. Chem.**, (1982), 47, 2873.

3. J.A. Schwindeman and P.D. Magnus, **Tetrahedron Lett.**, (1981), 4925.

4. S. Veeraraghavan, D. Bhattacharju and F.D. Popp, **J. Heterocyclic Chem.**, (1981), 18, 443.

5. J.H. Byers and T.A. Spencer, **Tetrahedron Lett.**, (1985), 713.

EXAMPLES:

$(CH_3)_2C(Cl)(CH_2)_3Cl \xrightarrow{Me_3SiCN,\ SnCl_4,\ CH_2Cl_2} (CH_3)_2C(CN)(CH_2)_3Cl$ (72%) ① [4]

2-methyloxetane + $(CH_3)_3SiCN \longrightarrow N{\equiv}C(CH_2)_2CH(CH_3)OSi(CH_3)_3$ (47%) ② [4]

$CH_3S(O)CH_2O$– (steroid-type bicyclic ether, H_3C) $\xrightarrow{Me_3SiCN,\ ZnI_2}$ $N{\equiv}CCH_2O$– (bicyclic, H_3C) (80%) ③ [4]

pyrimidine $\xrightarrow{Me_3SiCN,\ PhCOCl,\ AlCl_3,\ CH_2Cl_2}$ 1,3-bis(COPh)-2,4-dicyano-tetrahydropyrimidine (CN, COPh, COPh, CN) (58%) ④ [4]

(Sn-Bu vinylidene spiro cyclobutanone) $\xrightarrow[2.)\ n\text{-}Bu_4N^+F^-]{1.)\ Me_3SiCN,\ ZnI_2}$ (Sn-Bu octalone, =O) (59%) ⑤

REAGENT: 1,4-DIAZABICYCLO[2.2.2]OCTANE (**DABCO**) 

STRUCTURE:

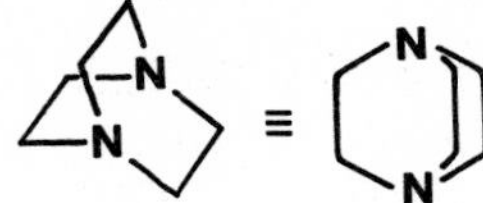

MW = 112.17
BP = 174°
MP = 158°

Fieser's Reagents for Organic Synthesis: Vol., pages

Vol.	pages	Vol.	pages	Vol.	pages	Vol.	pages
1	1203	**4**	119	**7**	86- 87	**10**	
2	99- 101	**5**	176- 77	**8**		**11**	153- 54
3		**6**		**9**		**12**	155- 56

March's Advanced Organic Chemistry: 922

MAJOR USES: ⚠1

1. Quenching agent for singlet oxygen
2. Basic catalyst
3. Dehydrohalogenation reagent

PREPARATION:

1. Commercially available

PRECAUTION:

1. Corrosive and hygroscopic.

NOTES:

1. The reagent readily sublimes at room temperature
2. The reagent is sometimes known as **triethylenediamine**, (TED).
3. After elimination of **HCl** there is an **imine-enamine** tautomerism.

REFERENCES:

1. H.-J. Liu, L.-K. Ho and H.K. Lai, **Can. J. Chem.**, (1981), 59, 1685.
2. A.J. Kolar and R.K. Olsen, **Synthesis**, (1977), 457.
3. P. Perlmutter and C.C. Teo, **Tetrahedron Lett.**, (1984), 5951.
4. H.-J. Liu and I.V. Oppong, **Can. J. Chem.**, (1982), 60, 94.
5. H.M.R. Hoffmann and J. Rabe, **Angew. Chem. Int. Ed.**, (1983), 22, 795.

EXAMPLES:

2-(COSEt)cyclohexanone $\xrightarrow{H_2CCHCOCH_3,\ DABCO}$ 2-(CH_2CH_2COMe)-2-(COSEt)cyclohexanone (100%) ① [2]

$ClNAc$ / $MeCHCOOMe$ $\xrightarrow[CH_2Cl_2]{DABCO}$ $CH_2{=}C(NHAc)COOMe$ (72-79%) ② [3] △3

$PhCH{=}NSO_2$-C_6H_4-Me + $CH_2{=}CHCO_2Et$ $\xrightarrow[80°]{DABCO}$ $H_2C{=}C(CO_2Et)$-CH(Ph)-$NHSO_2$-C_6H_4-Me (80%) ③ [2]

2-Cyclohexenone + $CH_2(COSEt)_2$ $\xrightarrow[1,2\text{-Dimethoxyethane, Argon, RT, 48 hrs, } H^+]{DABCO}$ 3-[$CH(COSEt)_2$]cyclohexanone (93%) ④ [2]

$PhCO_2$-$(CH_2)_3$-CHO + $HC{=}CH_2$ / CO_2Me $\xrightarrow[7\ days]{DABCO}$ $PhCO_2$-$(CH_2)_4$-$CH(OH)C(CO_2Me){=}CH_2$ (95%) ⑤ [2]

REAGENT: 1,5-DIAZABICYCLO[4.3.0]NONENE-5 , **(DBN)**

STRUCTURE:

MW = 124.19
BP = 95- 98° C (7.5 mm Hg)

Fieser's Reagents for Organic Synthesis: Vol., pages

1	189- 90	**4**	116- 19	**7**	86	**10**
2	98- 99	**5**	176	**8**		**11**
3		**6**	157	**9**		**12**

March's Advanced Organic Chemistry: 915

MAJOR USES:

1. Dehydrohalogenations
2. Base

PREPARATION:

1. Can be prepared from butyrolactam (2- pyrrolidone), see: **Ann.**, (1955), 596, 210.

2. Commercially available.

PRECAUTION:

1. Corrosive and moisture-sensitive.

NOTES:

REFERENCES:

1. L.Lombardo, L.N. Mander and J.V. Turner, **J. Am. Chem. Soc.**, (1980), 102, 6626.

2. M.B. Yunker and B. Fraser-Reid, **Can. J. Chem.**, (1976), 54, 3986.

3. W. Boland and K. Mertes, **Synthesis**, (1985), 705.

4. J.I. Levine and S.M. Weinreb, **J. Am. Chemn. Soc.**, (1983), 105, 1397.

5. M.R. Detty, **J. Org. Chem.**, (1980), 45, 924.

EXAMPLES:

Br, BzO, O, C, Me, CO_2Me, OH, O, O

DBN, THF, DMF

BzO, O, C, O, Me (90%) ① [1]

EtO, O, I, OBz

DBN

THF, 9 hrs, RT

EtO, OBz, O (86%) ② [1]

$EtC{\equiv}CCH_2Br$ + $HC{\equiv}CCO_2Et$

CuI, DBN / DMSO

$EtC{\equiv}CCH_2C{\equiv}CCO_2Et$ (90%) ③ [2]

N, O, CO_2Me

.75 eq DBN

16 hrs

[N, O, CO_2Me]

N, CO_2Me (76%) ④

O t-BuSiIMe$_2$

1.) CH_3CN, 0°

2.) DBN, THF, 65°, 48 hrs

OSiMe$_2$$t$-Bu (90%) ⑤

REAGENT: 1,8-DIAZABICYCLO[5.4.0]UNDEC-7-ENE (**DBU**) 

STRUCTURE:

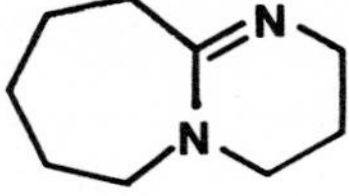

MW = 152.24
BP = 80- 83°
d = 1.018

Fieser's Reagents for Organic Synthesis: Vol., pages

1		**4**	16- 18	**7**	87- 88	**10**	
2	101	**5**	177- 178	**8**	141	**11**	155
3		**6**	158	**9**	132- 133	**12**	156

March's Advanced Organic Chemistry: 353, 361, 412, 915.

MAJOR USES:

1. Dehydrohalogenation
2. Organic base
3. Alkylations
4. Elimination reactions

PREPARATION:

1. The reagent can be prepared from caprolactam and acrylonitrile, see: H. Oediger and F. Moller, **Angew. Chem. Int. Ed.**, (1967), 6, 76.

2. This reagent can be purchased in 96% purity.

PRECAUTION:

Corrosive.

NOTES:

1. For a review, see: H. Oediger, F. Muller and K. Eiter, **Synthesis**, (1972), 591.

2. alpha-(p-Nitrophenyl)sulfonoxy = nosylate = **ONs**.

REFERENCES:

1. S. Ohta, A. Shimabayashi, M. Aono and M. Okamoto, **Synthesis**, (1982), 833.

2. L. Burton and J. White, **J. Am. Chem. Soc.**, (1981), 103, 3226.

3. H. Iio, M. Isobe, T. Kawai and T. Goto, **J. Am. Chem. Soc.**, (1979), 101, 6076.

4. R. Hoffman, B. Jankowski, C.S. Carr and E.N. Duesler, **J. Org. Chem.**, (1986), 51, 130.

5. D. Guillerm, M. Delarue, M. Jalali-Naini, P. Lemaitre and J.-Y. Lallemand, **Tetrahedron Lett.**, (1984), 1043.

EXAMPLES:

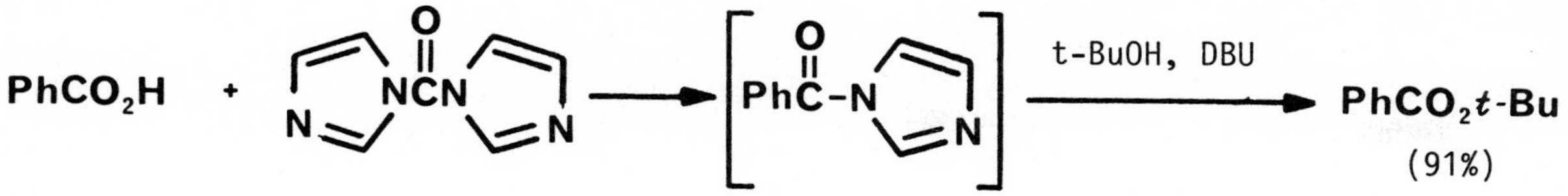

$PhCO_2H$ + [imidazole]NC(=O)N[imidazole] → [PhC(=O)-N(imidazole)] —t-BuOH, DBU→ $PhCO_2t$-Bu (91%) ①

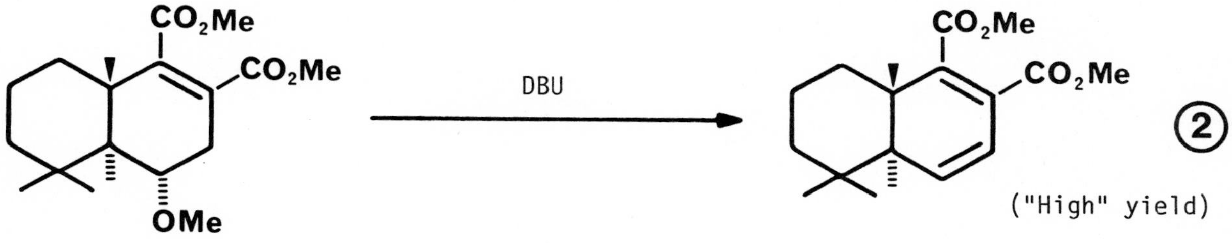

DBU ("High" yield) ② [4]

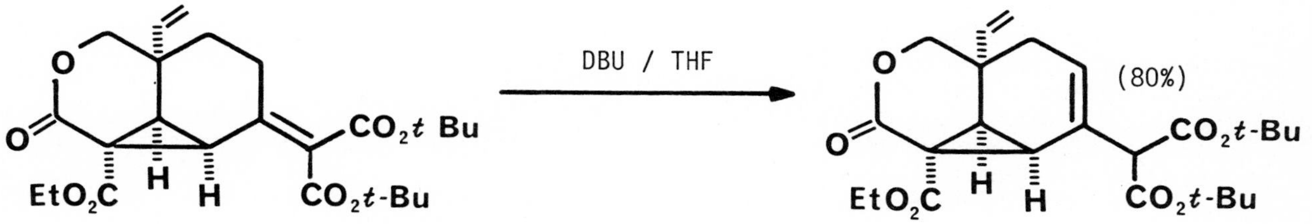

DBU / THF (80%) ③ [2]

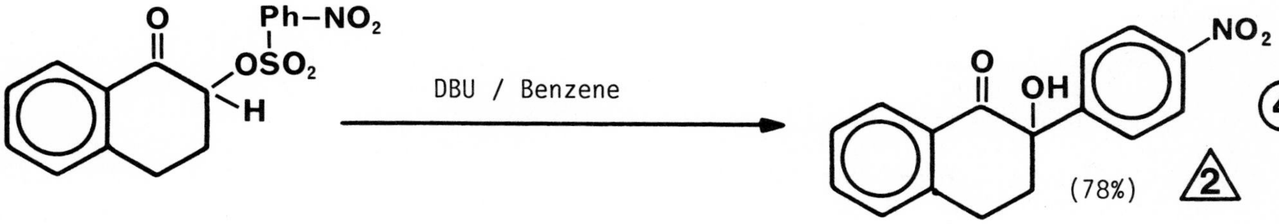

DBU / Benzene (78%) ④ △2

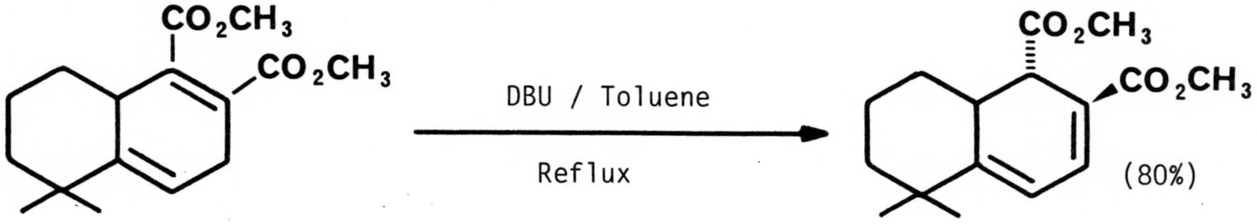

DBU / Toluene, Reflux (80%) ⑤ [2]

REAGENT: DIAZOMETHANE

STRUCTURE:

$$H_2C=\overset{\oplus}{N}=\overset{\ominus}{N} \leftrightarrow H_2\overset{\ominus}{C}-\overset{\oplus}{N}\equiv N \equiv CH_2N_2$$

MW = 42.04
BP = - 23°

Fieser's Reagents for Organic Synthesis: Vol., pages

1	191-195	**4**	120-122	**7**	88-89	**10**	
2	102-104	**5**	179-182	**8**		**11**	
3	74	**6**	159	**9**	133-135	**12**	157

March's Advanced Organic Chemistry: 172, 174, 175, 344, 355, 368, 441, 772, 866, 974, 976-78

MAJOR USES:

1. Esterification of acids
2. Methylation of alcohols
3. Ring expansion
4. **Arndt-Eistert reaction**
5. Cyclopropanation
6. Addition to ketenes

PREPARATION:

Diazomethane can be prepared from the following reagents:
a. N-nitroso-N-methylurea
b. Bis-(N-methyl-N-nitroso)terephthalamide
c. p-Toluenesulfonylmethylnitrosamine (**DIAZALD™**)
d. N,N'-dinitrosos-N,N'-dimethyloxamide
For details on these preparations, see: **Reagents**, Vol. 1, pages 191-192.

PRECAUTION:

1. The precursors to diazomethane are potentially carcinogenic

2. Diazomethane is toxic and highly explosive. Care must be exercised in its preparation to avoid overheating and the glassware used for its preparation must not have rough or sharp surfaces (avoid using ground glass joints). Use only with ventilation. Explosions may occur with alkali metals.

REFERENCES:

1. B.M. Trost, C.D. Shuey, F.DiNinno, Jr., ad S.S. McElvain, **J. Am. Chem. Soc.**, (1979), 101, 1284.

2. A.E. Greene, **Tetrahedron Lett.**, (1980), 3059.

3. F. Zutterman, H, DeWilde, R. Mijngneer, P. DeClercq and M. Vandewalle, **Tetrahedron**, (1979), 35, 2389.

4. T.V. Lee and J. Toczek, **Tetrahedron Lett.**, (1985), 473.

EXAMPLES:

1.) CO_2 / LDA

2.) CH_2N_2

NC —≡—H → NC —≡—CO_2Me (88%) ① [1]

CH_2N_2

—Cl → —Cl =O (NYA) ② [3]

1.) $LiCH_2CO_2Li$, DME

2.) CH_2N_2

OCH_2Ph, OH, O, CH_3O, H, H, O → OCH_2Ph, OH, O, CH_3O, H, H, OH, —CO_2CH_3 (70%) ③ [1]

CH_2N_2

O, O, CO_2H, CO_2H, H → O, O, CO_2Me, CO_2Me, H (100%) ④ [1]

REAGENT: DIBORANE

==

STRUCTURE:

MW = 27.69
(Not isolated)

$\equiv B_2H_6$

Fieser's Reagents for Organic Synthesis: Vol., pages

1	199- 207	**4**	124- 26	**7**	89	**10**	
2	106- 08	**5**	184- 86	**8**	141- 43	**11**	156- 57
3	76- 77	**6**	161- 62	**9**	136- 38	**12**	

March's Advanced Organic Chemistry: 703

MAJOR USES: 1

1. Reducing agent 2. Hydroboration

PREPARATION:

Two major methods of preparation include:(See: **Reagents**, Vol. 1, 199-200.)

1. $3\ NaBH_4 + 4\ BF_3 \longrightarrow 2\ B_2H_6 + 3\ NaBF_4$

2. $2\ NaBH_4 + Hg_2CL_2 \longrightarrow B_2H_6 + 2\ Hg + 2\ NaCl + H_2$

PRECAUTION:

1. Diborane is a flammable gas that can ignite in air.

2. The gas is highly toxic.

NOTES:

1. The reagent is an ineffective reducing agent for acid chlorides, nitro compounds, alkyl and aryl halides, esters and $-SO_2-$ groups.

REFERENCES:

1. P. Sammes and S. Smith, **Chem. Commun.**, (1982), 1143

2. W.C. Still and K.R. Shaw, **Tetrahedron Lett.**, (1981), 3725.

3. T. Nakata, G. Schmid, B. Vranesic, M. Okigawa, T. Smith-Palmer and Y. Kishi, **J. Am. Chem. Soc.**, (1978), 100, 2933.

4. E.J. Corey and R.H. Wollenberg, **Tetrahedron Lett.**, (1976), 4705.

5. S. Danishefsky, K. Vaughan, R.C. Gadwood and K. Tsuzuki, **J. Am. Chem. Soc.**, (1981), 103, 4136.

EXAMPLES:

B_2H_6

THF, -10°

$(CH_2)_3CH_3$ NCH$_3$ → HO NHCH$_3$ (78%) ① [1]

1.) B_2H_6

2.) H_2O_2 / OH^-

t-BuMe$_2$SiO → *t*-BuMe$_2$Si O OH OH (92%) ② [2]

B_2H_6, THF

OH → OH OH (NYA) ③ [2]

1.) B_2H_6 / THF

2.) H_2O_2 / OH^-

CO_2Et CO_2Et → HO CO_2Et CO_2Et (50-80%) ④ [2]

1.) B_2H_6

2.) H_2O_2, OH^-

$H_2C{=}CH$ O O MeO$_2$C O O → HOCH$_2$CH$_2$ O O MeO$_2$C O O (NYA) ⑤ [2]

REAGENT: 2,3-DICHLORO-5,6-DICYANO-1,4-BENZOQUINONE (**DDQ**)

STRUCTURE:

MW = 227.01
MP = 213 - 216°

Fieser's Reagents for Organic Synthesis: Vol., pages

1	215-219	**4**	130-134	**7**	96-97	**10**	135-136
2	112-117	**5**	193-194	**8**	153-156	**11**	166-167
3	83-84	**6**	168-170	**9**	148-151	**12**	174- 75

March's Advanced Organic Chemistry: 1053, 1055, 1077

MAJOR USES:

1. Oxidizing agent
2. Aromatization
3. Oxidative dimerization
4. Dehydrogenation
5. Cyclodehydrogenation
6. Dehydration
7. Cycloadditions
8. Oxidative coupling

PREPARATION:

1. The reagent can be prepared from benzoquinone (**Reagents**, Vol. 1, page 215.

2. Commercially available

PRECAUTION:

NOTES:

REFERENCES:

1. Y. Oikawa, T. Yoshioka and O. Yonemitsu, **Tetrahedron Lett.**, (1982), 885, 889.

2. K. Platt and F. Oesch, **Tetrahedron Lett.**, (1982), 163.

3. A. Kende and T. Blacklock, **Tetrahedron Lett.**, (1980), 3119.

4. R.W. Gammill, **Tetrahedron Lett.**, (1985), 1385.

5. L.A. Paquette, D.T. Belmont and Y.-L. Hsu, **J. Org. Chem.**, (1985), 50, 4667.

EXAMPLES:

DDQ, CH_2Cl_2, H_2O — OH (90%) + CHO / OCH_3 ① [1]

OH / OH — DDQ — O / O (55%) ② [1] [4]

O / CHOH / H — DDQ — O / CHO / H (70%) ③ [1]

O / OH / OMe — DDQ, Dioxane, Reflux — O / OH (46%) / OMe ④ [4]

$OSiMe_3$ — DDQ, Benzene — O (50%) ⑤

REAGENT: DICHLOROKETENE

STRUCTURE:

MW = 110.90 $Cl_2C{=}C{=}O$
(Not isolated)

Fieser's Reagents for Organic Synthesis: Vol., pages

1	221- 22	**4**	134- 35	**7**		**10**	139- 44
2	118	**5**		**8**	156	**11**	168- 70
3	87- 88	**6**		**9**	152- 54	**12**	176- 78

March's Advanced Organic Chemistry: See under "Ketene"

MAJOR USES:

1. Cycloadditions 2. Lactone formation 3. Cyclopentanone annulation

PREPARATION:

1. For a discussion of the in situ generation of this reagent, see: **Reagents**, Vol 1, 221- 22; Vol 2, 87.

PRECAUTION:

NOTES:

1. Dehalogenation of the alpha chloro acid chloride with zinc give the corresponing ketene derivative.

2. A dehydrohalogenation reaction gives the ketene.

3. Diazomethane ring expansion gives the five-membered ring and this becomes a general approach to the construction of cyclopentanones.

REFERENCES:

1. P.W. Jeffs, G. Molina, M.W. Cass and N.A. Cortese, **J. Org. Chem.**, (1982), 47, 3871.

2. W. Reid and O. Bellinger, **Synthesis**, (1982), 792.

3. J.P. Marino and M. Neisser, **J. Am. Chem. Soc.**, (1981), 103, 7687.

4. A.E. Greene, **Tetrahedron Lett.**, (1980), 3059.

5. G. Rosini, G.G. Spineti, E. Foresti and G. Pradella, **J. Org. Chem.**, (1981), 46, 2228.

EXAMPLES:

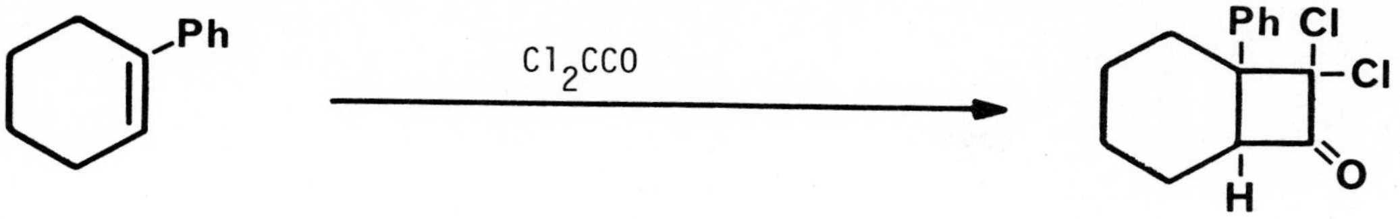

Ph

Cl_2CCO

Ph Cl

Cl

H O

① [1]

(NYA)

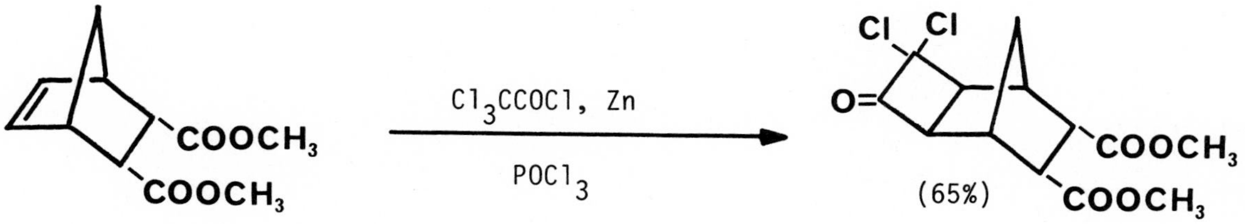

$COOCH_3$

$COOCH_3$

Cl_3CCOCl, Zn

$POCl_3$

Cl Cl

O=

(65%)

$COOCH_3$

$COOCH_3$

② [1] △1

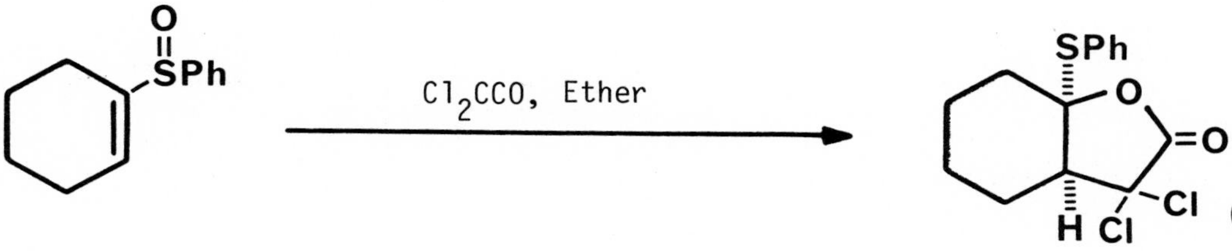

O

SPh

Cl_2CCO, Ether

SPh

O

=O

Cl

H Cl

(72%)

③ [1]

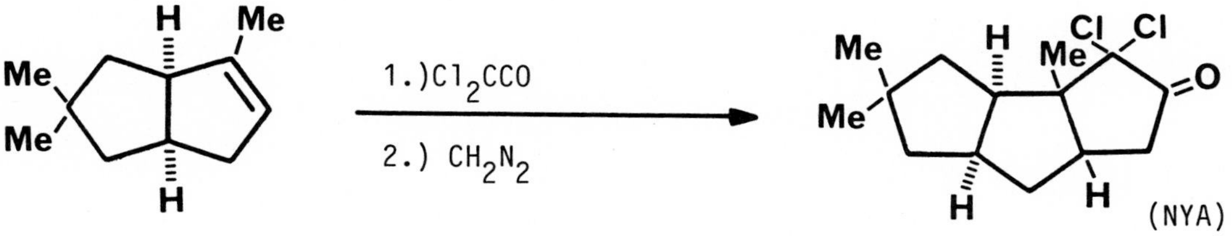

H Me

Me

Me

H

1.) Cl_2CCO

2.) CH_2N_2

Me H Cl Cl

Me

=O

Me

H H

(NYA)

④ [1] △2

S S

Me

$CHCl_2COCl$, Et_3N

Me Cl Cl O

S

S

(85%)

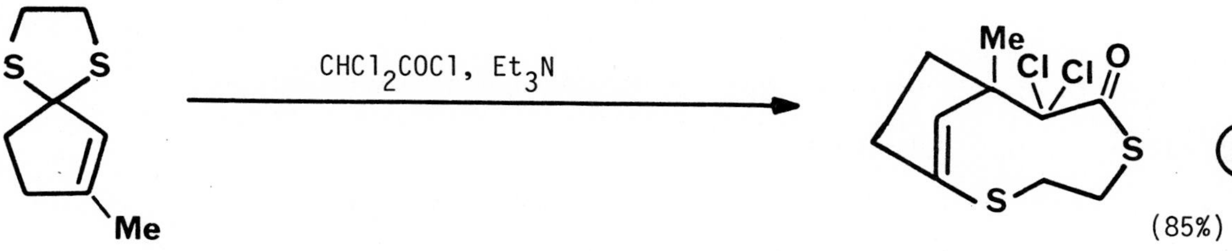

⑤ △3

REAGENT: DICOBALTOCTACARBONYL

==

STRUCTURE:

MW = 342.12 $Co_2(CO)_8$
MP = 51- 52° C

Fieser's Reagents for Organic Synthesis: Vol., pages

1	224- 25	**4**	139	**7**	99- 100	**10**	129- 30
2		**5**	204- 05	**8**	148- 50	**11**	162- 63
3	89	**6**	172	**9**	144- 45	**12**	

March's Advanced Organic Chemistry: 554, 597, 772, 724, 775.

MAJOR USES:

1. Oxo addition (CO addition)

PREPARATION:

1. The reagent can be prepared from Raney cobalt, cobalt carbonate or cobalt acetate (**Reagents**, Vol. 1, 224.

2. Commercially available.

PRECAUTION:

NOTES:

REFERENCES:

1. H. Alper and J.K. Currie, **Tetrahedron Lett.**, (1979), 2665.

2. S. Murai and N. Sonoda, **Angew. Chem. Int. Ed.**, (1979), 18, 809.

3. S.C. Shim, S. Antebi and H. Alper, **J. Org. Chem.**, (1985), 50, 147.

4. D.C. Billington and P.L. Paulson, **Organometallics**, (1982), 1, 1560.

5. T. Murai and S. Kato, **J. Am. Chemn. Soc.**, (1984), 106, 6093.

EXAMPLES:

Reaction 1: CH_3I, CO, $Co(CO)_8$ → product bearing $COCH_3$ (54%) ①

Reaction 2: $HSiMe(Et)_2$, CO / $Co_2(CO)_8$, 140° → product bearing $OSiMeEt_2$ (89%) ②

Reaction 3: CH_2SH / Me substituted arene; $Co_2(CO)_8$, EtOH / CO (60 atm), 190° → CH_2CO_2Et / Me substituted arene (75%) ③

Reaction 4: $CH_3C{\equiv}CCH_3$; $Co_2(CO)_8$ / THF → H_3C, CH_3, C≡C, $(CO)_3Co$, $Co(CO)_3$ (94%) ④

Reaction 5: epoxide with OMe + 2 $HSiEt_2Me$; $Co_2(CO)_8$ / CO, 1 atm, 25° / C_6H_6 → MeO, $OSiMeEt_2$, $OSiMeEt_2$ (75%) ⑤

REAGENT: 1,3-DICYCLOHEXYLCARBODIIMIDE (**DCC**)

STRUCTURE:

MW = 206.33
MP = 34- 35°C
BP = 122- 24° (6 mm Hg)

Fieser's Reagents for Organic Synthesis: Vol., pages

1	231-36	**4**	141	**7**	100- 01	**10**	142
2	126	**5**	206- 07	**8**	162- 63	**11**	173- 74
3	91	**6**	174	**9**	156- 57	**12**	

March's Advanced Organic Chemistry: 345, 349, 355, 357, 373, 441, 931, 1081.

MAJOR USES:

1. Dehydration 2. Lactonization 3. Heterocyclic synthesis

4. Esterification

PREPARATION:

1. Can be prepared by the oxidation of N,N'-dicyclohexylthiourea with mercuric oxide: E. Schmidt, F. Hitzler and E. Lahde, **Ber.**, (1938), 71, 1933

2. Commercially available in high purity.

PRECAUTION:

1. The reagent is moisture-sensitive, corrosive and a skin irritant.

2. For handling procedures, see: **Reagents**, Vol. 1, 231.

NOTES:

REFERENCES:

1. K. Findeisen, H. Heitzler and K. Dehnicke, **Synthesis**, (1981), 702.

2. J. Martinez and J. Laur, **Synthesis**, (1982), 979.

3. P. Knochel and D. Seebach, **Synthesis**, (1982), 1017.

4. K. Suda, F. Hino and C. Yijima, **Chem. Pharm Bull.**, (1985), 33, 882.

5. E.E. Ibrahim, A. Moshsen, M.E. Omar, N.S. Habib, O.M. Aboulwafa, **J. Heterocyclic Chem.**, (1982), 19, 761.

EXAMPLES:

$Me_2C(NHMe)-CN$ + DCC —DABCO→ $C_6H_{11}-NH-C(CN)=N-C_6H_{11}$ (82%) + $Me_2C=NMe$ ①

p-Nitrophenol + HCOOH —DCC, Ethyl acetate→ p-$NO_2C_6H_4$-O-CHO (72%) ②

t-Bu–CH(OH)–CH_2NO_2 —DCC, CuCl, Ether, RT→ t-Bu–CH=CH–NO_2 (94%) ③ [1]

i-Pr–CH(NHCOPh)–COOH —DCC, THF→ 4-i-Pr-2-phenyl-oxazol-5-one (80%) ④ [2]

Aminophenol steroid (H_2N, HO, Me, O) + PhN=C=S —DCC, EtOH→ PhNH–(benzoxazole) (89%) ⑤ [3]

REAGENT: DIIMIDE

==

STRUCTURE:

HN=NH

Fieser's Reagents for Organic Synthesis: Vol., pages

1	257- 58	**4**	154- 55	**7**		**10**	
2	139	**5**	220	**8**	172- 73	**11**	
3	99- 101	**6**	195	**9**		**12**	

March's Advanced Organic Chemistry: 177, 698, 812, 1106

MAJOR USES: △1 △2

1. Reducing agent

PREPARATION:

1. The reagent is generated in situ by oxidation of hydrazine, decomposition of sulfonyl or acyl hydrazides, or by the thermal decomposition of the salt of azodicarboxylic acid.

PRECAUTION:

NOTES:

1. The reagent will reduce double bonds between the same atoms. Thus, C=C, C C, =and N=N bonds will reduce while C=N or C=O bonds will not.

2. Carbon-carbon double bonds connected to electron-withdrawing groups will show reduced reactivity. Strained bonds will reduce faster than unstrained bonds.

REFERENCES:

1. M. Herin, P. Delbar, J. Remion, P. Sandra and A. Krief, **Tetrahedron Lett.**, (1979), 3107.

2. K. Mori, M. Ohki, A. Sato and M. Matsui, **Tetrahedron**, (1972), 28, 3739.

3. J.M. Hoffman, Jr. and R.H. Schlessinger, **Chem. Commun**, (1971), 1245.

4. C.E. Miller, **J. Chem. Ed.**, (1975), 42, 254.

5. P.B. Hulbert, E. Bueding and C.H. Robinson, **J. Med. Chem.**, (1973), 16, 72.

EXAMPLES:

HN=NH

(75%) ①

HN=NH

OH

OH (74%) ②

PhN=NPh —HN=NH→ PhNH—NHPh (95%) ③

HN=NH

(76%) ④

O_2N–(furan)–CH=CHCO$_2$H —HN=NH, HOAc→ O_2N–(furan)–CH$_2$CH$_2$CO$_2$H (89%) ⑤

REAGENT: DIISOBUTYLALUMINUM HYDRIDE (**DIBAL, DIBAH**)

===

STRUCTURE:

$(CH_3)_2CHCH_2-Al(H)-CH_2CH(CH_3)_2$

MW = 142.22
BP = 116- 118°C (1 mm Hg)

Fieser's Reagents for Organic Synthesis: Vol., pages

1	260- 62	**4**	158- 61	**7**	111- 13	**10**	149
2	140- 42	**5**	224- 25	**8**	173- 74	**11**	185- 88
3	101- 02	**6**	198- 201	**9**	171- 72	**12**	191- 93

March's Advanced Organic Chemistry: 394, 397, 694, 809, 816, 1052, 1105, 1108.

MAJOR USES:

1. A versatile reducing agent.

PREPARATION:

1. The reagent can be prepared from triisobutylaluminum and heptane [W. Gensler and J. Bruno, **J. Org. Chem.**, (1963), 28, 1254]. See also J. Eisch and W. Kaska, **J. Am. Chem. Soc.**, (1966), 88, 2213.

2. The reagent is commercially available.

PRECAUTION:

1. THF cannot be used as solvent because it forms a complex with the reagent.

2. The reagent is moisture sensitive, and is reactive to air. Must be maintained under an inert atmosphere.

NOTES:

1. **DIBAH** can reduce an ester or a nitrile to an aldehyde. In the presence of n-BuLi an "ate" complex is formed that allows for selective reduction in polyfunctional molecules. See: **Reagents**, Vol. 12, 191- 193.

REFERENCES:

1. H. Yamamoto and K. Maruoka, **J. Am. Chem. Soc.**, (1981), 103, 4186.

2. P.A. Bartlett and J.L. Adams, **J. Am. Chem. Soc.**, (1980), 102, 337.

3. A.E. Greene, C. LeDrian, P. Crabbe, **J. Am. Chem. Soc.**, (1980), 102, 7583.

4. W.H. Parsons, R.H. Schlessinger and M.L. Quesada, **J. Am. Chem. Soc.**, (1980), 102, 889.

EXAMPLES:

$CH_3(CH_2)_5CHO$

DIBAL

NH_2 NH_2

HN NH $CH_3(CH_2)_5$

NH_2 NH $CH_3(CH_2)_6$ (94%) ①

MeO_2C COOH

DIBAL

Toluene

H O COOH (77%) ② ⚠1

t-BuSiMe$_2$ SMe MeS-C-COOn-Bu MeO- O

DIBAL

t-BuSiMe$_2$ SMe MeS-C-CHO MeO- O ("High yield") ③ ⚠1

MeO H Me– –COOMe O COOMe

1.) DIBAL

2.) H^+

3.) MnO_2

MeO H Me– –CHO O CH_2OH (65%) ④ ⚠1

REAGENT: DIISOPINOCAMPHEYLBORANE 1

STRUCTURE:

MW = 286.36

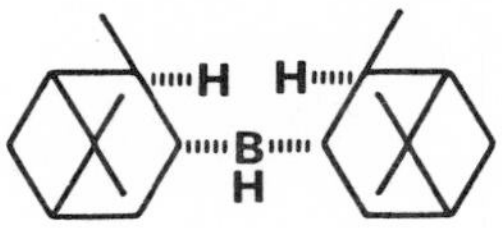

Fieser's Reagents for Organic Synthesis: Vol., pages

1	262- 63	**4**	161- 62	**7**		**10**	
2		**5**		**8**	174	**11**	188
3		**6**	202	**9**		**12**	193

March's Advanced Organic Chemistry: 107, 705

MAJOR USES:

1. Asymmetric hydroboration

PREPARATION:

1. The reagent can be prepared by the hydroboration of alpha pinene [see: H.C. Brown, M. Desai and P. Judhav, **J. Org. Chem.**, (1982), 47, 5065. See also: H.C. Brown and B. Singaram, **J. Org. Chem.**, (1984), 49, 945.]

PRECAUTION:

1. Use the same precautions as with diborane.

NOTES:

1. **Chem. Abstract.** refers to this as Tetra-3-pinanylborane.

2. Monoisopinocampheylborane undergoes chemical reaction with similar diastereoselectivity.

REFERENCES:

1. R.K. Varma, M. Koreeda, D. Yagen, K. Nakanishi and E. Caspi, **J. Org. Chem.**, (1975), 40, 3680.

2. J.J. Partridge and G. Zweifel, **J. Am. Chem. Soc.**, (1961), 83, 486.

3. J.J. Partridge, N.K. Chadha and M.R. Uskokovic, **J. Am. Chem. Soc.**, (1973), 95, 532.

4. H.C. Brown, P.K. Jadhav and A.K. Mandal, **J. Org. Chem.**, (1982), 47, 5074.

EXAMPLES:

THPO

(-) DIPCB

CH_2OH

H

THP = Tetrahydropyranyl

1

HO

(83% Optical purity, NYA)

CH_2CO_2Me

(+) DIPCB

H_2O_2

CH_2CO_2Me

OH

2

(45%)

Me

1.) DIPCB / THF, -78°

2.) NaOH, H_2O_2, 0°

Me

OH

3

(33%)

Me

BH_2

Ph

+

-25°

24 hrs

NaOH, H_2O_2

Ph

OH

4

2

(92%)

REAGENT: N,N-DIMETHYLHYDRAZINE

==

STRUCTURE:

MW = 60.10
BP = 62- 64°/ 753 mm
d = 0.791

Me_2NNH_2

Fieser's Reagents for Organic Synthesis: Vol., pages

1	289- 90	**4**		**7**	126- 30	**10**	
2	154- 55	**5**	254	**8**	192- 93	**11**	200- 01
3	117	**6**	223	**9**	184- 85	**12**	205

March's Advanced Organic Chemistry: See under "hydrazines"

MAJOR USES: 1 2

1. Hydrazone synthesis

PREPARATION:

1. Commercially available.

PRECAUTION:

1. Highly toxic and a suspected carcinogen.

NOTES:

1. The dimethylhydrazones are readily converted to nitriles (See for example, ref. 5)

2. The hydrazones can be removed by m-chloroperoxybenzoic acid (**MCPBA**), see:
Reagents, 12.

3. The starting material was made by the reaction of N,N-dimethylhydrazine and the corresponding carbonyl compound.

REFERENCES:

1. B. Serckx-Poncin, A.-M. Hesbain-Frisque and L. Ghosez, **Tetrahedron Lett.**, (1982), 3261.

2. R.W.M. Aben and H.W. Scheeren, **Tetrahedron Lett.**, (1985), 1889.

3. F. Yoneda and T. Nagamabu, **J. Chem. Soc., Perkin I**, (1976), 1547.

4. G.R. Newkome and D.L. Fishel, **J. Org. Chem.**, (1966), 31, 677.

5. R.F. Smith and L.E. Walker, **J. Org. Chem.**, (1962), 27, 4372.

EXAMPLES:

Me, CH_2, N, NMe_2 → ($H_2C{=}CH{-}CN$) → Me, N, CN, NMe_2 (53%) ① ⚠3

$-CHO$, $-OMe$, Me OMe → (H_2NNMe_2) → $-CH{=}N{-}NMe_2$, $-OMe$, Me OMe (80%) ②

O, Me, N, NO, O, N, H, NHMe + PhCHO → (H_2NNMe_2) → O, Me, N, N, Ph, O, N, H, N (71%) ③

COMe, Cl → (H_2NNMe_2, EtOH) → $MeC{=}NNMe_2$, Cl (88%) ④

CHO → (Me_2NNH_2, C_6H_6 Reflux) → $CH{=}NNMe_3^{\oplus}I^{\ominus}$ (92%) ⑤

REAGENT: DIMETHYLOXOSULFONIUM METHYLIDE 

STRUCTURE:

$$H_3C-\overset{\overset{\displaystyle O}{\|}}{\underset{\underset{\displaystyle CH_3}{|}}{S}}=CH_2$$

Fieser's Reagents for Organic Synthesis: Vol., pages

1	315-18	**4**	197- 99	**7**	133	**10**	168- 69
2	171- 73	**5**	254- 57	**8**	194- 96	**11**	
3	125- 27	**6**		**9**	186- 87	**12**	213- 14

March's Advanced Organic Chemistry: 599, 773, 864

MAJOR USES:

1. Methylations 2. Epoxide synthesis 3. Cyclopropane synthesis

PREPARATION:

1. Can be prepared from trimethylsulfonium iodide and NaH.

PRECAUTION:

NOTES:

1. Also known as **dimethylsulfoxonium methylide**
2. Dimethylsulfonium methylide has similar behavior.

REFERENCES:

1. E.J. Corey, M. Tius, J. Das, **J. Am. Chem. Soc.**, (1980), 102, 7612.
2. W.C. Still, **J. Am. Chem. Soc.**, (1979), 101, 2493.
3. T. Kunieda and B. Witkop, **J. Org. Chem.**, (1970), 35, 3981.
4. A.G. Hortmann and R.L. Harris, **J. Am. Chem. Soc.**, (1971), 93, 2471.
5. Y. Sugimura, N. Soma and Y. Kishida, **Tetrahedron Lett.**, (1971), 91.

EXAMPLES:

$Me_2S(O)CH_2$

(NYA, 5:1) ① [2]

$Me_2S(O)CH_2$

(75%) ② [2] △2

$Me_3S^+(O)Cl^-$

NaH

$[Me_2S^+(O)C^-H_2]$

THF / N_2

(90%) ③ [1]

$Me_3S^+(O)I^-$

NaI

$[Me_2S(O)CH_2]$

DMSO, RT, 17 hrs

$PhC{\equiv}CC(O)CMe_3$

(63-76%) ④

$Me_2S(O)CH_2$

THF

(95%) ⑤ [3]

REAGENT: DIMETHYLSULFONIUM METHYLIDE

STRUCTURE:

$$CH_3\overset{\overset{CH_2}{\|}}{S}CH_3$$

MW = 76.17

Fieser's Reagents for Organic Synthesis: Vol., pages

1	314-15	**4**	196- 97	**7**		**10**	
2	169- 71	**5**		**8**	198- 99	**11**	213- 14
3	119- 23	**6**		**9**	188	**12**	

March's Advanced Organic Chemistry: 773, 864

MAJOR USES:

1. Epoxide formation from carbonyl groups 2. Ring formation 3. Base

PREPARATION:

1. Most often prepared in situ from reaction of trimethylsulfonium salts and a base.

PRECAUTION:

NOTES:

1. The reaction first results in an epoxide formed at the carbonyl position.

2. A silicon analog.

REFERENCES:

1. L.C. Garver and E.E. van Tamelen, **J. Am. Chem. Soc.**, (1982), 104, 867.

2. E. Borredon, M. Delmas and A. Gaset, **Tetrahedron Lett.**, (1982), 5283.

3. R. Okazaki, Y. Negishi and N. Inamoto, **J. Org. Chem.**, (1984), 49, 3819.

4. M. Chaykowsky, E.J. Modest and S.K. Sengupta, **J. Heterocyclic Chemistry**, (1977), 14, 661.

5. K. Ackermann, J. Chapuis, D.E. Horning and J.M. Muchowski, **Can. J. Chem.**, (1969), 47, 4327.

EXAMPLES:

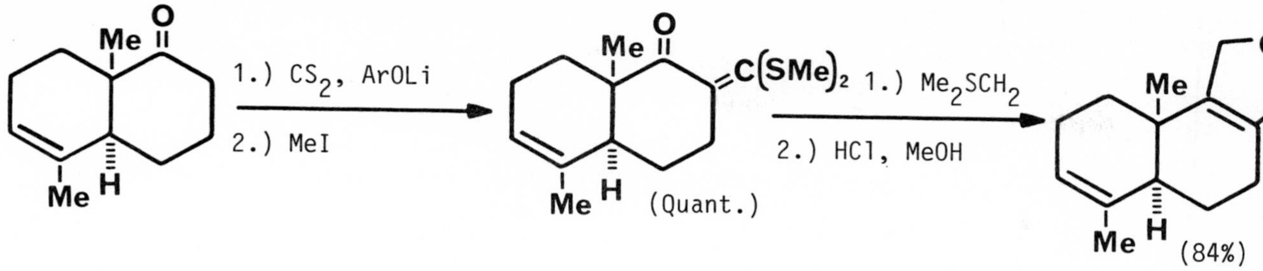

1.) CS_2, ArOLi

2.) MeI

(Quant.)

1.) Me_2SCH_2

2.) HCl, MeOH

(84%)

① [1] △1

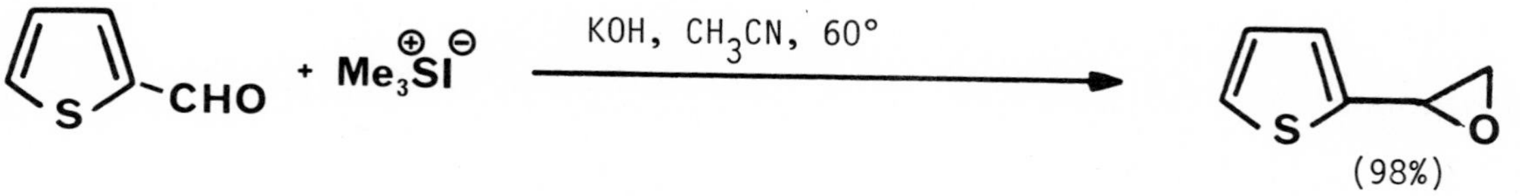

$Me_3S^{\oplus}I^{\ominus}$

KOH, CH_3CN, 60°

(98%)

② △2

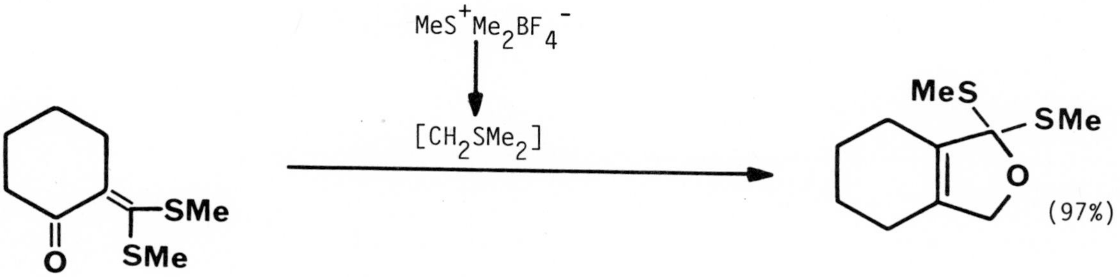

$MeS^+Me_2BF_4^-$

$[CH_2SMe_2]$

(97%)

③ [2] △1

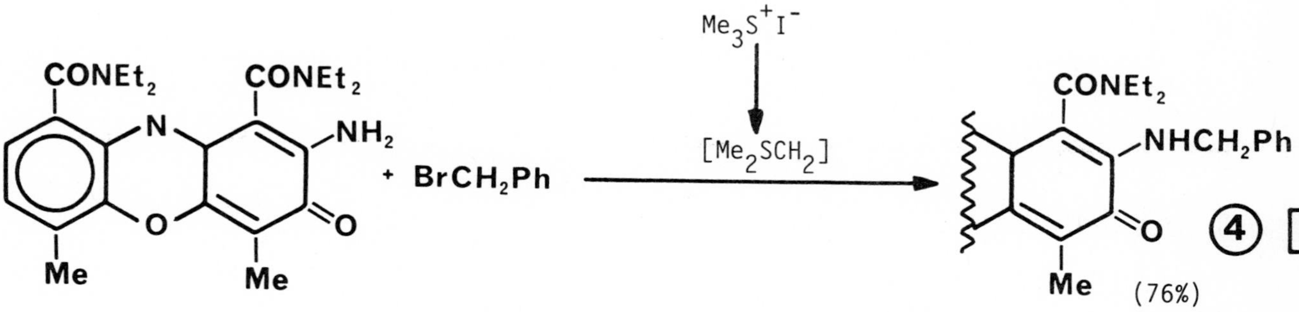

$CONEt_2$ $CONEt_2$ NH_2 + $BrCH_2Ph$

$Me_3S^+I^-$

$[Me_2SCH_2]$

$CONEt_2$ $NHCH_2Ph$

(76%)

④ [3]

Me_2SCH_2

DMSO

(90%)

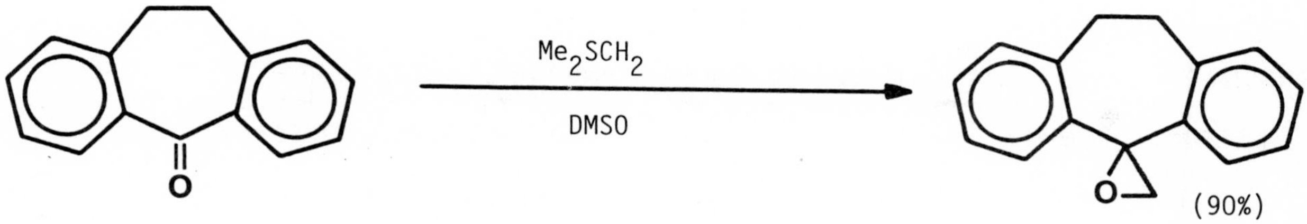

⑤ [1]

REAGENT: DIPHENYLDISELENIDE (PHENYL DISELENIDE)

STRUCTURE:

Ph-Se-Se-Ph

MW = 312.13
MP = 61- 63°C

Fieser's Reagents for Organic Synthesis: Vol., pages

1		**4**		**7**	136- 37	**10**	170
2		**5**	272- 76	**8**		**11**	
3		**6**	235	**9**	199	**12**	

March's Advanced Organic Chemistry: 536

MAJOR USES: ⚠1

1. Isomerizations 2. Allylic alcohol formation 3. Alkene synthesis

PREPARATION:

1. Can be prepared from phenyl Grignard and selenium [see, for example, H. Reich, M. Cohen and P. Clark, **Org. Synth.**, (1979), 59, 141, and references cited.].

2. Commercially available in high purity.

PRECAUTION:

1. The reagent is very toxic and should be used with caution.

NOTES:

1. Reduction of the reagent with $NaBH_4$ produces the highly nucleophilic phenylselenium anion.

2. Via a selenoxide elimination reaction.

REFERENCES:

1. A. Barrett, D. Barton and G. Johnson, **Synthesis**, (1978), 741.

2. L. Snyder and J. Mlochowski, **Synthesis**, (1984), 439.

3. M.D. Erion and J.E. McMurry, **Tetrahedron Lett.**, (1985), 559.

4. D.N. Brattesani and C.H. Heathcock, **Tetrahedron Lett.**, (1974), 2279.

5. K.B. Sharpless and R.F. Lauer, **J. Am. Chem. Soc.**, (1973), 95, 2697.

EXAMPLES:

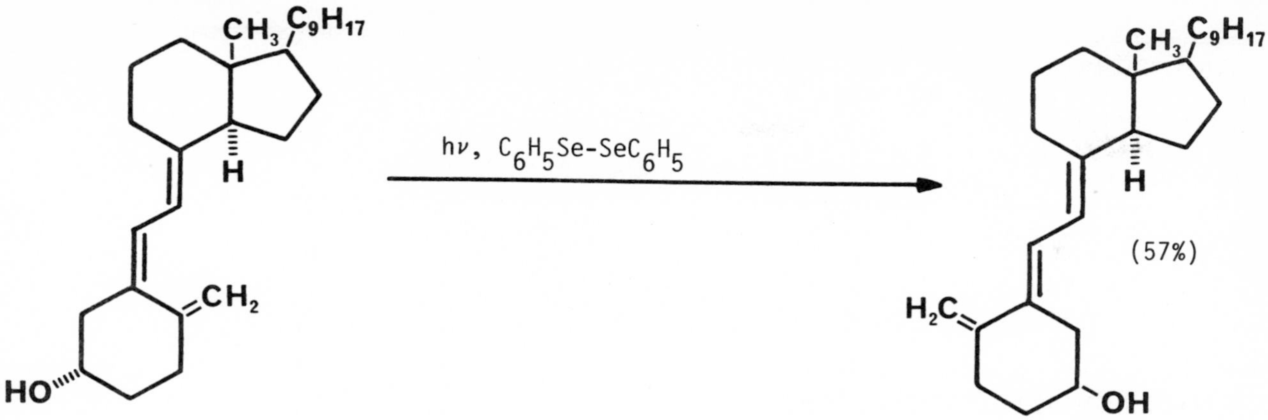

$h\nu$, $C_6H_5Se-SeC_6H_5$ (1) [1]

(57%)

(2)

PhSeSePh → (H_2NNH_2 / NaOH) → 2 [PhSeNa] → (CH_2Cl_2) → $PhSeCH_2SePh$

(96%)

1.) Ph_2Se_2, $h\nu$, Benzene

2.) Sodium periodate

(3) [3]

(89%) [triangle 2]

(4)

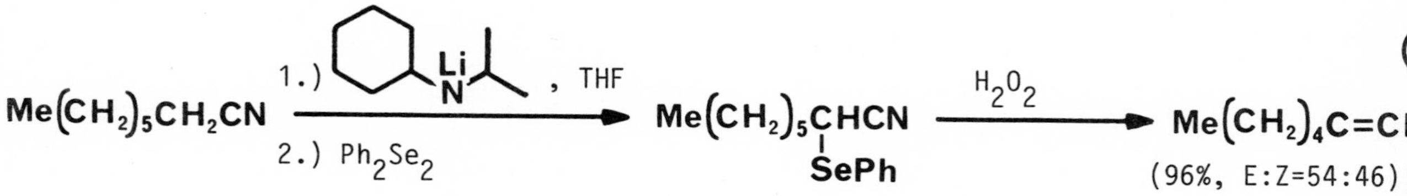

$Me(CH_2)_5CH_2CN$ → (1.) LiN(cyclohexyl)(isopropyl), THF; 2.) Ph_2Se_2) → $Me(CH_2)_5CH(SePh)CN$ → (H_2O_2) → $Me(CH_2)_4C{=}CHCN$

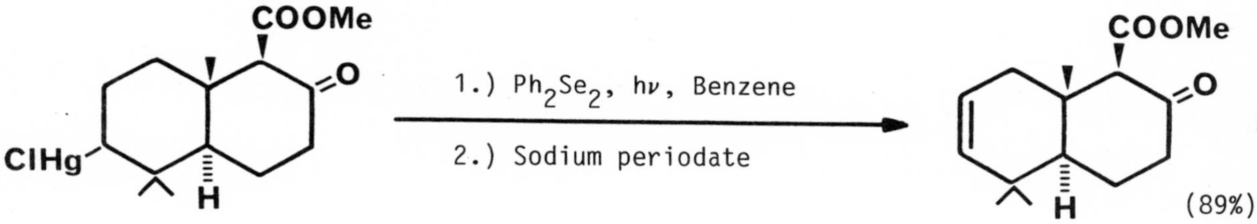

(96%, E:Z=54:46) [3]

[triangle 2]

1.) Ph_2Se_2, n-BuOH, $NaBH_4$

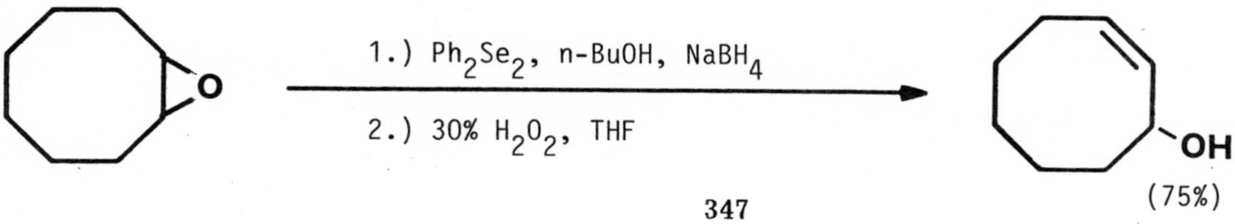

2.) 30% H_2O_2, THF

(5) [2]

(75%)

REAGENT: DIPHENYLPHOSPHORYL AZIDE (**DPPA**)

==

STRUCTURE:

MW = 275.20
BP =157°C (0.17 mm Hg)
d = 1.277

$(PhO)_2P(=O)N_3$

Fieser's Reagents for Organic Synthesis: Vol., pages

1	**4** 210- 11	**7** 138	**10** 173
2	**5** 280	**8** 211- 12	**11** 222
3	**6**	**9**	**12**

March's Advanced Organic Chemistry: 380

MAJOR USES:

1. Amine synthesis 2. Heterocyclic synthesis 3. Source of azide

---

PREPARATION:

1. The reagent can be prepared from diphenylphosphorylchloridate and sodium azide in acetone [see: T. Shiori, K. Ninomiya and S. Yamada, **J. Am. Chem. Soc.**, (1972), 94, 6203].

2. Commercially available.

PRECAUTION:

1. An irritant and is toxic.

NOTES:

1. This can be considered a modified **Hofmann degradation**.

REFERENCES:

1. Y. Hamada and T. Shioiri, **Tetrahedron Lett.**, (1982), 235.

2. J.W. Lyga, **Synth. Commun.**, (1986), 16, 163.

3. B. Danieli, G. Lesma, G. Palmisand and S. Tollari, **Synthesis**, (1984), 353.

4. J.H. Rigby and N. Balasubrahamanian, **J. Org. Chem.**, (1984), 49, 4569.

5. S. Mori, T. Aoyama and T. Shioiri, **Tetrahedron Lett.**, (1984), 429.

EXAMPLES:

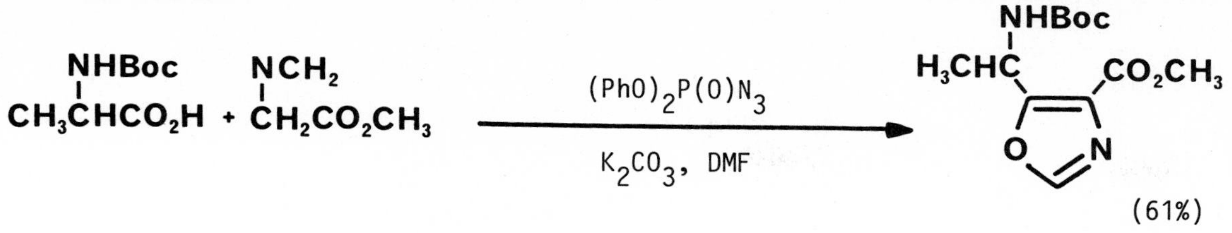

NHBoc, CH_3CHCO_2H + NCH_2, $CH_2CO_2CH_3$ —$(PhO)_2P(O)N_3$, K_2CO_3, DMF→ NHBoc, H_3CHC, CO_2CH_3, O, N (61%) ① [2]

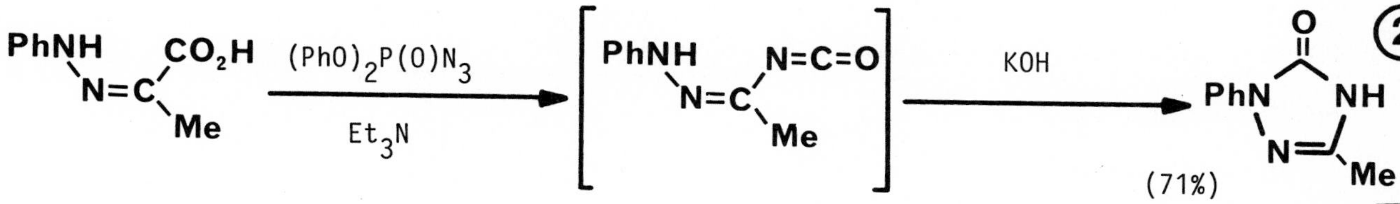

PhNH, CO_2H, N=C, Me —$(PhO)_2P(O)N_3$, Et_3N→ [PhNH, N=C=O, N=C, Me] —KOH→ O, PhN, NH, N, Me (71%) ② [2]

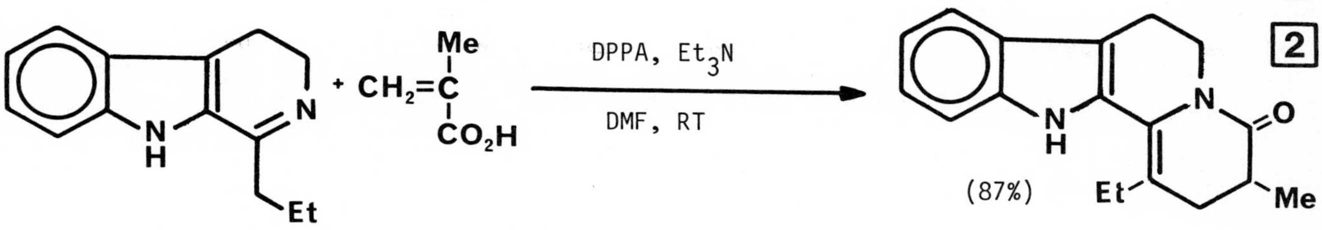

N, H, N, Et + CH_2=C(Me)CO_2H —DPPA, Et_3N, DMF, RT→ N, H, N, O, Et, Me (87%) ③ [2]

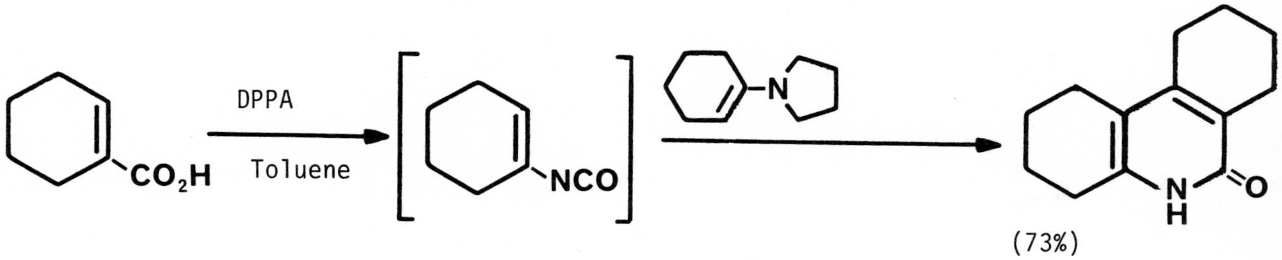

CO_2H —DPPA, Toluene→ [NCO] —N→ N, H, O (73%) ④ [2]

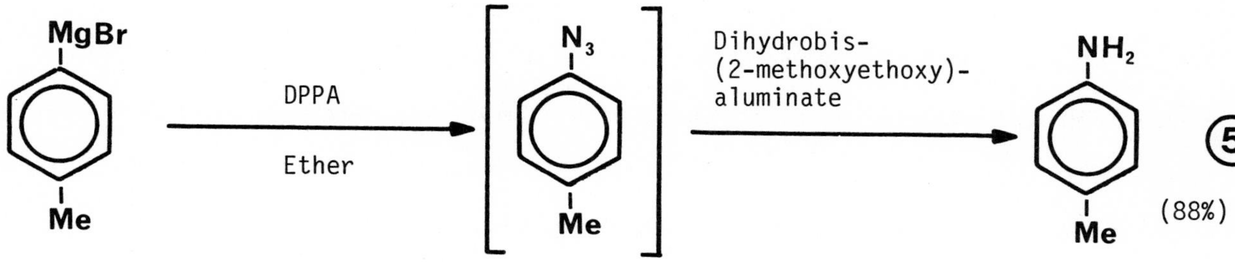

MgBr, Me —DPPA, Ether→ [N_3, Me] —Dihydrobis-(2-methoxyethoxy)-aluminate→ NH_2, Me (88%) ⑤ [3]

REAGENT: DISIAMYLBORANE (BIS-3-METHYL-2-BUTYLBORANE), Si_2BH

STRUCTURE:

MW = 154.10
MP = 35- 40°C

Me Me
Me_2CHCH $CHCHMe_2$
B
H
$\equiv Sia_2BH$

Fieser's Reagents for Organic Synthesis: Vol., pages

1	57- 59	**4**	37	**7**		**10**	40
2		**5**	39- 41	**8**	41	**11**	226
3	22	**6**		**9**		**12**	

March's Advanced Organic Chemistry: 703- 706, 1085

MAJOR USES: 1

1. Selective hydroboration

PREPARATION:

The reagent is prepared by reacting 3-methyl-2-butene with diborane.

PRECAUTION:

1. The reagent is toxic, and highly flammable.

NOTES:

1. The reagent does not reduce carboxylic acids, sulfonic acids, amides, esters, acid chlorides, acid anhydrides or sulfones. Oxides and nitriles are slowly reduced. The reducing ability is essentially the same as diborane; however, the greater steric bulk of the reagent gives it increased ability to reactive in stereo- or regioselective manner.

2. **PCC** = Pyridinium chlorochromate

REFERENCES:

1. H. Brown, S. Kulkarni and C. Rao, **Synthesis**, (1980), 151.

2. C.A. Brown, M.C. Desai and P.K. Jadhav, **J. Org. Chem.**, (1986), 51, 162.

3. E. Urbina, A. Guerrero, L. Cuellar and R. Contreras, **Synthesis**, (1983), 113.

4. G. Zweifel, A. Horng and J.E. Plamondon, **J. Am. Chem. Soc.**, (1974), 96, 316.

5. H. Yatagai, Y. Yamamoto and K. Maruyama, **Chem. Commun.**, (1978), 702.

EXAMPLES:

1.) Sia_2BH

2.) PCC

CH_2 → CHO (67%) ① ②

BMB, THF

H — $C{\equiv}C{-}C_9H_{19}$ → H — $CH{=}CH{-}C_9H_{19}$ (70%) ②

H, $OSiMe_3$ + $Me_3N{\rightarrow}O$ → $[Sia]_2BH$ → H, $OSiMe_3$, O (98%) ③

1.) Sia_2BH, THF

2.) NaOH, H_2O_2

$-C{\equiv}CCH(OEt)_2$ → $-CH_2\overset{O}{\overset{\|}{C}}CH_2OEt$ (80%) ④

⑤

$n\text{-}C_6H_{13}C{\equiv}CEt$ → BMB, THF → [C_6H_{13}, H, C=C, Et, $B(Sia)_2$] → $Pd(OAc)_2$, N_2 → $n\text{-}C_6H_{13}$, H, C=C, Et, H (70%)

REAGENT: DISODIUM TETRACARBONYLFERRATE (COLLMAN'S REAGENT)

==

STRUCTURE: 1

$Na_2Fe(CO)_4$

MW = 213.88

Fieser's Reagents for Organic Synthesis: Vol., pages

1		**4**	461- 65	**7**	341	**10**	174- 75
2		**5**	624- 25	**8**	216- 17	**11**	
3	267- 68	**6**	550- 52	**9**	205- 07	**12**	

March's Advanced Organic Chemistry: 420

MAJOR USES:

1. Carbonyl group insertion 2. Reductions

PREPARATION:

1. Can be prepared by reacting iron pentacarbonyl with 1% sodium amalgam under an inert atmosphere.

2. Commercially available.

PRECAUTIONS:

1. Very toxic and a suspected carcinogen.

NOTES:

1. Also known as **tetracarbonylferrate(II)**.

REFERENCES:

1. J.P. Collman, S.R. Winter and D.R. Clark, **J. Am. Chem. Soc.**, (1972), 94, 1788.

2. J.P. Collman and N.W. Hoffman, **J. Am. Chem. Soc.**, (1973), 95, 2689.

3. J.E. McMurry and A. Andrus, **Tetrahedron Lett.**, (1980), 4687.

4. R.G. Fincke and T.N. Sorell, **Org. Synth.**, (1979), 59, 102.

5. T. Mitsuda, Y. Watanabe, M. Tanaka, K. Yamamoto and Y. Takegami, **Bull. Chem. Soc.,Jpn.**, (1972), 45, 305.

EXAMPLES:

$n\text{-}C_8H_{17}Br \xrightarrow[\text{N-Methyl-2-pyrrolidone}]{Na_2Fe(CO)_4} [n\text{-}C_8H_{17}Fe(CO)_4]^{\ominus} \xrightarrow{EtI} n\text{-}C_8H_{17}C(O)Et$ (80%) ① [1]

$Me(CH_2)_7CH_2Br \xrightarrow[THF]{Na_2Fe(CO)_4} Na[Fe(CO)_4(CH_2)_8Me] \xrightarrow[THF]{C_6F_5C(O)Cl} n\text{-}C_6F_5C(O)(CH_2)_8Me$ (99%) ② [1]

1-(vinyl)-1-(tosyloxymethyl)cyclohexane ($=CH_2$, OTs) $\xrightarrow[THF]{Na_2Fe(CO)_4}$ spiro[4.5]decanone (=O) (65%) ③ [1]

Hexahydrophthalic anhydride $\xrightarrow[\text{2.) } H^+]{\text{1.) } Na_2Fe(CO)_4}$ 2-formylcyclohexanecarboxylic acid (CHO, CO_2H) (61%) ④ [2]

Phthaloyl chloride (COCl, COCl) $\xrightarrow{Na_2Fe(CO)_4}$ biphthalidylidene (O=, O, =O) (23%) ⑤ [2]

REAGENT: ETHYL VINYL ETHER

STRUCTURE:

MW = 72.10
BP = 33°C

Et-O-HC=CH_2

Fieser's Reagents for Organic Synthesis: Vol., pages

1	386- 88	**4**	234- 35	**7**		**10**	
2	198	**5**		**8**		**11**	235- 36
3		**6**		**9**		**12**	

March's Advanced Organic Chemistry: 418

MAJOR USES:

1. Protecting group 2. Trans vinylation 3. Synthesis of aldehydes by **Claisen Rearrangement**

PREPARATION:

1. Commercially available.

2. The reagent can be prepared by reacting acetaldehyde, ethanol and HCl [C. Hurd and D. Botteron, **J. Am. Chem. Soc.**, (1946), 68, 1200].

PRECAUTION:

NOTES:

1. See **Claisen Rearrangement**

REFERENCES:

1. L. Paquette, H. Schostarez and G. Annis, **J. Am. Chem. Soc.**, (1981), 103, 6526.

2. S. Karlson, P. Froyen and L. Skattebol, **Acta Chem. Scand.**, (1976), B 30, 664.

3. K. Mikami, N. Kishi and T. Nakai, **Chem. Letters**, (1981), 1721.

4. H. Yamamoto and H.L. Sham, **J. Am. Chem. Soc.**, (1979), 101, 1609.

5. R.K. Boeckman, Jr., D.M. Blum and S.D. Arthur, **J. Am. Chem. Soc.**, (1979), 101, 5060.

EXAMPLES:

①

CH$_2$OH → (CH$_2$=CH-O-Et, Hg(OAc)$_2$) → H$_2$C=CHOCH$_2$ (82%) → (145-50°) → CHO, H$_2$C, CH$_2$ (69%)

[2] [3] [△1]

OH → (EtOCHCH$_2$, Hg(OAc)$_2$) → (69%) ② [2]

[2]

OH → (EtOCHCH$_2$, Hg(OAc)$_2$) → O → (△) → CHO (56%) ③ [3] [△1]

OH → (1.) CH$_2$CHOEt, Hg^{++}; 2.) △) → CHO (78%) ④ [3] [△1]

HO, COOMe → (1.) CH$_2$CH-OEt, Hg^{++}; 2.) 160°) → CHO, COOMe (60-65%) ⑤ [3]

REAGENT: FLUOROSULFONIC ACID

STRUCTURE:

MW = 100.08

$F-S(=O)_2-OH$

Fieser's Reagents for Organic Synthesis: Vol., pages

1	396- 97	**4**		**7**		**10**
2	199- 200	**5**	310- 11	**8**		**11**
3		**6**	262- 3	**9**		**12**

March's Advanced Organic Chemistry: 219- 220, 475

MAJOR USES:

1. Dehydration 2. Cyclization 3. With SbF_5 can make stable carbonium ions. 4. Acid catalyst

PREPARATION:

1. Commercially available

PRECAUTION:

1. Corrosive, moisture-sensitive.

2. Toxic

NOTES:

1. **Pinacol rearrangement**

REFERENCES:

1. M. Suzuki, H. Hart, E. Dunkelblum and W. Li, **J. Am. Chem. Soc.**, (1977), 99, 5083.

2. A. Rosowsky and K.K.N. Chen, **J. Org. Chem.**, (1974), 39, 1248.

3. J. Kagen and D.A. Agdeppa, Jr., **Helv.**, (1972), 55, 2255.

4. A.V. Fokin, Y.N. Studnev, A.I. Rapkin, V.G. Chilikin and O.V. Verenikin, **Bull. Ac. Sci., USSR**, (1983), 1306.

5. S.M. Kupchan, A.J. Liepa, V. Kameswaran and R.F. Bryan, **J. Am. Chem. Soc.**, (1973), 95, 6861.

EXAMPLES:

HO / O (bicyclic ether) → FSO_2OH / CH_2Cl_2 / N_2, -78° → O (90%) ① [1]

HN / H_2N / N / H / N / N / Ph → FSO_2OH, TFA / ½ hr, RT → HN / H_2N / N / H / N / N (74%) ② [2]

$CH_3-C(OH)(Ph)-C(OH)(Ph)-COOEt$ → FSO_2OH / 0°, 2 min → $CH_3-C(Ph)_2-C(=O)-COOEt$ (76%) ③ [4] △1

$O_2NSO_3F + CF_2{=}CF_2$ → FSO_2OH / Freon-113 → $O_2NCF_2CF_2OSO_2F$ (71%) ④ [4]

MeO / MeO / N–OMe / MeO / MeO → VOF_3, FSO_2OH, TFA / CH_2Cl_2, -30° → MeO / MeO / N–OMe / MeO / MeO / MeO (43%) ⑤ [2]

REAGENT: FORMALDEHYDE

STRUCTURE:

MW = 30.03
BP = -21°C

$$H\overset{O}{\overset{\|}{C}}H$$

Fieser's Reagents for Organic Synthesis: Vol., pages

1	397- 402	**4**	238- 39	**7**	158- 60	**10**	186
2	200- 01	**5**		**8**	231- 32	**11**	240
3		**6**	264- 67	**9**	224- 25	**12**	252

March's Advanced Organic Chemistry: 784 (See March Index for numerous references)

MAJOR USES:

1. Reducing agent 2. Condensations 3. **Mixed aldol** condensations.

PREPARATION:

1. Commercially available as a 37% aqueous solution.

2. Dry gas may be obtained by heating solid **paraformaldehyde** and directing the gas into the reaction flask.

PRECAUTION:

1. Irritating to the mucous membranes.

2. A carcinogen.

NOTES:

1. This process uses excess paraformaldehyde.

REFERENCES:

1. L. Overman and K. Ball, **J. Am. Chem. Soc.**, (1981), 102, 1851.

2. P.A. Grieco, M. Nishizawa, T. Oguri, S.D. Burke and N. Marinovic, **J. Am. Chem. Soc.**, (1977), 99, 5573.

3. Y. Tamura, M. Okada, O. Kitao and Z.-I. Yoshida, **Tetrahedron Lett.**, (1984), 5797.

4. G. Schneider, I. Vincze, L. Hackler and G. Dombi, **Synthesis**, (1983), 665.

5. H. Loibner, A. Pruckner and A. Stutz, **Tetrahedron Lett.**, (1984), 2535.

EXAMPLES:

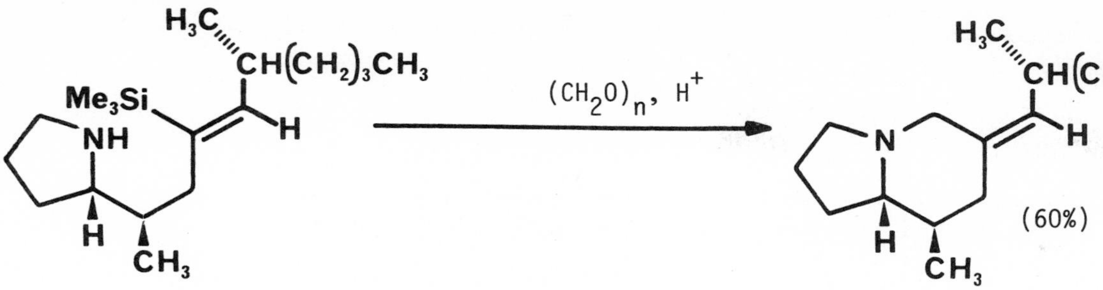

H_3C
$CH(CH_2)_3CH_3$
Me_3Si
NH
H
H
CH_3
$(CH_2O)_n$, H^+
N
(60%)

(1) [2] △1

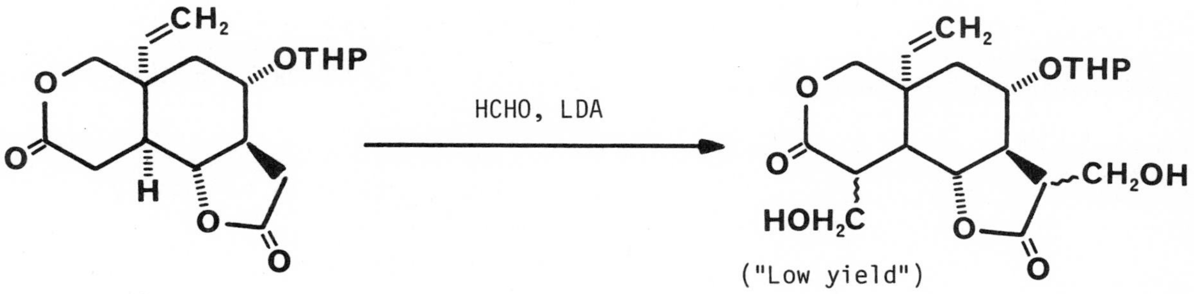

CH_2
OTHP
O
O
H
O
O
HCHO, LDA
CH_2OH
HOH_2C
("Low yield")

(2) [3]

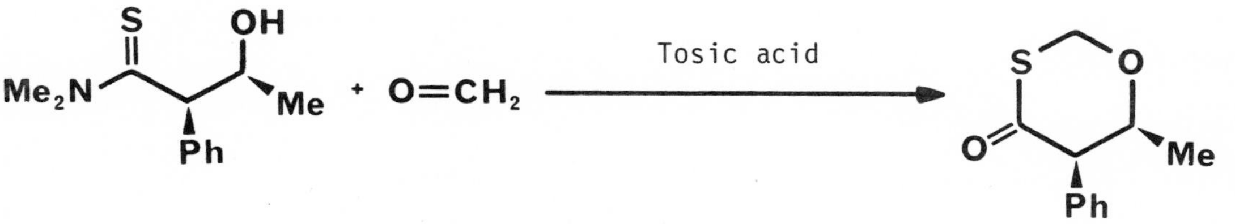

S
OH
Me_2N
Me
Ph
$O{=}CH_2$
Tosic acid
S
O
O
Me
Ph

(95%)

(3) [2]

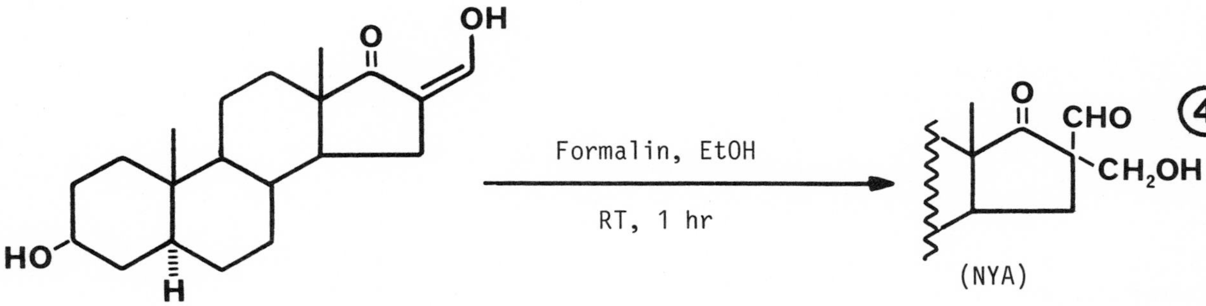

O
OH
HO
H
Formalin, EtOH
RT, 1 hr
O
CHO
CH_2OH
(NYA)

(4)

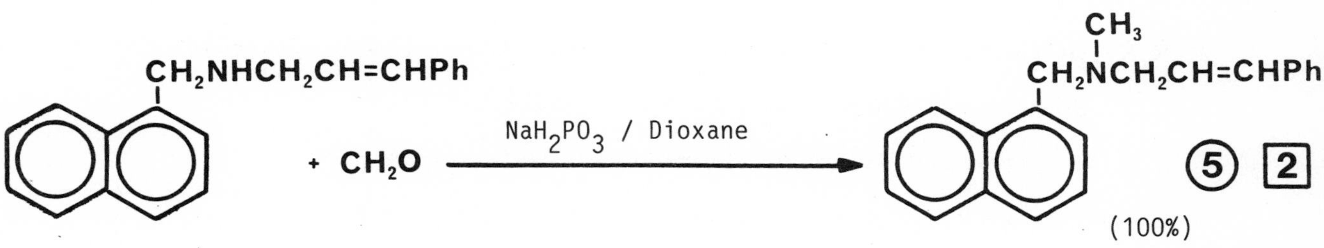

$CH_2NHCH_2CH{=}CHPh$
CH_2O
NaH_2PO_3 / Dioxane
CH_3
$CH_2NCH_2CH{=}CHPh$
(100%)

(5) [2]

REAGENT: FORMIC ACID

===

STRUCTURE:

MW = 46.03
BP = 100- 101°
pK_a = 3.77

$HC(=O)-OH$

<u>Fieser's Reagents for Organic Synthesis</u>: **Vol., pages**

1	404- 07	**4**	239- 40	**7**	160	**10**	
2	202- 03	**5**	316- 19	**8**	232	**11**	243
3	147	**6**		**9**	226- 27	**12**	

March's <u>Advanced Organic Chemistry</u>: 555, 720, 788, 791, 799,987, 1063, 1103

MAJOR USES:

1. Acid 2. Reduction 3. Conversion of vinyl chlorides to carbonyl groups

---

PREPARATION:

1. Commercially available

PRECAUTION:

1. Corrosive

NOTES:

1. **DCC** is used for the esterification reaction.

2. **Koch- Haaf reaction**.

REFERENCES:

1. J. Martinez and J. Laur, **Synthesis**, (1982), 979.

2. R.K. Geivandov and E.I. Kovshev, **Zh. Org. Chim**, (1980), <u>16</u>, 2615.

3. R.E. Bowman, **J. Chem. Soc., <u>Perkin 1</u>**, (1983), 897.

4. J.W. Scheeren and J. Lange, **Tetrahedron Lett.**, (1984), 1609.

5. G. Mehta and K.S. Rao, **Tetrahedron Lett.**, (1984), 1839.

EXAMPLES:

HO, Cl, Cl, Cl (2,4,5-trichlorophenol) + HCOOH → (DCC, Ethyl acetate) → HC(O)O–C$_6$H$_2$Cl$_3$ ① △1 (75%)

Et / OH (bicyclo[2.2.2]octanol) + HCOOH → (Conc. H_2SO_4, 25°) → Et / COOH ② △2 (81%)

O, NH_2 → ($OHC\text{-}CO_2H$ / HCO_2H, Et_3N) → O, CHO, $NHCH_2CO_2H$ ③ (85%)

Me, O, =O, O, Ph, HO, Me → (20% HCO_2H, 40°, 15 hrs) → HO, O, =O, Ph, HO, Me ④ [1] (90%)

Cl, =O → ($NaBH_4$ / $CeCl_3$, MeOH) → Cl, OH → (HCO_2H) → O ⑤ [3] (65%)

REAGENT: GRIGNARD REAGENTS ⚠1

STRUCTURE:

R–Mg–X

Fieser's Reagents for Organic Synthesis: Vol., pages

1	415	**4**		**7**	163- 4	**10**	189- 94
2	205	**5**	321	**8**	235- 38	**11**	245- 49
3		**6**	269- 70	**9**	229- 33	**12**	

March's Advanced Organic Chemistry: 805

MAJOR USES:

1. Reactions with carbonyl groups and carboxyl groups, imines, and nitriles.

2. The **Grignard reagent** can also function as a base.

PREPARATION:

1. Generally prepared from the reaction with an organohalogen compound and magnesium in an ether solvent.

2. Many are commercially available.

PRECAUTION:

1. Reactions should be carried out in anhydrous conditions and without protic solvents.

NOTES:

1. See: **Grignard Reaction**.

REFERENCES:

1. H. Ishikawa, T. Mukaiyama and S. Ikeda, **Bull. Chem. Soc. Jpn**, (1981), 54, 776.

2. K. Hattori, K. Maruoak and H. Yamamoto, **Tetrahedron Lett.**, (1982), 3395.

3. L.M. Weinstock, R.B. Currie and A.V. Lovell, **Synth. Commun.**,(1981), 11, 943.

4. P. Canonne, M. Akssira and G. Lemay, **Tetrahedron Lett.**, (1981), 2611.

5. B. Miller and J. Haggerty, **J. Org. Chem.**, (1986), 51, 174.

EXAMPLES:

$CH_3CH(OEt)(OPh)$ + $PhCH_2MgBr$ —($TiCl_4$)→ $CH_3CH(OEt)(CH_2Ph)$ (51%) ①

Cyclohexanone O-mesyloxime (N–OMs) —1.) CH_3MgI; 2.) $HC{\equiv}CH_2MgBr$→ 2-methyl-2-($CH_2C{\equiv}CH$)azepane (N–H) (66%) ②

Bicyclic anhydride —2 $Me_2CHMgBr$, Ether→ lactone with Me_2CH and $CHMe_2$ (77%) ③ [1]

Cyclohexadienone (O, OAc, Me, Me, Br, Br, Me) —$Me_2CHMgBr$, Ether→ $OCHMe_2$-substituted aryl (Me, Me, Br, Br, Me) (28%) + OH-substituted aryl (Me, Me, Br, Br, Me) (72%) ④ [1]

$PhCH_2CH_2MgBr$ + $EtOC(=O)-C(=O)OEt$ —THF, -10°→ $PhCH_2CH_2C(=O)-C(=O)OEt$ (55%) ⑤ [1]

REAGENT: HEXAMETHYLPHOSPHOROUS TRIAMIDE (**HMPT**)
(TRIS(DIMETHYLAMINO))PHOSPHINE ⚠1

STRUCTURE:

MW = 163.21
BP = 55- 58°C

$[(CH_3)_2N]_3P$

Fieser's Reagents for Organic Synthesis: Vol., pages

1	425	**4**		**7**		**10**	199
2	208- 10	**5**		**8**		**11**	253- 54
3	149- 53	**6**	279- 80	**9**	235- 36	**12**	239- 41

March's Advanced Organic Chemistry: 316, 591, 616, 915, 933, 934, 987, 1112.

MAJOR USES:

1. Dehydrohalogenations 2. Displacement reactions 3. Alkylations

4. Reductions 5. A good aprotic solvent

PREPARATION:

1. The reagent can be prepared from PCl_3 and Me_2NH. Also commercially available.

PRECAUTION:

1. Possible carcinogen

2. Skin irritant and highly flammable

NOTES:

1. This reagent is sometimes confused with hexamethylphosphoric triamide, **HMPA**, $[(Me)_2N]_3PO$.

REFERENCES:

1. M. Hirama, **Tetrahedron Lett.**, (1981), 1905.

2. G.A. Crafft and T.L. Siddall, **Tetrahedron Lett.**, (1985), 4867.

3. Y. Chapleur, **Chem. Commun.**, (1984), 449.

4. D.N. Harpp, D.K. Ash and R.A. Smith, **J. Org. Chem.**, (1980), 45, 5155.

5. H. Fauduet and R. Burgada, **Synthesis**, (1980), 642.

EXAMPLES:

Cyclohexenone + $CH_2=CHCH_2SO_2Ph$ → (n-BuLi, THF, HMPT; -78°) → 3-[1-(phenylsulfonyl)allyl]cyclohexanone, SO_2Ph (89%) ① [3]

$Ph-CH_2-SH$ → (1.) t-BuOCl; 2.) $P(NMe_2)_3$; 3.) NH_4PF_6) → $Ph-CH_2-S-P(NMe_2)_3^{\oplus}$ $^{\ominus}PF_6$ → (Br^-) → $Ph-CH_2-Br$ (98%) ② [2]

Lactone (bis-acetonide) → (HMPT / CCl_4; THF, -30°, Argon) → $=CCl_2$ (92%) ③ [4]

$PhCH_2S\text{-}S\text{-}SCH_2Ph$ → (HMPT; Ether, RT, 2 hrs) → $PhCH_2S\text{-}SCH_2Ph$ (100%) ④ [4]

$PhC(=O)-C(=O)OMe$ + $P(NMe_2)_3$ → (CH_2Cl_2; N_2, -45°) → $PhC-OP(NMe_2)_3$ / $MeOC-O$ → ($H_2C=CHCO_2Me$) → Ph, CO_2Me, $MeOC(=O)$ cyclopropane (60%) ⑤ [3]

REAGENT: HYDRAZINE

==

STRUCTURE:

MW = 32.05

MP = 1.4°C

BP = 113.5°C

H_2NNH_2

<u>Fieser's Reagents for Organic Synthesis</u>: **Vol., pages**

1	434- 35	**4**	248	**7**	170- 71	**10**	
2	211	**5**	327- 29	**8**	245	**11**	255
3	153	**6**	280- 81	**9**	236- 37	**12**	241

March's <u>**Advanced Organic Chemistry**</u>: 177, 370, 375, 378, 446, 594, 693, 696, 1096- 98, 1103, 1117.

MAJOR USES:

1. Reductions   2. Protecting groups   3. Heterocyclic synthesis

----

PREPARATION:

1. Anhydrous hydrazine can be prepared from the more available 95% hydrazine with barium pernitride (see: Y. Okamoto and J.C. Goswami, **Inorg. Chem.**, (1966), <u>5</u>, 1281).

2. Anhydrous hydrazine is also commercially available.

3. For a useful review of this reagent, see: **Aldichimica Acta**, (1980), <u>13</u>, 33.

PRECAUTION:

1. The reagent is suspected to be a carcinogen, and is highly toxic. Has been known to explode during distillation.

NOTES:

1. **Wolff- Kishner** reduction.

REFERENCES:

1. H.-J. Hansen, H.-R. Sliwaka and W. Hug, **helv.**, (1979), <u>62</u>, 1120.

2. K.C. Nicolaou, W.E. Barnette and R.L. Magolda, **J. Am. Chem. Soc.**, (1979), <u>101</u>, 766.

3. I. Reichelt and H.-U. Reissig, **Synthesis**, (1984), 786.

4. E.C. Taylor and J.E. Macor, **J. Heterocyclic Chem.**, (1985), <u>22</u>, 409.

5. J. Bosch, N. Amat and A. Domingo, **Heterocycles**, (1984), <u>22</u>, 561.

EXAMPLES:

$H_2NNH_2 \cdot H_2O$, K_2CO_3, 200°

(68%) ① [1] △1

H_2NNH_2

Et

(90%) ② [3]

Me

$OSiMe_3$

OMe

+ H_2NNH_2

MeOH

RT, 16 rhs

(93%) ③ [3]

i-Pr

OMe

NMe_2

+ H_2NNH_2

Abs EtOH / N_2

3 hrs

(86%) ④ [3]

1.) $H_2NNH_2 \cdot H_2O$

2.) KOH

(79%) ⑤ [3]

REAGENT: HYDROGEN PEROXIDE

==

STRUCTURE:

MW = 34.02 HOOH

d = 1.110

Fieser's Reagents for Organic Synthesis: Vol., pages

1	457- 71	**4**	253- 55	**7**	174	**10**	201- 03
2		**5**	337- 39	**8**	247- 48	**11**	
3		**6**	286	**9**	241- 42	**12**	242- 45

March's Advanced Organic Chemistry:

MAJOR USES: 1

1. A powerful oxidizing agent

PREPARATION:

1. Commercially available in concentrations from 30 to 100%. The most common concentration is the 30% material.

PRECAUTION:

1. The reagent is potentially explosive and should be handled with care. Treat all peroxide compounds with extreme care.

2. Avoid contact with skin and eyes.

3. A strong oxidizing agent.

NOTES:

1. For a procedure to increase the concentration of the hydrogen peroxide, see: J. Blum, Y. Pickholtz and H. Hart, **Synthesis**, (1972), 195.

REFERENCES:

1. R.M. Wilson and J.W. Rekers, **J. Am. Chem. Soc.**, (1981), 103, 206.

2. S. Uemura, S. Fukuzawa and K. Ohe, **Tetrahedron Lett.**, (1985), 921.

3. C.W. Jefford, G. Bernardinelli, J.-C. Rossier, S. Kohmoto and J. Boukouvalas, **Tetrahedron Lett.**, (1985), 615.

4. R. Bergman and G. Magnusson, **J. Org. Chem.**, (1986), 51, 212.

5. C.A. Maryanoff, R.C. Stanzione, J.M. Plampin and J.E. Mills, **J. Org. Chem.**, (1986), 51, 1882.

EXAMPLES:

H_2O_2, BF_3, Et_2O (73%) ①

H_2O_2 (70%, 86:14) ②

H_2O_2 / Et_2O, Amberlyst-15 (90%) ③

H_2O_2, NaOH (90%) ④

Ph-NH-C(=S)-NH_2 → H_2O_2 → Ph-N=C(SO_3H)(NH_2) (85%) ⑤

REAGENT: HYDROXYLAMINE and HYDROXYLAMINE HYDROCHLORIDE

==

STRUCTURE:

$H_2NOH \cdot HCl$

MW = 69.49
MP = 155- 157° (Dec)

Fieser's Reagents for Organic Synthesis: Vol., pages

1	478- 481	**4**		**7**	176- 177	**10**	206- 207
2		**5**		**8**		**11**	257- 258
3		**6**		**9**	245	**12**	

March's Advanced Organic Chemistry: 370, 375, 600, 689, 694, 805

MAJOR USES:

1. Nitrile preparation from aldehydes
2. Isoxazole synthesis
3. Oxime preparation

PREPARATION:

Commercially available in 99% purity

PRECAUTION:

Moisture sensitive and Corrosive.

NOTES:

REFERENCES:

1. F.W. Lichtenhaler and P. Jarglis, **Tetrahedron Lett.**, (1980), 1425.

2. D.J. Brunelle, **Tetrahedron Lett.**, (1981), 3699.

3. D. Hagiwara, K. Sawada, T. Ohnami and M. Hashimoto, **Chem. Pharm. Bull.**, (1982), 30, 3061.

4. J. Ojima, T. Nakada, M. Nakamura and E. Ejiri, **Tetrahedron Lett.**, (1985), 635.

5. E.E. Sugg, J.F. Griffin and P.S. Portoghese, **J. Org. Chem.**, (1985), 50, 5032.

EXAMPLES:

$NH_2OH \cdot HCl$ / Pyridine, 25°

① [3] (89%)

$CH_3CCH_2CCH_3$ → H_2NOH

(85%)

② [2]

H_2NOH, CH_2Cl_2, DMF

(88%) ③ [3]

H_2NOH

④ [3] (44%)

$NH_2OH \cdot HCl$, Pyridine

⑤ [3] (99%)

REAGENT: IODINE

STRUCTURE:

MW = 253.81

I_2

BP = 184.4° **MP** = 113.5°

d = 4.930

Fieser's Reagents for Organic Synthesis: Vol., pages

Vol.	Pages	Vol.	Pages	Vol.	Pages	Vol.	Pages
1	495-500	**4**	258-260	**7**	179-181	**10**	210-211
2	220-222	**5**	346-347	**8**	256-260	**11**	261-267
3	159-160	**6**	293-295	**9**	248-249	**12**	252- 56

March's Advanced Organic Chemistry: See comprehensive listing in March

MAJOR USES:

1. Iodination
2. Oxidation
3. Dehydration
4. Catalysis
5. Iodohydrin preparation
6. Iodolactonization
7. Iodocyclization
8. Iodoamination

PREPARATION:

1. Iodine is commercially-available in very high purity.

PRECAUTION:

1. Iodine is corrosive and is a lachrymator.

REFERENCES:

1. A. Bongini, G. Gardillo, M. Orena, G. Porzi and S. Sandri, **J. Org. Chem.**, (1982), 47, 4626.

2. V.-F. Wang, T. Izawa, S. Kobayashi and M. Ohno, **J. Am. Chem. Soc.**, (1982), 104, 6465.

3. S. Danishefshy, P.F. Schuda, T. Kitahara and S.J. Etheredge, **J. Am. Chem. Soc.**, (1977), 99, 6066.

4. S. Takano and S. Hatakeyama, **Heterocycles**, (1982), 19, 1243.

5. Y. Tamura, S. Kawamura and Z. Yoshida, **Tetrahedron Lett.**, (1985), 2885.

EXAMPLES:

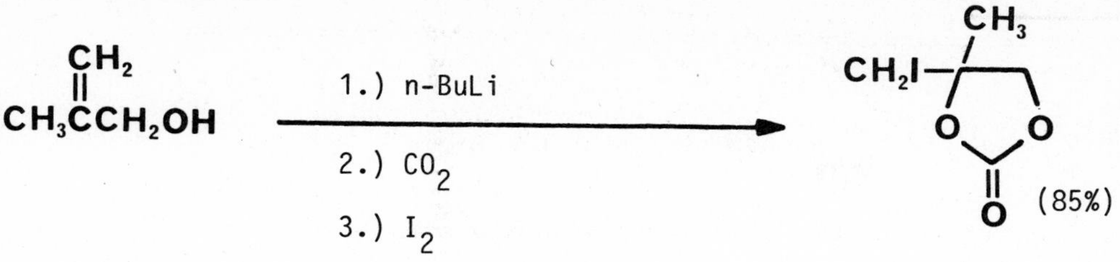

$CH_2{=}C(CH_3)CH_2OH$ → 1.) n-BuLi 2.) CO_2 3.) I_2 → CH_2I/CH_3 cyclic carbonate (85%) ① [6]

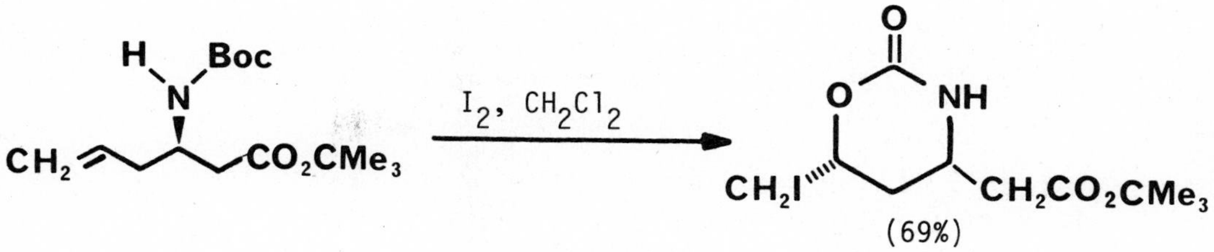

I_2, CH_2Cl_2 → (69%) ② [6]

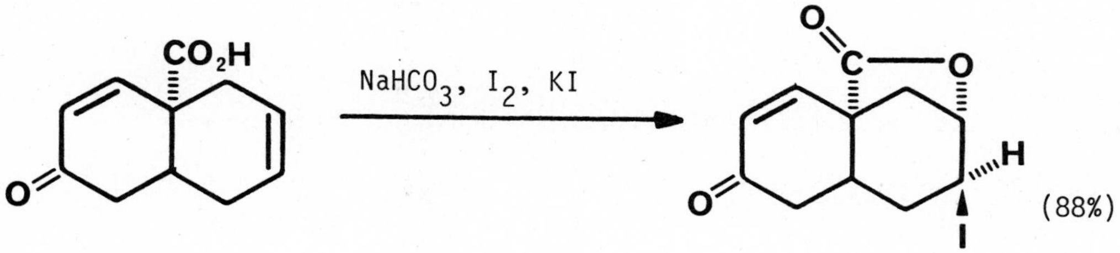

$NaHCO_3$, I_2, KI → (88%) ③ [6]

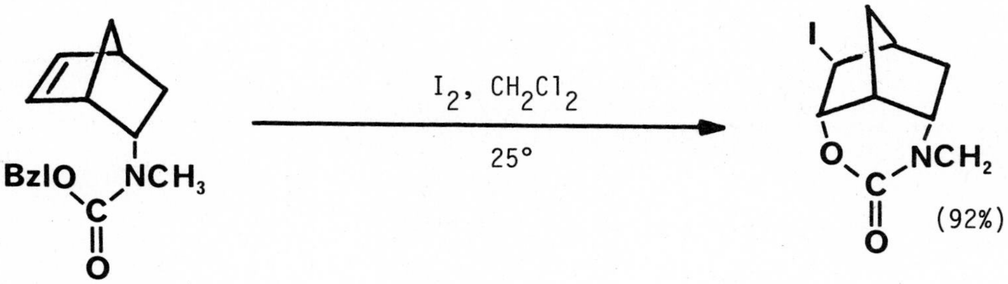

I_2, CH_2Cl_2, 25° → (92%) ④ [6]

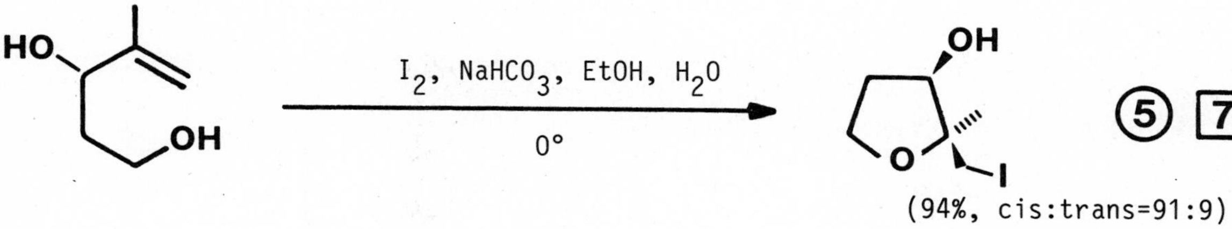

I_2, $NaHCO_3$, EtOH, H_2O, 0° → (94%, cis:trans=91:9) ⑤ [7]

REAGENT: IODOTRIMETHYLSILANE

STRUCTURE:

MW = 200.10 ISiMe$_3$
BP = 106°C

Fieser's Reagents for Organic Synthesis: Vol., pages

1	**4**	**7**	**10** 216- 19
2	**5**	**8** 261- 63	**11** 271- 75
3	**6**	**9** 251- 56	**12** 259- 63

March's Advanced Organic Chemistry: 331, 334, 385, 547, 920, 1055

MAJOR USES:

1. Sulfoxide deoxygenation 2. Silyl-enol ether synthesis 3. Reductive rearrangements 4. **Nazarov**-type cyclizations 5. Ring opening 6. Conversion of primary nitro compounds to aldehydes

PREPARATION:

1. From chlorotrimethylsilane and LiI, see: M. Lissel and K. Dreksler, **Synthesis**, (1983), 459.

2. Commercially available

PRECAUTION:

1. Flammable and corrosive.

NOTES:

REFERENCES:

1. J. P. Marino ad R.J. Linderman, **J. Org. Chem.**, (1981), 46, 3696.

2. K. Saski, T. Kushida, M. Iyoda and M. Oda, **Tetrahedron Lett.**, (1982), 2117.

3. S.P. Brown, B.S. Bal and H.W. Pinnick, **Tetrahedron Lett.**, (1981), 4891.

4. N.G. Kundu, **Synth. Commun.**, (1981), 787.

5. V. Nair and S.D. Chamberlain, **J. Org. Chem.**, (1985), 50, 5069.

EXAMPLES:

CO_2Et — Me_3SiI, CCl_4, 25° → [OSiMe$_3$, CO_2Et, I] — H_2O → CO_2Et (48%) ① [2]

Me_3SiI → (96%) ② [3]

CO_2Et — 1.) $ISiMe_3$ 2.) K_2CO_3, THF, Heat → (86%) ③

$PhCH_2OCH_2$, CH_2OCH_2Ph — Me_3SiI → HN, N H (100%) ④

NH_2, $OSiMe_2$-t-Bu, t-Bu-Me_2SiO, $OSiMe_2$-t-Bu — Me_3SiI, N_2, CH_2I_2, n-Pentyl nitrate → I (70%) ⑤

REAGENT: ION- EXCHANGE RESINS 

STRUCTURE: These are polymeric materials and have no unique structures.

Molecular weights and physical properties depend on the particular resin.

Fieser's Reagents for Organic Synthesis: Vol., pages

1	511- 17	**4**	266- 67	**7**	182	**10**	220- 21
2	227- 28	**5**	255- 56	**8**	263- 64	**11**	276
3		**6**	302- 04	**9**	256- 57	**12**	

March's Advanced Organic Chemistry: Not indexed

MAJOR USES:

1. Catalyst   2. Source of acid or base

PREPARATION:

1. Commercially available

PRECAUTION:

NOTES:

1. Resins are generally known by their trade names.

2. The resins are polystyrene matrix cross-linked with divinylbenzene. Both cation and anion forms are available.

3. Mixed acid/base resin = **REXYN** (Fisher Scientific).

4. **THPO-** = tetrahydropyranyloxy

REFERENCES:

1. J. Stowell and H. Hauck, Jr. ,**J. Org. Chem.**, (1981), 46, 2428.

2. S. Danishefsky, T. Kitahara, P.F. Shuda and S.J. Etheridge, **J. Am. Chem. Soc.**, (1977), 99, 6066.

3. G.W.J. Fleet and P.W. Smith, **Tetrahedron Lett.**, (1985), 1469.

4. C.W. Jefford, G. Bernardinelli, J.-C. Rossier, S. Kohmoto and J. Boukouvalas, **Tetrahedron Lett.**, (1985), 65.

5. F.-T. Luo and E.-I. Negishi, **J. Org. Chem.**, (1985), 50, 4762.

EXAMPLES:

1

Rexyn, H^+

Rexyn, OH^-

$OCH(CH_2)_2CC_2H_5$

$(CH_2)_2CC_2H_5$

CH_3 ① △1 △3

(48%)

$HOCH_2CH_2OH$, $MgSO_4$

Dowex 50W-X8, H^+

(66%)

② 1

2

③

Dowex 50W-X8

MeOH, Reflux

$PhCH_2O$ OMe OH

(92%, 2.5:1)

H_2O_2 / EtOH

Amberlyst-15, 25°

OH

(42%) (36%)

④ 1

THPO CO_2H

1.) MeI, K_2CO_3 DMF

2.) Amberlite IR-120 MeOH

HO CO_2Me

(89%)

⑤ 1 △4

REAGENT: ISOPROPENYLACETATE

STRUCTURE:

$$H_2C=C(OC(=O)CH_3)CH_3$$

MW = 100.12
BP = 94°

Fieser's Reagents for Organic Synthesis: Vol., pages

1	524- 26	4	273	7		10	
2		5		8		11	
3		6	308	9		12	

March's Advanced Organic Chemistry: 352

MAJOR USES:

1. Formation of enol acetates

PREPARATION:

1. Commercially available in high purity.

PRECAUTION:

1. A mild skin irritant, and flammable.

NOTES:

REFERENCES:

1. P.A. Grieco, N. Nishizawa, T. Oguri, S.D. Burke and N. Marinovic, **J. Am. Chem. Soc.**, (1977), 99, 5773.

2. J. Wolinski and R.B. Login, **J. Org. Chem.**, (1970), 35, 3205.

3. N. Pravdic, I. Franjic-Mihalic and B. Danilov, **Carbohydrate Research**, (1975), 45, 302.

4. M. Kosugi, M. Suzuki, I. Hagiwara, K. Goto, K. Siatoh and T. Migita, **Chem. Lett.**, (1982), 939.

5. F. Merenyi and M. Nilsson, **Acta Chem. Scand.**, (1964), 18, 1368.

EXAMPLES:

OMe, OAc, O=, AcO, COOMe

Me Me
$H_2C=C-O-C=O$

TsOH

OMe, OAc, AcO, AcO, COOMe

(78%) ①

CHO + OAc

Tosic acid

54 hrs

[OAc] + CO_2Me, CO_2Me → CO_2Me, CO_2Me ②

(80%)

CH_2OAc, O, OAc, AcO, Cl, NH_2 + OAc

Tosic acid

Reflux, 24 hrs

CH_2OAc, O, OAc, AcO, NAc_2, H ③

(80-86%)

Br, Me + Bu_3SnOMe

$H_2C=C(Me)OC(O)Me$, Pd (Cat.)

Toluene, 100°, 5 hrs

O, CH_2CMe, Me + $BuSnBr$ ④

(91%)

O, O, O + OAc

$AlCl_3$, $ClCH_2CH_2Cl$

O, H, CMe, O, O ⑤

(55%)

REAGENT: LEAD TETRAACETATE

STRUCTURE:

$Pb(O\overset{O}{\overset{\|}{C}}CH_3)_4$

MW = 443.39

Fieser's Reagents for Organic Synthesis: Vol., pages

1	537- 63	**4**	278- 82	**7**	185- 88	**10**	228
2	234- 38	**5**	365- 70	**8**	269- 72	**11**	
3	168- 71	**6**	313- 17	**9**	265- 69	**12**	270- 72

March's Advanced Organic Chemistry: See Index

MAJOR USES:

1. Oxidizing agent 2. Diol cleavage 3. Acetoxylation

PREPARATION:

1. Can be prepared from Pb_3O_4 , HOAc and acetic anhydride.

2. Commercially available.

PRECAUTION:

1. Moisture sensitive.

2. Toxic and a skin irritant.

NOTES:

1. Via:

OH OH O → O CH CH O O → O CH C–OH O O → OH O O

REFERENCES:

1. E.J. Corey and A. Gross, **Tetrahedron Lett.**, (1980), 1819.

2. S. Danishefsky, M. Hirama, K. Gombatz, T. Harayama, E. Berman and P. Schuda, **J. Am. Chem. Soc.**, (1979), 101, 7020.

3. E.J. Corey, R.L. Danheiser, S. Chandrasekaran, G.E. Keck, B. Gopalom, S.D. Larsen, P. Siret and J.-L. Gras, **J. Am. Chem. Soc.**, (1978), 100, 8034.

4. S. Danishefsky, T. Kitahara, P.F. Schuda and S.J. Etheridge, **J. Am. Chem. Soc.**, (1977), 99, 6066.

5. R.B. Gammill, **Tetrahedron Lett.**, (1985), 1385.

EXAMPLES:

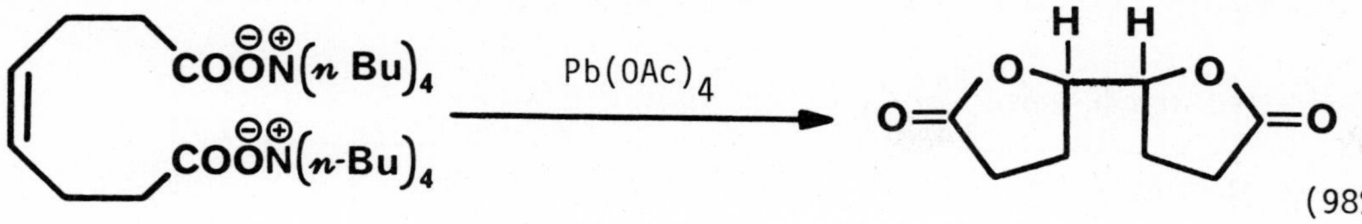

$Pb(OAc)_4$ (98%) ①

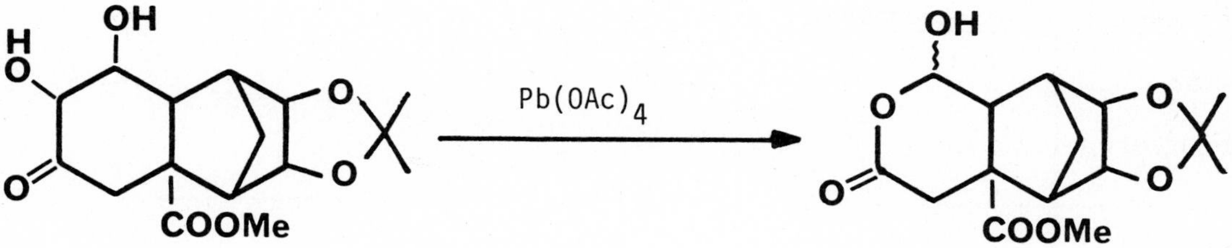

$Pb(OAc)_4$ (100%) ② [2] △1

$Pb(OAc)_4$ (Excess of 64%) ③ [2]

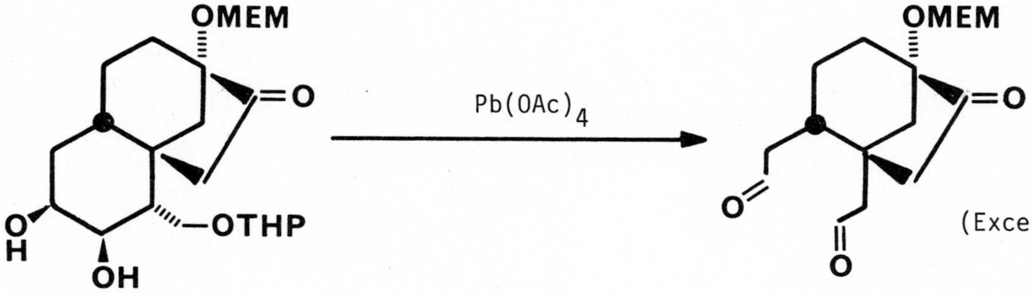

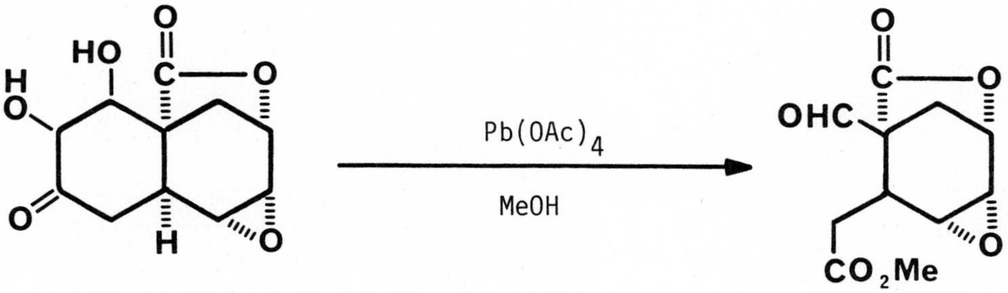

$Pb(OAc)_4$, MeOH (86%) ④ [2]

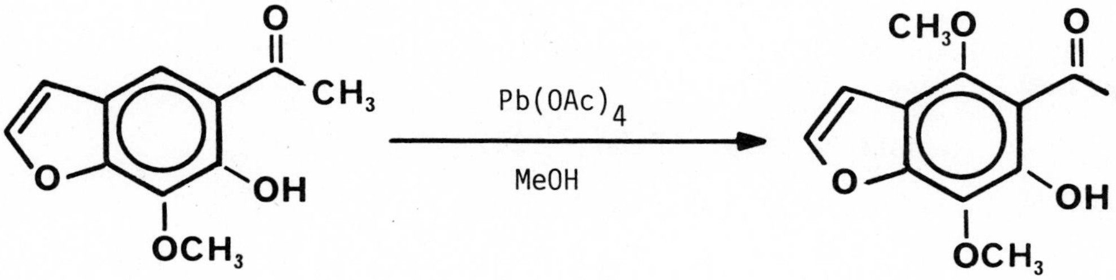

$Pb(OAc)_4$, MeOH (61%) ⑤ [3]

REAGENT: LINDLAR CATALYST

STRUCTURE:

MW = 429.64 $Pd-CaCO_3-PbO$

<u>Fieser's</u> <u>Reagents for Organic Synthesis</u>: **Vol., pages**

1	566- 67	**4**	283	**7**		**10**	
2		**5**		**8**		**11**	
3	171- 72	**6**	319	**9**	270- 71	**12**	272

March's <u>Advanced Organic Chemistry</u>: 695

MAJOR USES:

1. Selective hydrogenation

PREPARATION:

1. See: H. Lindlar and R. Dubois, **Org. Synth**, Coll. Vol. 5, (1975), 880.

2. Commercially available.

3. See: J. Rajaram, A.P.S. Narula, H.P.S. Chawla and S. Dev., **Tetrahedron,** (1983), <u>39</u>, 2315 for methods of preparing reproducible catalysts.

PRECAUTION:

NOTES:

1. **THP** = Tetrahydropyranyloxy

REFERENCES:

1. W. Rousch, H. Gillis and S. Hall, **Tetrahedron Lett.**, (1980), 1023.

2. M. Rosenberger and C. Neukom, **J. Am. Chem. Soc.**, (1980), <u>102</u>, 5425.

3. H.R. Buser, P.M. Guerin. M. Toth, G. Sziocs, A. Schmid, W. Franke and H. Arn, **Tetrahedron Lett.**, (1985), 403.

4. F.-T. Luo and E.-P. Negishi, **J. Org. Chem.**, (1985), <u>50</u>, 4762.

5. A. Claesson and C. Bogentoft, **Synthesis**, (1973), 539.

EXAMPLES:

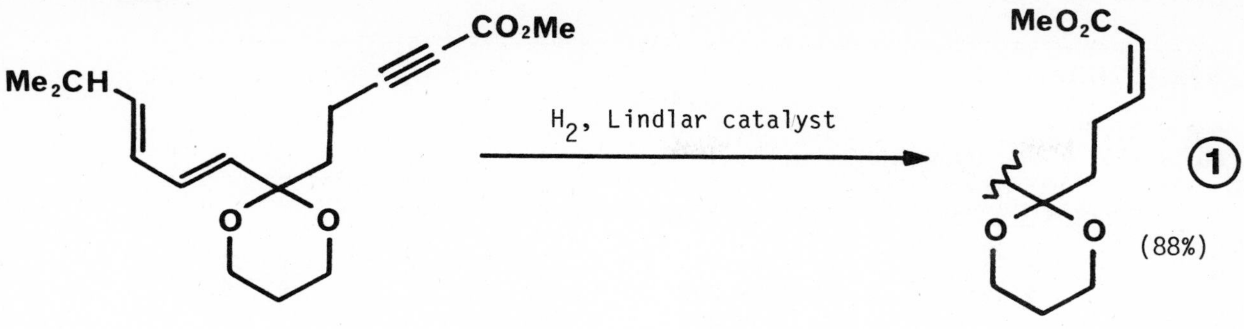

O
CO₂Me
H₂, Lindlar
O
(50%)
2

3

$CH_3(CH_2)_8C{\equiv}CCH_2C{\equiv}CH_2CH_2OTHP \xrightarrow{H_2,\ Lindlar}$ ⌇-$CH{=}CHCH_2CH{=}CHCH_2CH_2OTHP$

(82%) △1

4

HO ≡ CO_2Me — 1.) H_2, Lindlar; 2.) Acetyl chloride, Pyridine, THF → AcO = CO_2Me

(79%)

5

$OCMe_3$
n-Pr$\overset{|}{C}$H$C{\equiv}CCH_2OH$ — Lindlar; Pet ether, 60-85° → *n*-Pr$CH{=}CHCH_2CH_2OH$

(68%)

REAGENT: LITHIUM ALUMINUM HYDRIDE (**LAH**)

STRUCTURE:

$LiAlH_4$

MW = 37.95
MP = 125° (decomposition)

Fieser's Reagents for Organic Synthesis: Vol., pages

1	581-595	**4**	291- 293	**7**	196	**10**	236- 237
2	242	**5**	382- 389	**8**	286- 289	**11**	289- 292
3	176- 177	**6**	325- 326	**9**	274- 277	**12**	272- 275

March's Advanced Organic Chemistry: See index

MAJOR USES:

1. A powerful reducing agent.

2. Source of hydride for nucleophilic displacement reactions.

PREPARATION:

1. Commercially available in solid form and in high purity.

2. Can be obtained in solution (non-protic solvents such as THF, ether, diglyme, etc.).

PRECAUTION:

1. The reagent is very sensitive to water, and care must be taken to make sure that the reagent is handled under anhydrous conditions. Reacts with protic solvents.

3. The reagent is pyrophoric.

NOTES:

1. The hydride reacts with the carbonyl group of the ester to release the hydroxyl function.

2. Hydolysis leaves an acid which can be reduced to an alcohol.

REFERENCES:

1. P.T. Lansbury and J. Vacca, **Tetrahedron Lett.**, (1982), 2623.

2. V. Schurig, U. Leyer and D. Wistuba, **J. Org. Chem.**, (1986), 51, 242.

3. C. Ortiz and R. Greenhouse, **Tetrahedron Lett.**, (1985), 2831.

4. F. Cotteneau, N. Maigrot and J.P. Mazaleyrat, **Tetrahedron Lett.**, (1985), 421.

5. E.J. Trybulski, R.I. Fryer, E. Reeder, S. Vitone, and L. Todaro, **J. Org. Chem.**, (1986), 51, 2191.

EXAMPLES:

H, Me₃CO, OAc, CHCN — $LiAlH_4$, THF → H, Me₃CO, OH, CH_2CN (95%) ① [1] △1

H_2C—CHCHEt (epoxide O), Me — $LiAlH_4$ / Et_2O → HO, $CH_3CHCHEt$, Me (66%) ② [2]

N, CN — 1.) KOH, EtOH, H_2O 2.) LAH, THF, 25° → N, CH_2OH (94%) ③ [1] △2

⊕N, Me, Me, ⊖Br — LAH / THF → NMe_2, Me (61%) ④ [2]

O, N, Cl, Cl — LAH / THF → HO, N, Cl, Cl (80%) ⑤ [1]

REAGENT: LITHIUM- AMMONIA

==

STRUCTURE: Li,NH_3

Fieser's Reagents for Organic Synthesis: Vol., pages

1	601- 03	**4**	288- 90	**7**	195	**10**	234- 36
2	245	**5**	379- 81	**8**	282- 84	**11**	286- 87
3	178- 82	**6**	322- 23	**9**	273- 74	**12**	

March's Advanced Organic Chemistry: 394, 397, 406, 693, 695, 699- 702, 822, 1097.

MAJOR USES: △1

1. Reductions
2. Reductive alkylations
3. **Birch Reductions**
4. Solvent/ reactant for metal amide formation.

PREPARATION:

1. Lithium (or other active metal) is added to anhydrous liquid ammonia.

PRECAUTION:

NOTES:

1. One must be careful not to add iron salts, since they will catalyze the formation of the metal- amide.

REFERENCES:

1. A. Furst, L. Labler and W. Meier, **Helv.**, (1981), 64, 1870.

2. G. Subba Roa, H. Ramanathan and K. Raj, **Chem. Commun.**, (1980), 315.

3. G. Stork and E. Logusch, **J. Am. Chem. Soc.**, (1980), 102, 1218.

4. J.H. Hutchinson and T. Money, **Tetrahedron Lett.**, (1985), 1819.

5. A.V. Rama Rao, E. Rajarathnam Reddy, G.V.M. Sharma, P. Yadagiri and I.S. Yadaw, **Tetrahedron Lett.**, (1985), 465.

EXAMPLES:

1.) Li, NH_3
2.) NH_4Cl

(49%) ① [1]

1.) Li, NH_3, -78°
2.) $H_3C\text{-}CH\text{=}CH\text{-}CO_2CH_3$

COOH

HOOC, $CHCH_2COOCH_3$, CH_3

(70%) ② [3]

1.) Li, NH_3-THF
2.) $EtCH_2Br$

$EtCH_2$ H H

(65%) ③ [2]

CH_2 H OCH_3

Li / NH_3
EtOH

CH_2 H_3CO H

(57%) ④ [1]

H—≡—⁀⁀OH

$LiNH_2$, Liq. NH_3
$Me(CH_2)_4Br$

—≡—⁀⁀OH

(74%) ⑤ [4]

REAGENT: LITHIUM BOROHYDRIDE (LITHIUM TETRAHYDROBORATE)

STRUCTURE:

$LiBH_4$

MW = 21.79
MP = 284°C (dec)

Fieser's Reagents for Organic Synthesis: Vol., pages

1	603	**4**	296	**7**		**10**	
2		**5**		**8**		**11**	293
3		**6**		**9**		**12**	276

March's Advanced Organic Chemistry: 999, 1100- 01

MAJOR USES:

1. Reducing agent

PREPARATION:

1. Can be prepared from sodium borohydride and lithium chloride [H.C. Brown, E. Mead and B. Subba Rao, **J. Am. Chem. Soc.**, (1955), 77, 6209].

2. A solution can be conveniently prepared from $NaBH_4$ and LiBr [H.C. Brown, Y. Choi and S. Narasimhan, **Inorg. Chem.**, (1981), 20, 4454].

3. Commercially available

PRECAUTION:

1. Decomposes in moist air.

NOTES:

REFERENCES:

1. T.V. Lee and J. Toczek, **Tetrahedron Lett.**, (1985), 473.

2. U. Stache, K. Radscheit, W. Fritsch, W. Haede, H. Kohl and H. Ruschig, **Ann.**, (1971), 750, 149.

3. H.C. Brown and S. Narasimham, **J. Org. Chem.**, (1982), 47, 1604.

4. H.C. Brown, S. Narasimham and Y.M. Choi, **J. Org. Chem.**, (1982), 47, 4702.

5. H.C. Brown and S. Narasimham, **Organometallics**, (1982), 1, 762.

EXAMPLES:

LiBH$_4$ → H_3O^+ → (60%) ①

LiBH$_4$, Pyridine, 0° → (18%) ②

LiBH$_4$ $(MeO)_3B$ (Cat.), Ether, Reflux → (100%) ③

LiBH$_4$, Et_2O, Toluene, 100°, 1 hr → (91%) ④

2 [cyclohexene] + $MeCO_2Et$ — LiBH$_4$, Ethyl acetate / Ether, 25°, 1 hr → (94%) ⑤

REAGENT: LITHIUM DIISOPROPYLAMIDE (**LDA**)

STRUCTURE:

$$\begin{array}{ccccc} H_3C & & & & CH_3 \\ | & & & & | \\ HC & - & N & - & CH \\ | & & | & & | \\ H_3C & & Li & & CH_3 \end{array}$$

MW = 101.19

Fieser's Reagents for Organic Synthesis: Vol., pages

1	611	**4**	298- 302	**7**	204- 07	**10**	241- 43
2	249	**5**	400- 06	**8**	292	**11**	296- 99
3	184- 85	**6**	334- 39	**9**	280-83	**12**	277- 99

March's Advanced Organic Chemistry: 391, 421

MAJOR USES:

1. Strong base, often used for generating kinetic enolates.

PREPARATION:

1. The reagent is prepared by the reaction of n-butyllithium on diisopropylamine , and is stored in hexane. See: H.O. House, W.V. Phillips, T.S.B. Sayer and C.-C. Yan, **J. Org. Chem.**, (1978), 43, 700.

2. For a convenient preparation of molar amounts of this reagent, see: M.T. Reetz and W.F. Maier, **Ann.**,(1980), 1471.

3. Also commercially available.

PRECAUTION:

1. In the solid state is flammable and corrosive.

NOTES:

1. A dehydrohalogenation reaction.

REFERENCES:

1. C.H. Heathcock, M.C. Pirrung, C.T. Buse, J.P. Hagen, S.D. Young, and J.E. Sohn, **J. Am. Chem. Soc.**, (1979), 101, 7077.

2. E. Negishi, A.O. King and J.J.M. Tour, **Org. Synth.**, (****)

3. D. Liotta, M. Saindane and D. Brothers, **J. Org. Chem.**, (1982), 47, 1598.

4. M.T. Reetz and H. Muller-Starke, **Tetrahedron Lett.**, (1984), 3301.

5. S. Raucher and P. Klein, **J. Org. Chem.**, (1986), 51, 123.

EXAMPLES:

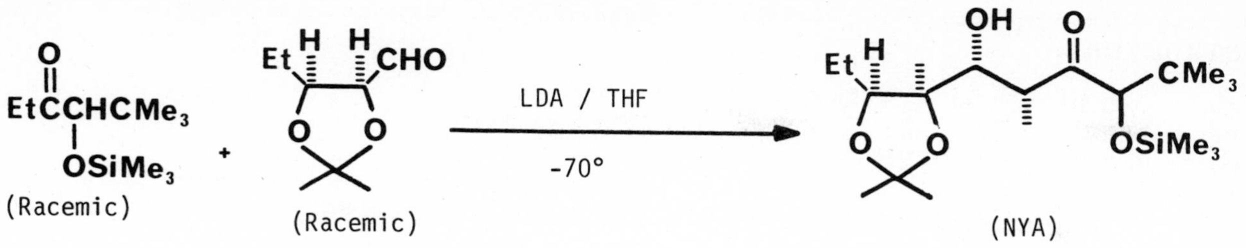

EtC(O)CH(OSiMe$_3$)CMe$_3$ (Racemic) + Et/CHO dioxolane (Racemic) → LDA / THF, -70° → (NYA) ①

1.) LDA, THF
2.) ClP(O)(OEt)$_2$
3.) LDA
C≡CH (72%) ② △1

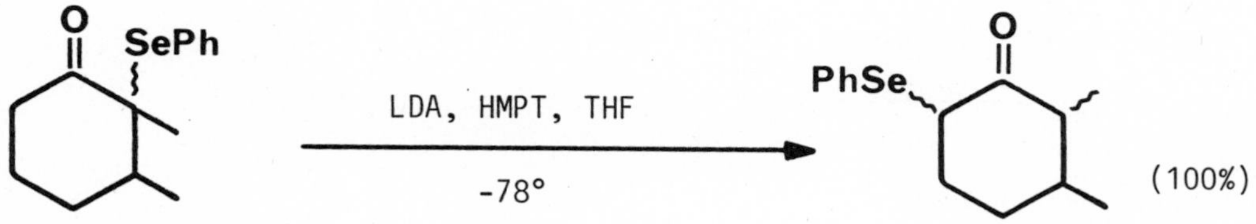

SePh; LDA, HMPT, THF, -78°; PhSe (100%) ③

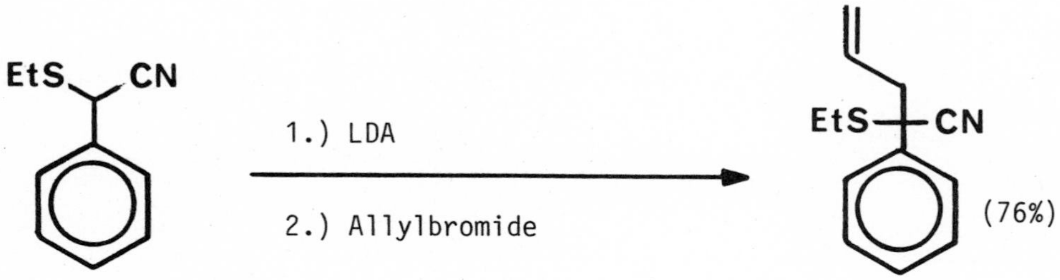

EtS, CN; 1.) LDA 2.) Allylbromide; EtS—CN (76%) ④

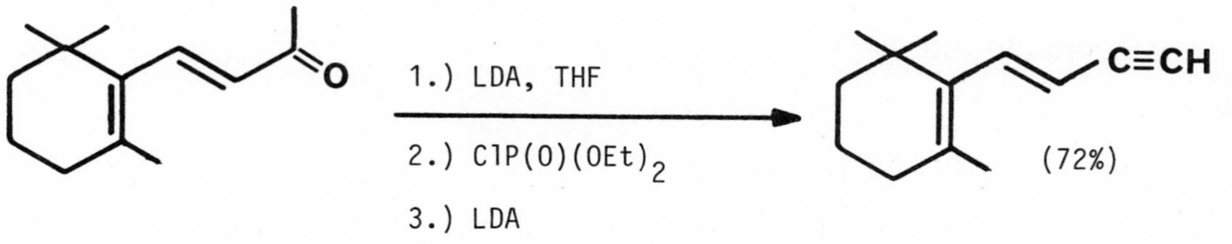

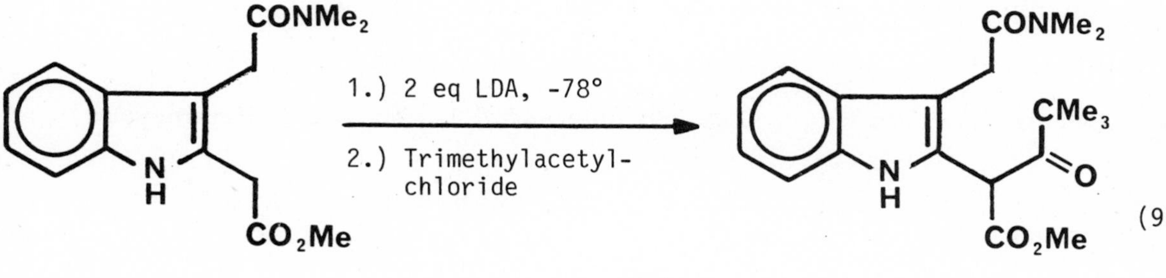

CONMe$_2$, N H, CO_2Me; 1.) 2 eq LDA, -78° 2.) Trimethylacetyl-chloride; CONMe$_2$, CMe$_3$, O, N H, CO_2Me (95%) ⑤

REAGENT: LITHIUM TRIETHYLBOROHYDRIDE (**SUPER-HYDRIDE**) ⚠1

STRUCTURE:

$LiBEt_3H$

MW = 105.94
FP = -17°C
d = 0.920

<u>**Fieser's Reagents for Organic Synthesis**</u>: **Vol., pages**

1		**4**	313- 14	**7**	215- 16	**10**	249- 50
2		**5**		**8**	309- 310	**11**	304- 05
3		**6**	348- 49	**9**	286	**12**	289

March's <u>Advanced Organic Chemistry</u>: 390, 392, 394, 1095, 1099, 1101

MAJOR USES:

1. Reducing agent 2. Nucleophile

PREPARATION:

1. Can be prepared from LiH and triethylborane in THF.

2. Commercially available in solution as "Super-hydride"

PRECAUTION:

1. Flammable and moisture sensitive.

2. Reactions should be carried out under an inert atmosphere.

NOTES:

1. For a Review, see: S. Krishnamurthy, **Aldrichimica Acta**, (1974), <u>7</u>, 55.

REFERENCES:

1. A.G. Kelly and J.S. Roberts, **Chem. Commun.**, (1980), 228.

2. R.O. Hutchins and K. Learn, **J. org. Chem.**, (1982), <u>47</u>, 4380.

3. F. Hansske and M.J. Roberts, **J. Am. Chem. Soc.**, (1983), <u>105</u>, 6736.

4. R.C. Cambie, P.S. Rutledge, G.A. Strange and P.D. Woodgate, **Heterocycles**, (1982), <u>19</u>, 1501.

5. R. Baker, P.D. Ravenscroft and C.J. Swain, **Chem. Commun.**, (1984), 74.

EXAMPLES:

1.) $LiEt_3BH$

2.) TsCl, C_5H_5N

(51%) ① [2]

C_7H_{15}CH=CHCH$_2$OPh → C_7H_{15}CH=CHCH$_3$

$LiEt_3BH$, Pd(0), Ph_3P

Dry THF, Argon

(71%, 97% trans) ② [1]

$LiEt_3BH$

Dry DMSO, THF, N_2, 21°

14 hrs

(98%) ③ [2]

$LiB(Et)_3H$, THF

Dry Ethyl ether, N_2

Dark, 24 hrs

(100%) ④

$LiEt_3BH$

THF

(88%) ⑤ [1]

REAGENT: MANGANESE DIOXIDE

==

STRUCTURE:

MW = 86.94 MnO_2

MP = 535° (decomp)

Fieser's Reagents for Organic Synthesis: Vol., pages

1	637 - 42	**4**	317- 18	**7**		**10**	
2	257- 63	**5**	422- 24	**8**	312	**11**	311
3	191- 94	**6**	357	**9**		**12**	

March's Advanced Organic Chemistry: 788, 1051-53, 1057, 1062, 1081, 1092

MAJOR USES:

1. Oxidizing agent 2. Dehydrogenation 3. Coupling

4. Particularly noted for its ability to be a mild oxidizing agent for allylic alcohols

---

PREPARATION:

1. For methods of preparing fresh, reactive MnO_2, see: a. I.M. Goldman, **J. Org. Chem.**, (1969), 34, 1979. b. L.A. Carpino, **J. Org. Chem.**, (1970), 35, 3971.

2. The reagent is commercially available in very high purity.

PRECAUTION:

Oxidizing agent, Irritant.

NOTES:

REFERENCES:

1. S. Mashraqui and P. Keehn, **Synth. Commun.**, (1982), 637.

2. L. Lombardo, L.N. Mander and J.V. Turner, **J. Am. Chem. Soc.**, (1980), 102, 6626.

3. A.E. Greene, C. LeDrian and P. Crabbe, **J. Am. Chem. Soc.**, (1980), 102, 7583.

4. R. Bergman and G. Magnusson, **J. Org. Chem.**, (1986), 51, 212.

5. B. Errazuriz, R. Tapia and J.A. Valderrama, **Tetrahedron Lett.**, (1985), 819.

EXAMPLES:

MnO_2, Benzene, Δ — (62%) ① [2]

MnO_2, CH_2Cl_2 — HO, OH, CO_2Me → O, OH, CO_2Me (50-60%) ② [4]

MnO_2 — HO, CO_2*n*-Bu → O, CO_2*n*-Bu (70%) ③ [4]

MnO_2, Hexane — OH → O (81%) ④ [4]

MnO_2 / HNO_3, CH_2Cl_2 — OMe, OMe → O, O (85%) ⑤ [1]

REAGENT: MERCURY (II) ACETATE (MERCURIC ACETATE)

STRUCTURE:

MW = 318.68
MP = 179- 182°

$(CH_3CO_2)_2Hg$

Fieser's Reagents for Organic Synthesis: Vol., pages

Vol.	pages	Vol.	pages	Vol.	pages	Vol.	pages
1	644- 652	**4**	319- 323	**7**	222- 223	**10**	252- 253
2	264- 267	**5**	424- 427	**8**	315- 316	**11**	315- 317
3	194- 196	**6**	358- 359	**9**	291	**12**	298- 303

March's Advanced Organic Chemistry: 1054, 1084, 1085

MAJOR USES:

1. Oxidation
2. Dehydrogenation
3. Alkyne hydration
4. Alkene additions
5. Bromination
6. Cyclization
7. Isomerization
8. Solvomercuration
9. Oxymercuration
10. Oxidative cleavage
11. Vinyl chloride hydrolysis
12. Reductive coupling of allenes
13. Aminomercuration

PREPARATION:

Commercially available in high purity

PRECAUTION:

As with many mercury-containing compounds, this reagent is toxic. The reagent is also light-sensitive.

NOTES:

1. Removal of the mercury leaves a radical. Usually, this captures a hydrogen and results in non-stereospecific reduction of the mercurial. However, in the presence of unsaturated aldehydes, ketones and esters, 1,4-addition of the radical readily takes place (**Giese reaction**).

REFERENCES:

1. Y. Saitoh, Y. Moriyama, H. Hirota, T. Takahashi and Q. Khudng-Huu, **Bull. Chem. Soc. Japan**, (1981), 54, 488.

2. B. Giese, K. Heuck and U. Luning, **Tetrahedron Lett.**, (1981), 2155.

3. S. Danishefsky, S. Chackalamannil and B.-J. Uang, **J. Org. Chem.**, (1982), 47, 2231.

4. S. Wolff and W.C. Agosta, **Tetrahedron Lett.**, (1985), 703.

5. L.A. Paquette, D.T. Belmont and Y.-L. Hsu, **J. Org. Chem.**, (1985), 50, 4667.

EXAMPLES:

1.) $Hg(Oac)_2$, CH_3OH

2.) $NaBH_4$

$(CH_2)_{10}CH_3$ NH₂ H H

$(CH_2)_{11}CH_3$ (76%) ① 13

$Hg(OAc)_2$, CH_3OH, NaCl

OMe HgCl (NYA) ② 8

1.) $Hg(OAc)_2$

2.) $NaBH(OMe)_3$, CH_2Cl_2

Ph AcO (70%) ③ 9 △1

$Hg(OAc)_2$

CMe₃ (93%) ④ 6

$Hg(OAc)_2$

$BF_3 \cdot Et_2O$

Cl (91%) ⑤ 11

REAGENT: METHANESULFONYL CHLORIDE (MESYL CHLORIDE)

==

STRUCTURE:

CH_3SO_2Cl

MW = 114.55
BP = 60° @ 12 mm

Fieser's Reagents for Organic Synthesis: Vol., pages

1	662- 64	**4**	326- 7	**7**		**10**	
2	286- 89	**5**	435- 36	**8**		**11**	322
3		**6**	362- 63	**9**		**12**	

March's Advanced Organic Chemistry: 350

MAJOR USES:

1. Preparing mesylate derivatives
2. Beckmann elimination
3. **Beckmann fragmentation**
4. Dehydrations

PREPARATION:

This reagent is commercially available. For a synthesis from the sulfonic acid and thionyl chloride, see: P.J. Hearst and C.R. Noller, **Org. Synth. Coll.** 4, 571, (1963)

PRECAUTION:

Very corrosive and highly toxic. A skin irritant.

NOTES:

REFERENCES:

1. J.A. Glinski and L.H. Zalkow, **Tetrahedron Lett.**, (1985), 2857.

2. X. Creary and M.E. Mehrsheikh-Mohammadi, **J. Org. Chem.**, (1986), 51, 7.

3. S. Castillon, A. Dessinges, R. Faghih, G. Lukacs, A. Olesker and T.T. Thang, **J. Org. Chem.**, (1985), 50, 4913.

4. M.G. Constantino, P.M. Donate and N. Petragnani, **J. Org. Chem.**, (1986), 51, 387.

5. S. Raucher and P. Klein, **J. Org. Chem.**, (1986), 51, 123.

EXAMPLES:

HO, H, O-C(=O)-Ph → MeSO$_2$O, H, O-C(=O)-Ph

$MeSO_2Cl$, Et_3N, CH_2Cl_2

-2°, 1½ hrs

(93%) ① [1]

HO, CH-P(=O)(OEt)$_2$, CF$_3$ → MsO, CH-P(=O)(OEt)$_2$, CF$_3$

MsCl, CH_2Cl_2

Et_3N

(86%) ② [1]

CH_2OH, O, F, HO, OBzl, N_3 → CH_2OMs, O, F, HO, OBzl, N_3

MsCl

Pyridine

(87%) ③ [1]

C≡C-C≡CH, OH → C≡C-C≡CH

MsCl

Pyridine

(63%) ④ [4]

HO, O=C-CH_2CCl_3, O-N, NH, O, CO_2CH_3 → MeSO$_2$O, NH

MsCl

(95%) ⑤ [1]

REAGENT: METHYL CHLOROFORMATE

==

STRUCTURE:

$Cl-CO_2CH_3$

MW = 94.5
BP = 70 -72° C

Fieser's Reagents for Organic Synthesis: Vol., pages

1	**4**	**7** 236	**10**
2	**5**	**8**	**11** 336- 37
3	**6** 376	**9**	**12**

March's Advanced Organic Chemistry:

MAJOR USES:

1. Carbomethoxylations
2. Trapping enolates

PREPARATION:

Commercially available in high purity.

PRECAUTIONS:

1. Flammable

2. Vapors toxic and irritating to the eyes.

NOTES:

REFERENCES:

1. S. Danishefsky, M. Kahn and M. Silvestri, **Tetrahedron Lett.**, (1982), 703.

2. M.E. Kuehne and P.J. Seaton, **J. Org. Chem.**, (1985), 50, 4790.

3. S. Raucher and P. Klein, **J. Org. Chem.**, (1986), 51, 123.

4. H. Kobler, R. Munz and G. Al Gasser, **Ann.**,(1978), 1937.

5. E. Piers and M. Soucy, **Can. J. Chem.**, (1974), 52, 3563.

EXAMPLES:

$MeSCH_2S(O)Me$ → n-BuLi, THF, -70° → $MeS(O)\text{-}\overset{\ominus}{C}H\text{-}SMe$

$(EtO)_2P(O)C(CO_2Et)=CH_2$ + $MeS(O)\text{-}\overset{\ominus}{C}H\text{-}SMe$ → [$(EtO)_2P(O)\overset{\ominus}{C}(CO_2Et)\text{-}CH_2CH(SMe)S(O)Me$]

PhCHO → $MeS(O)(MeS)CH\text{-}CH_2C(CO_2Et)=CHPh$ (62%) → $MeSCH=CH\text{-}C(CO_2Et)=CHPh$ (56%) ①

1-Naphthyl-MgBr → $MeS(O)CH_2SMe$, THF → 1-Naphthyl-$CH(SMe)_2$ → SO_2Cl_2, SiO_2, H_2O → 1-Naphthyl-CHO (93%) ②

$MeO\text{-}C_6H_4\text{-}CHO$ → MMTS / Triton B, THF, 70°, Reflux → $MeO\text{-}C_6H_4\text{-}CH=C(SMe)S(O)Me$ (82%) ③

2-Bromopyridine → MMTS, NaH, DME, 30° → 2-Pyridyl-$CH(SMe)S(O)Me$ → 200° → 2-Pyridyl-CHO (74%) ④

REAGENT: METHYL METHYLTHIOMETHYLSULFOXIDE (**MMTS**)

STRUCTURE:

MW = 124.23
BP = 92- 93° @ 2.5 mm Hg

$CH_3\overset{O}{\overset{\|}{S}}CH_2SCH_3$

Fieser's Reagents for Organic Synthesis: Vol., pages

1	**4**	341- 42	**7**		**10**	
2	**5**	456- 57	**8**	344- 45	**11**	
3	**6**	390- 92	**9**	314	**12**	

March's Advanced Organic Chemistry: Not indexed

MAJOR USES:

1. Synthesis of ketones and aldehydes or derivatives (thioenol ether, or thioketal).

PREPARATION:

1. The reagent can be prepared by the oxidation of formaldehyde dimethyl mercaptal with 30% hydrogen peroxide in acetic acid (K. Ogura and G.I. Tsuchihashi, **Bull. Chem. Soc. Jpn.**, (1972), 45, 2203).

2. Commercially available.

PRECAUTION:

1. The reagent has a very disagreeable odor.

NOTES:

REFERENCES:

1. T. Minami, K. Niskimura and I. Hirao, **J. Org. Chem.**, (1982), 47, 2360.

2. H. Hojo, R. Mazuda, T. Saeki, K. Fujimori and S. Tsutsumi, **Tetrahedron Lett.**, (1978), 1303.

3. K. Ogura, Y. Ito and G. Tsuchihashi, **Bull. Chem. Soc. Jpn.**, (1979), 52, 2013.

4. G.R. Newkome, J.M. Robinson and J.D. Sauer, **Chem. Commun.**, (1974), 410.

EXAMPLES:

1.) $BrMg(CH_2)_3CH{=}CH_2$, CuI
2.) MeO_2CCl

① [3] (90%)

$ClCO_2Me$, Na_2CO_3
CH_2Cl_2, N_2

② [1] (64%)

$ClCO_2Me$, CH_2Cl_2
Et_3N

③ [1] (89%)

④

Et_3N + MeO_2CCl → 1.) CH_2Cl_2, -15° 2.) Δ, 1 hr → $Et_3\overset{\oplus}{N}MeCl^{\ominus}$ + CO_2 (100%)

⑤ [1]

$(n\text{-}Bu)_2CuLi$, $ClCO_2Me$
Dry ether, Argon, -78°

(86%, 98:2)

REAGENT: MOLECULAR SIEVES

STRUCTURE: General Formula and Pore Diameter

3 A, $K_9Na_3[(AlO_2)_{12}(SiO_2)_{12}]\cdot 27\ H_2O$, 3 Å

4 A, $Na_{12}[(AlO_2)_{12}(SiO_2)_{12}]\cdot 27\ H_2O$, 4 Å

5 A, $Ca_{4.5}Na_3[(AlO_2)_{12}]\cdot 30\ H_2O$, 5 Å

13 X $Na_{86}[(AlO_2)_{86}(SiO_2)_{106}]\cdot X\ H_2O$, 10 Å

Fieser's Reagents for Organic Synthesis: Vol., pages

1	703- 05	**4**	345	**7**		**10**	273
2	286- 87	**5**	465	**8**		**11**	350
3	206	**6**	411	**9**	316- 17	**12**	

March's Advanced Organic Chemistry: Not Indexed

MAJOR USES:

1. Drying agent 2. Useful in driving equilibrium reactions to completion by removal of small products.

PREPARATION:

1. Commercially available in a variety of pore sizes.

PRECAUTION:

1. Hygroscopic.

NOTES:

1. ROH = [structure: OH, Me, Me, Me]

REFERENCES:

1. G.A. Taylor, **J. Chem. Soc.**Perkin I, (1981), 3132.

2. L. Lombardo, L.N. Mander and J.V. Turner, **J. Am. Chem. Soc.**, (1980), 102, 6626.

3. J.-N. Denis, A.E. Greene, A.A. Serra and M.J. Luche, **J. Org. Chem.**, (1986), 51, 46.

4. B.P. Mundy and W.G. Bornmann, **Synth. Commun.**, (1978), 227.

5. J.H. Markgraf, E.W. Greeno, M.D. Miller, W.J. Zaks, and G.A. Lee, **Tetrahedron Lett.**, (1983), 241.

EXAMPLES:

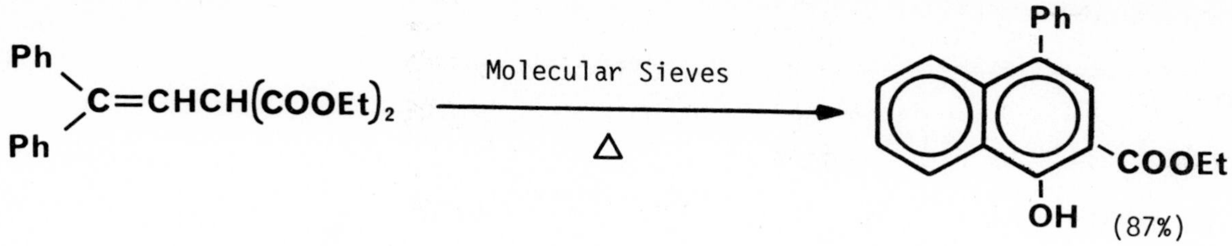
Ph
C=CHCH(COOEt)$_2$
Ph
Molecular Sieves
Δ
Ph
COOEt
OH
(87%)

(1) 2

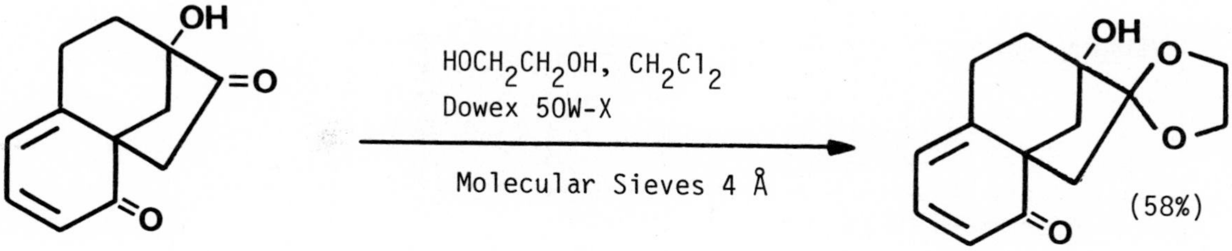
OH
=O
=O
HOCH$_2$CH$_2$OH, CH$_2$Cl$_2$
Dowex 50W-X
Molecular Sieves 4 Å
OH
O
O
=O
(58%)

(2) 2

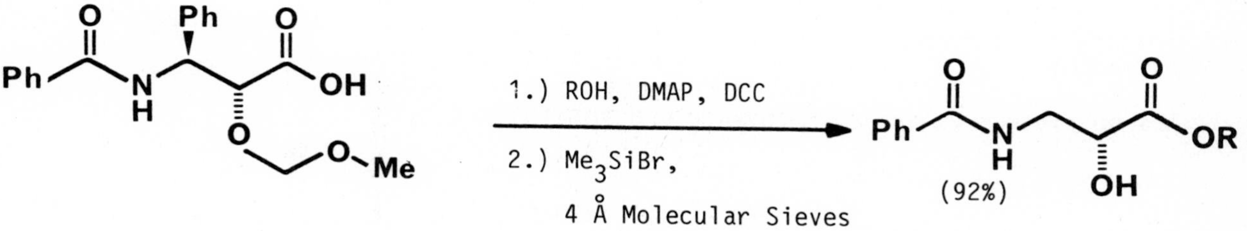
O
Ph
O
Ph
N
H
OH
O
O
Me
1.) ROH, DMAP, DCC
2.) Me$_3$SiBr,
4 Å Molecular Sieves
O
O
Ph
N
H
OR
(92%)
OH

(3) 2 1

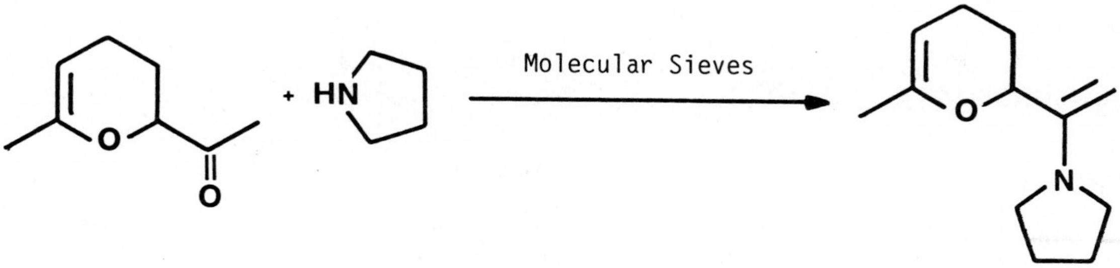
O
O
+ HN
Molecular Sieves
O
N

(4) 2

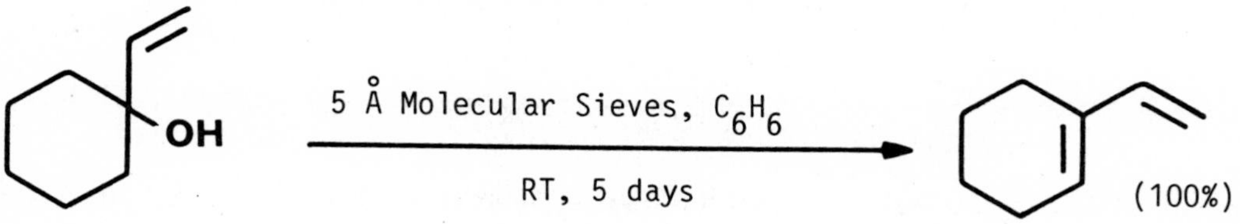
OH
5 Å Molecular Sieves, C$_6$H$_6$
RT, 5 days
(100%)

(5) 1

REAGENT: NICKEL (II) ACETYLACETONATE
BIS(2,4-PENTANEDIONATO)NICKEL(II)

STRUCTURE:

MW = 256.93
MP = 229.30°C
d = 1.455

Fieser's Reagents for Organic Synthesis: Vol., pages

Vol.	pages	Vol.	pages	Vol.	pages	Vol.	pages
1		**4**		**7**	250	**10**	42
2		**5**	471	**8**		**11**	58- 59
3		**6**	417	**9**	51- 52	**12**	

March's Advanced Organic Chemistry: Not Indexed

MAJOR USES:

1. Catalyst for alkylations, and condensations

PREPARATION:

1. Can be prepared from 2,4-pentanedione and $NiCl_2 \cdot 6\ H_2O$

2. Commercially available as the hydrate.

PRECAUTION:

1. A suspected carcinogen.

NOTES:

1. $\mathbf{C_2N_2}$ = Dicyanogen (**NC-CN**).

REFERENCES:

1. J.H. Nelson, P.N. Howells, G.C. DeLullo and G.L. Landon, **J. Org. Chem.**, (1980), 45, 1246.

2. T. Hayashi, Y. Katsuro, Y. Okamoto and K. Kumada, **Tetrahedron Letters**, (1981), 4449.

3. J.-L. Luce, C. Petrier, J.-A. Lansard and A.E. Greene, **J. Org. Chem.**, (1983), 48, 3837.

4. J.B. Corain, M. Basato and H.F. Klein, **Angew. Chem. Int. Ed.**, (1981), 20, 972.

5. M. Julia and J.-N. Verpeaux, **Tetrahedron Letters**, (1982), 2457.

EXAMPLES:

$CH_3COCH_2COCH_3$ + $CH_2=CHCOCH_3$ —(1% $Ni(acac)_2$, Dioxane)→ $(CH_3CO)_2CHCH_2CH_2COCH_3$ (90%) ①

1-Naphthyl–$OP(O)(OEt)_2$ —($AlEt_3$, $Ni(acac)_2$)→ 1-Ethylnaphthalene (95%) ②

Li + PhBr —($ZnBr_2$)→ [Ph_2Zn] + $Me_2C=CHCOMe$ —($Ni(acac)_2$)→ $Me_2C(Ph)CH_2COMe$ (98%) ③

2 $MeCOCH_2COMe$ —(+ 2 C_2N_2, $Ni(acac)_2$; Dichloroethane, 20°, 140 hrs)→ 4-Me-5-(COMe)-6-($CONH_2$)-2-[$H_2N-C=C(COMe)_2$]pyrimidine (100%) ④

$PhSO_2CH(MgBr)Ph$ —(2% $Ni(acac)_2$; THF, Reflux, 6 hrs)→ PhCH=CHPh (70%, E:Z=80:20) ⑤

REAGENT: ORGANOCOPPER REAGENTS

STRUCTURE: $RCuCNLi$ Me_2CuLi $R_2CuCNLi_2$

$RCu \cdot BF_3$ Me_5Cu_3Li Ph_2PCu

Fieser's Reagents for Organic Synthesis: Vol., pages

1		**4**	127- 8	**7**	92, 212	**10**	282- 90
2	151	**5**	187- 88	**8**	334- 35	**11**	365- 73
3	79, 106-12	**6**	163	**9**	328- 34	**12**	245- 50

March's Advanced Organic Chemistry: Look under organocopper reagents and Lithium dialkylcopper reagents.

MAJOR USES:

1. Conjugate additions
2. Displacement of leaving groups

PREPARATION:

There are many variations in the organocopper reagents- both in their reactivity and preparation. See Lipshutz's Review: **Tetrahedron**, (1984), 40, 5005.

PRECAUTION:

Reactions should be carried out under anhydrous conditions and under an inert atmosphere.

NOTES:

REFERENCES:

1. D.L.J. Clive, V. Farina and P. Beaulieu, **J. Org. Chem.**, (1982), 47, 2572.

2. B.H. Lipshutz, R.S. Wilhelm and D.M. Floyd, **J. Am. Chem. Soc.**, (1981), 103, 7672.

3. Y. Yamamoto, S. Yamamoto, H. Yatagai, Y. Ishhiara and K. Maruyama, **J. Org. Chem.**, (1982), 47, 119.

4. M.J. O'Donnell and J.-B. Falmagne, **Tetrahedron Lett.**, (1985), 699.

5. D.H. Hua and A. Verma, **Tetrahedron Lett.**, (1985), 547.

EXAMPLES:

[Reaction 1: 2-cyclohexylidenepropanal (cyclohexane ring =C(CHO)CH$_3$) → 1-methylcyclohexyl-CH(CH$_3$)CHO + 1,2-Adduct] ① [1]

$Me_5Cu_3Li_2$ ....(88%, 99.5:0.5)

Me_2CuLi ....(91%, 80:20)

[Reaction 2: Cyclopentyl–Br + $(n\text{-Bu})_2Cu(CN)Li_2$ —THF→ Cyclopentyl–n-Bu (86%)] ② [2]

[Reaction 3: 1-cyclohexenyl–C(=O)–n-Bu → 2-(n-Bu)cyclohexyl–C(=O)–n-Bu] ③ [1]

$n\text{-BuCu}\cdot BF_3$ ....(90%)

$(n\text{-Bu})_2CuLi$ ....(74%)

$Ph_2C{=}N\text{-}CH(OAc)\text{-}CO_2Et$ —$Bu_2Cu(CN)Li_2$, THF→ $Ph_2C{=}N\text{-}CH(n\text{-Bu})\text{-}CO_2Et$ (46%) ④ [2]

[Reaction 5: 4-acetoxy-2-azetidinone (O=C, NH, $OCCH_3$ with C=O) + 2 (1-cyclohexenyl)$_2$CuLi —Ether / DMS→ 4-(1-cyclohexenyl)-2-azetidinone (95%)] ⑤ [2]

REAGENT: OSMIUM TETROXIDE

==

STRUCTURE:

MW = 254.20
MP = 39.5-40°
PB = 130°

OsO_4

Fieser's Reagents for Organic Synthesis: Vol., pages

1	759- 67	**4**		**7**		**10**	290
2	301	**5**		**8**		**11**	
3		**6**		**9**		**12**	358

March's Advanced Organic Chemistry: 628, 732, 1071

MAJOR USES:

Although a general oxidizing agent, OsO_4 has found primary use as a hydroxylating agent

---

PREPARATION:

Available commercially. Can be obtained as solution in a variety of solvents.

PRECAUTION:

Vapor poisonous. Irritant.

NOTES:

1. Via the silylenol ether

2. **TMNO** = Trimethylamine N-oxide monohydrate

3. **NMMO** = N-Methylmorpholine N-oxide

REFERENCES:

1. J.P. McCormick, W. Tomasic and M.W. Johnson, **Tetrahedron Lett.**, (1981), 607.

2. S. Danishefsky, P.F. Schuda, T. Kitihara and S.J. Etheredge, **J. Am. Chem. Soc.**, (1977), 99, 6066.

3. F.M. Hauser and S.R. Ellenberger, **J.Org. Chem.**, (1986), 51, 50.

4. S. Knapp, G. Sankar Lal ad D. Sahai, **J. Org. Chem.**, (1986),51, 380.

5. V. Nair and S.D. Chamberlain, **J. Org. Chem.**, (1985), 50, 5069.

EXAMPLES:

$CH_3(CH_2)_6COCH_3$ → 1.) $HN[SiMe_3]_2$, $ISiMe_3$; 2.) OsO_4-NMMO → $CH_3(CH_2)_5CH(OH)COCH_3$ (69%) ① [1] △1 △3

OsO_4; Barium chlorate, THF (84%) ② [1]

OsO_4 / TMNO (56%) + (38%) ③ [1] △2

1.) OsO_4, NMMO; 2.) MeOH, TsOH (68%) ④ [1] △3

1.) OsO_4, Pyridine; 2.) $NaHSO_4$, Pyridine, H_2O (78%) ⑤ [1]

REAGENT: OZONE

STRUCTURE:

O_3

MW = 48.0
BP = -111.9°

Fieser's Reagents for Organic Synthesis: Vol., pages

1	773- 77	**4**	363- 4	**7**	269- 71	**10**	295- 6
2		**5**	491- 95	**8**	374- 77	**11**	387- 88
3		**6**	436- 41	**9**	341- 43	**12**	365- 66

March's Advanced Organic Chemistry: 627, 634, 785, 1060, 1066-71, 1087- 89

MAJOR USES: ⚠1

Ozonolysis of alkenes

PREPARATION:

Ozone is prepared in the laboratory by a special piece of equipment that allows dry oxygen to pass between two electrodes at high voltage. See: **Reagents**, Vol. 1, page 773.

PRECAUTION:

The blue-colored gas is explosive, as are the intermediates formed after reacting an alkene with ozone. Special care must be exercised to be sure that the intermediates have been properly decomposed. Exposure to ozone can be harmful.

NOTES:

1. Ozone can be adsorbed on silica gel where it can be used for reaction. Up to 4- 5% ozone (by weight) can be adsorbed.

2. The recovered cyclopentanone is readily recycled by aldol conditions.

REFERENCES:

1. S. Schreiber, R. Claus and J. Reagan, **Tetrahedron Lett.**, (1982), 3867.

2. J. Wrobel and J. Cook, **Synth. Commun.**, (1980), 333.

3. A.M. Caminade, M. LeBlank, F. El Khatib and M. Koenig, **Tetrahedron Lett.**, (1985), 2889.

4. J.H. Hutchinson and T. Money, **Tetrahedron Lett.**, (1985), 1819.

5. T.V. Lee and J. Toczek, **Tetrahedron Lett.**, (1985), 473.

EXAMPLES:

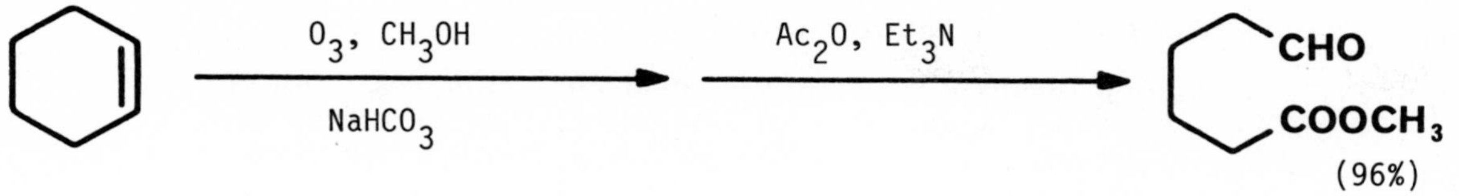

①

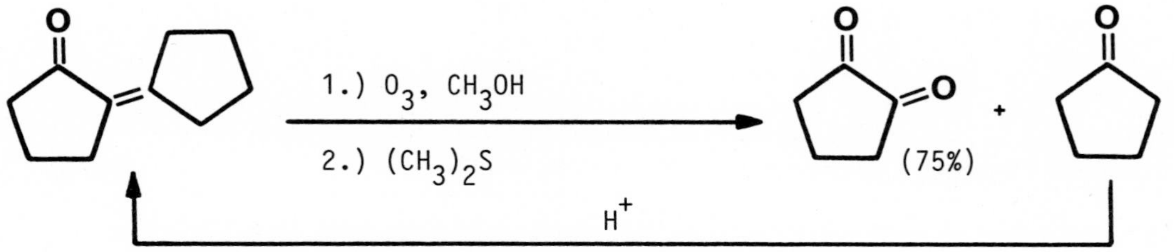

② △2

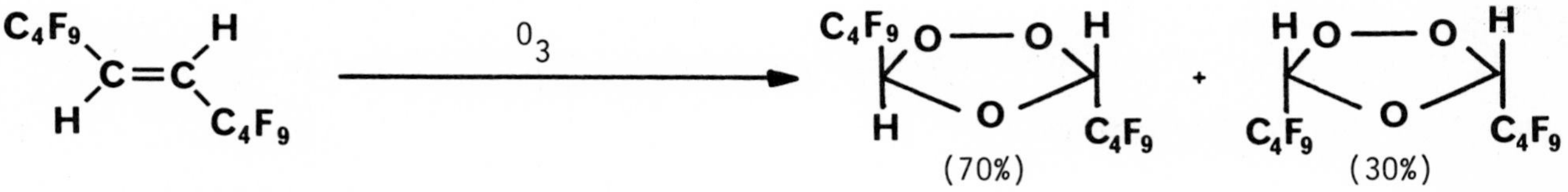

③

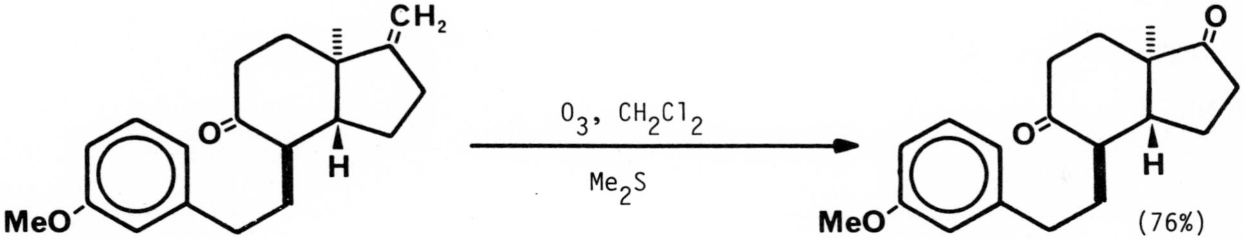

④

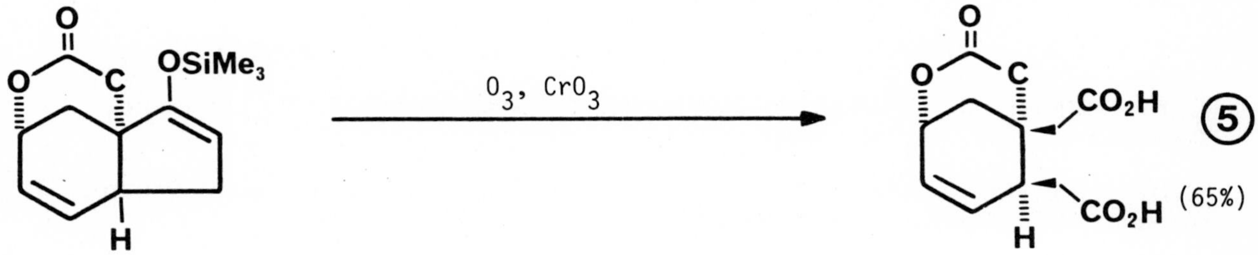

⑤

REAGENT: PALLADIUM

==

STRUCTURE:

Pd/C, Al, Al_2O_3, $BaCO_3$, $BaSO_4$, $CaCO_3$

The palladium catalyst is often adsorbed onto surfaces such as: Carbon, Aluminum oxide, Barium carbonate, Barium sulfate and Calcium carbonate. Generally the palladium content will vary from 1 to 10%.

MW = 106.40

<u>Fieser's Reagents for Organic Synthesis</u>: Vol., pages

1	778- 82	**4**	368- 69	**7**	275- 77	**10**	299- 300
2	303	**5**	499	**8**	382- 83	**11**	392- 93
3		**6**	445- 46	**9**	351- 52	**12**	

March's <u>Advanced Organic Chemistry</u>: 692, 1056, 1097.

MAJOR USES:

1. Hydrogenation catalyst
2. Dehydrogenation catalyst
3. Decarbonylation
4. Reductive cleavages
5. Deoxygenation

PREPARATION:

1. Various preparations are commercially available.
2. See also **Lindlar catalyst.**

PRECAUTION:

1. The catalyst on carbon is highly flammable.
2. The catalyst on barium carbonate is toxic.

NOTES:

REFERENCES:

1. H. Urbach and R. Henning, **Tetrahedron Lett.**, (1985), 1839.

2. R.B. Gammill, **Tetrahedron Lett.**, (1985), 1385.

3. G.W.J. Fleet and P.W. Smith, **Tetrahedron Lett.**, (1985), 1469.

4. D. Guillerm, M. Delarme, M. Jalalinaimi, P. Lemaitre and J.-Y. Lallemand, **Tetrahedron Lett.**, (1984), 1043.

5. G.P. Boldrini, D. Savoia, E. Tagliavini, C. Trombini and A. Umani-Ronchi, **J. Organomettalic Chem.**,(1984), <u>268</u>, 97.

EXAMPLES:

Pd/C, H_2, 30°; EtOH, H_2SO_4 — CO_2Et, N, CH_2Ph → H, H, N H, CO_2Et (58%) ① [1] [4]

H_2, Pd·C; K_2CO_3 / EtOH — OAc, O, OAc, OAc → O, OAc, OAc (83%) ② [1]

1.) Pd-Black, H_2, EtOH; 2.) $PhCH_2OCOCl$, Ether, aq $NaHCO_3$ — O, O, OMe, O, N_3, CH_2Ph → OCH_2Ph, MeO, O, CH_2OH, N, $OCOCH_2Ph$ (43%) ③

1.) H_2/Pd/C; 2.) $LiAlH_4$ — CO_2Me, CO_2Me → CH_2OH, CH_2OH, H (93%) ④ [1]

$PhSO_2Na$; Pd-Graphite, PPh_3 — O, OCMe → SO_2Ph (100%) ⑤

REAGENT: PALLADIUM ACETATE

==

STRUCTURE:

MW = 224.49 $Pd(CH_3CO_2)_2$
MP = 205° C

<u>**Fieser's Reagents for Organic Synthesis**</u>**: Vol., pages**

1	778	**4**	365	**7**	274	**10**	297
2	303	**5**	496- 97	**8**	378- 82	**11**	389- 91
3		**6**	442- 43	**9**	344- 49	**12**	

March's <u>**Advanced Organic Chemistry:**</u> 484, 636, 643, 1001, 1075.

MAJOR USES:

1. Oxidations 2. Oxidative coupling 3. Cyclopropanations

PREPARATION:

1. The reagent can be prepared from the reaction of palladium(II)nitrate and acetic acid.

2. Commercially available.

PRECAUTION:

NOTES:

REFERENCES:

1. A.S. Kende, B. Roth and P.J. San Filippo, **J. Am. Chem. Soc.**, (1982), <u>104</u>, 1784.

2. J. Tsuji, H. Nagashima and K. Sato, **Tetrahedron Lett.**, (1982), 3085.

3. T. Itahara, **Chem. Commun.**,(1981), 859.

4. M. Suda, **Synthesis**, (1981), 714.

5. A. Matsuura, Y. Ito and T. Matsuura, **J. Org. Chem.**, (1985), <u>50</u>, 5002.

EXAMPLES:

$(CH_2)_3CH{=}CH_2$ / $OSi(CH_3)_3$ —[$Pd(OAc)_2$, CH_3CN]→ =CH_2, =O (40%) + $-CH_3$, =O (40%) + CH_3, =O (18%) ① [1]

OH, CH_3, Br, O, O —[$Pd(OAc)_2$, PPh_3; K_2CO_3]→ O, CH_3, O, O (75%) ② [1]

O, O + p-$PhMe_2$ —[$Pd(OAc)_2$; HOAc]→ CH_3, O, CH_3, O (78%) ③ [2]

$CH{=}CH_2$ + CH_2N_2 —[$Pd(OAc)_2$, Ether; 5°]→ H, H (77%) ④ [3]

$PhCH{=}CH_2$ + O, H_3C OOH —[$Pd(OAc)_2$ / N_2]→ $PhCCH_3$ (O) (67%) + O, H_3C OH (79%) ⑤

REAGENT: PALLADIUM (II) CHLORIDE

==

STRUCTURE:

MW = 177.31 $PdCl_2$
MP = 678.8°

Fieser's Reagents for Organic Synthesis: Vol., pages

1	782	**4**	368- 70	**7**	277	**10**	300- 02
2	303- 05	**5**	500- 03	**8**	384	**11**	393- 95
3		**6**	447- 50	**9**	352	**12**	

March's Advanced Organic Chemistry: 1054, 1083- 85

MAJOR USES:

1. Catalyst for oxidations, couplings, alkylations, cyclizations.

PREPARATION:

1. Commercially available.

PRECAUTION:

1. Highly toxic.

NOTES:

REFERENCES:

1. J. Muzart and J.-P. Pete, **Chem. Commun.**,(1980), 257.

2. L.S. Hegedus, G.F. Allen and D.J. Oleson, **J. Am. Chem. Soc.**, (1980), 102, 3583.

3. S.B. Fergusson and H. Alper, **Chem. Commun.**, (1984), 1349.

4. R.C. Larock, L.W. Harrison and M.H. Hsu, **J. Org. Chem.**,(1984), 39, 3662.

5. M.F. Semmelhack and C. Bodurow, **J. Am. Chem. Soc.**, (1984), 106, 1496.

EXAMPLES:

(1) hν — (32%)

(2) $PdCl_2$ — (91%)

$CH_2{=}CHCCH_3$ (O over C)

(3) $C_8H_{17}CH{=}CH_2$ + $HOCH_2CH_2OH$ + CO — $PdCl_2$, THF / HCl; $CuCl_2$, Ethylene glycol → $C_8H_{17}CH(CH_3)COCH_2CH_2OH$ (C=O) (69%)

(4) ClHg … Cl — $PdCl_2$ / LiCl, Acetonitrile → [PdCl … HO … Cl] — Reflux, NH_4Cl → (71%)

(5) OH — $CuCl_2$, $PdCl_2$; CO / MeOH → CO_2Me … O (50%)

REAGENT: PHASE-TRANSFER CATALYSTS 

STRUCTURE:

$CH_3N[(CH_2)_7CH_3]_3Cl$ (Aliquat 336), Bu_4NBr, Bu_4NHSO_4

There are two major types of phase transfer catalysts- (1) Quaternary ammonium or phosphonium salts, and (2) Crown ethers and cryptans

---

Fieser's Reagents for Organic Synthesis: Vol., pages

Vol.	pages	Vol.	pages	Vol.	pages	Vol.	pages
1		**4**		**7**	4, 380	**10**	305-06
2		**5**	460	**8**	387-91	**11**	403-07
3		**6**	10, 404	**9**	356-61	**12**	

March's Advanced Organic Chemistry: 320-22

MAJOR USES:

1. Catalysts for coupling reactions 2. Alkylations 3. Condensations

---

PREPARATION:

PRECAUTION:

NOTES:

1. Several phase-transfer reagents are commercially available:

a. Addgen 464 b. Aliquat 336 c. Tetrabutylammonium bromide

d. 18-Crown ether e. Tetrabutylammonium hydrogensulfate

2. For Reviews, see: G. Gokel and W. Weber, **J. Chem. Educ.**, (1978), 35, 350

3. **HDTAB** = Hexadecyltrimethylammonium bromide

REFERENCES:

1. M. Lissel, **Ann.**, (1982), 1589.

2. E. Dehmlow and A. Sharmout, **Ann.**, (1982), 1750.

3. L. Ghosez, J. Antoine, E. Deffense, M. Navarro, V. Libert, M. O'Donnel, W. Bruder, K. Willey and K. Wojciechowski, **Tetrahedron Lett.**, (1982), 4255.

4. M. Lissel, **J. Chem. Res.(M)**, (1982), 2946.

5. F.D. Mills, **Synth. Commun.**, (1986), 905.

EXAMPLES:

(1) 3-Methylbenzene-1,2-dithiol (4-methyl, SH, SH) + Cl_2CHCO_2Et —[Aliquat 336; K_2CO_3, $PhCH_3$]→ 2-(CO_2Et)-1,3-benzodithiole bearing CH_3 (63%)

(2) $(CH_3)_2C(OH)\text{-}C{\equiv}CH$ + $(CH_3)_2C{=}O$ —[$Bu_4N^+Br^-$, NaOH, $PhCH_3$; H_2O, 30 min]→ $(CH_3)_2C(OH)\text{-}C{\equiv}C\text{-}C(OH)(CH_3)_2$ (100%)

(3) 4-ClPhCH=N-CH_2CO_2Et + EtI —[$Bu_4N^+HSO_4^-$]→ Et-CH(NH_2)-CO_2H (76%)

(4) 2-Chlorocyclohexanone + Ph-SH —[Aliquat 336; Na_2CO_3, PhMe, H_2O]→ 2-(S-Ph)cyclohexanone (91%)

(5) $CH_3C(=O)CH(C(=O)\text{-}O\text{-}Et)CH_2CH_2CH_3$ —[HDTAB; KOH, N_2, H_2SO_4]→ $CH_3C(=O)(CH_2)_3CH_3$ (75%) [3]

REAGENT: PHOSPHOROUS PENTACHLORIDE

STRUCTURE:

PCl_5

MW = 208.24
MP = 178- 81° C
d = 1.600

Fieser's Reagents for Organic Synthesis: Vol., pages

1	866- 70	**4**	388- 89	**7**	290	**10**	
2		**5**	534	**8**		**11**	
3		**6**		**9**		**12**	

March's Advanced Organic Chemistry: Not indexed

MAJOR USES:

1. Acid chloride preparation
2. Dehydration
3. Chlorination
4. **Beckmann Rearrangement** catalyst

PREPARATION:

1. Commercially available.

PRECAUTION:

1. Moisture sensitive and corrosive.

NOTES:

REFERENCES:

1. J. Ojima, T. Nakada, E. Ejira and N. Nakamura, **Tetrahedron Lett.**, (1985), 639.

2. E.E. Sugg, J.F. Griffin and P.S. Portoghese, **J. Org. Chem.**,(1985), 50, 5032.

3. H. Singh, S.K. Aggarwac and N. Malhotra, **Synthesis**, (1983), 791.

4. F. Pochat, **Synthesis**, (1984), 146.

5. F. Camps, J. Coll, A. Messeguer and M.A. Pericas, **Tetrahedron Lett.**, (1979), 3901.

EXAMPLES:

NOH; PCl_5; THF, RT, 5 hrs; O; H N; (41%) ① [4]

Me Ph; NOH; PCl_5, CH_2Cl_2, 10°; Me Ph; Cl; Cl; N H; O; (76%) ② [4]

N; N; PCl_5; Xylene, Reflux; [N Cl; N Cl]; H_2O; Reflux; H N O; N H O; (60%) ③ [3]

N CN; HN O; Cl; PCl_5; $CHCl_3$, Reflux; N CN; N Cl; Cl; NaSH, H_2O; DMSO; N S N; Cl; (92%) ④ [3]

O; MeO; O; PCl_5, CCl_4; RT, 8 hrs; O; MeO; Cl; Cl; (82%) ⑤ [3]

REAGENT: PHOSPHOROUS (V) SULFIDE (PHOSPHOROUS PENTASULFIDE)

STRUCTURE:

P_4S_{10}

MW = 444.54
MP = 286-90°C
d = 2.090

Fieser's Reagents for Organic Synthesis: Vol., pages

1	870- 71	**4**	389	**7**		**10**	320
2		**5**	534- 35	**8**	401	**11**	428
3	226- 28	**6**		**9**	374	**12**	

March's Advanced Organic Chemistry: 794

MAJOR USES:

1. Dehydration 2. Thionation 3. Sulfoxide reduction

PREPARATION:

1. Commercially available in 99% purity.

PRECAUTION:

1. The solid is flammable and moisture sensitive.

NOTES:

1. Via:

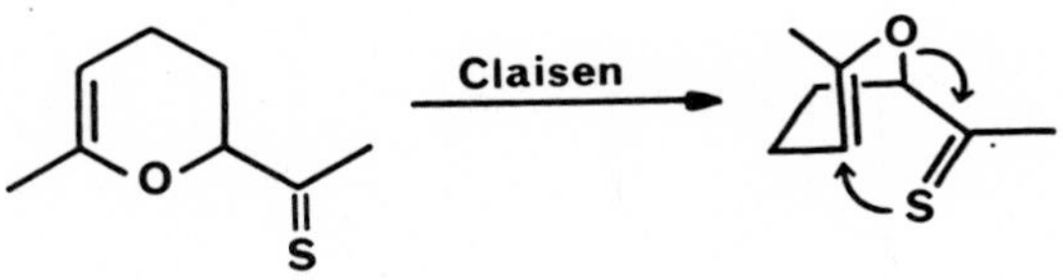

Claisen

REFERENCES:

1. J. Kuipers, B. Lammerink, I. Still and B. Zwanenberg, **Synthesis**, (1981), 295.

2. K.B. Lipkowitz, S. Scarpone, B.P. Mundy and W.G. Bornmann, **J. Org. Chem.**, (1979), 44, 486.

3. S. Raucher and P. Klein, **J. Org. Chem.**, (1986), 51, 123.

4. M. Yamamoto, S. Iwasa, K. Takatsuki and K. Yamada, **J. Org. Chem.**, (1986), 51, 346.

5. H. Alper, J. Currie and R. Sachdeva, **Angew. Chem., Int. Ed.**, (1978), 17, 689.

EXAMPLES:

$Ph_2C{=}S{=}O$ —[P_4S_{10}, CH_2Cl_2; 6 hrs]→ $Ph_2C{=}S$ (58%) ① [2]

—[P_2S_5; Δ]→ (NYA) ② [2] ⚠1

—[P_2S_5; Ultrasound]→ (72%) ③ [2]

—[1.) P_2S_5; 2.) MeI]→ (87%) ④ [2]

Cl-C_6H_4-NHC(=O)C(Me)=CH_2 —[P_4S_{10}, $NaHCO_3$; MeCN]→ Cl-C_6H_4-NHC(=S)CH(Me)CH_2SH (56%) ⑤ [2]

REAGENT: PHOSPHORYL CHLORIDE (PHOSPHOROUS OXYCHLORIDE)

==

STRUCTURE:

MW = 153.33

MP = 1.25° C

BP = 105.8° C

$POCl_3$

Fieser's Reagents for Organic Synthesis: Vol., pages

1	876- 82	**4**	390	**7**	292- 93	**10**	
2	330- 31	**5**	390	**8**	401- 02	**11**	**429**
3	228	**6**		**9**	374- 75	**12**	

March's Advanced Organic Chemistry: 372, 382, 487, 495, 539, 593, 732, 933

MAJOR USES:

1. Dehydration of alcohols and amides 2. Molecular rearrangements

3. Generation of imminium salts 4. Catalyst for aromatic substitution reactions.

PREPARATION:

1. Commercially available

PRECAUTION:

1. Extremely irritating to skin, eyes and mucous membranes. Avoid inhalation.

NOTES:

REFERENCES:

1. S. Fujita, K. Koyama and Y. Inagaki, **Synthesis**, (1982), 68.

2. H. Bates and H. Rapoport, **J. Am. Chem. Soc.**, (1979), 101, 1259.

3. O. Meth-Cohn, **Tetrahedron Lett.**, (1985), 1901.

4. B. Maurer and A. Hauser, **Tetrahedron Lett.**, (1984), 1061.

5. R. Gleiter and J. Uschmann, **J. Org. Chem.**, (1986), 51, 370.

EXAMPLES:

POCl$_3$, CH$_3$CONMe$_2$ (60-90%) B ① [2]

POCl$_3$ 105° (NYA) ② [3]

POCl$_3$; H$_2$C=CHOAc (93%) ③ [3]

POCl$_3$,DMF, 15-25° (93%) ④ [4]

Me$_2$CO / POCl$_3$ N$_2$, RT, 15 hrs (90%) ⑤

REAGENT: POLYPHOSPHORIC ACID (**PPA**)

STRUCTURE:

$$HO-\overset{\overset{O^{\ominus}}{|}}{\underset{\underset{HO}{|}}{P^{\oplus}}}-O-\left[-\overset{\overset{O^{\ominus}}{|}}{\underset{\underset{OH}{|}}{P^{\oplus}}}-O-\right]_n-\overset{\overset{O^{\ominus}}{|}}{\underset{\underset{OH}{|}}{P^{\oplus}}}-OH$$

Fieser's Reagents for Organic Synthesis: Vol., pages

1	894- 95	**4**	395- 97	**7**	294- 95	**10**	321- 22
2	334- 36	**5**	540- 42	**8**		**11**	
3	231- 33	**6**	474- 75	**9**		**12**	399- 400

March's Advanced Organic Chemistry: 486, 494, 985, 988.

MAJOR USES: △1

1. Dehydration 2. Cyclization 3. Hydrolysis 4. Isomerizations

5. Acid catalyst

PREPARATION:

1. Commercially available.

2. For a preparation of "super" PPA, see: D. Bailey, C. DeGrazia, H. Lape, R. Frering, D. Fort and T. Skulan, **J. Med. Chem.**, (1973), 16, 151.

PRECAUTION:

1. An irritant and is corrosive.

2. Very hygroscopic

NOTES:

1. For a discussion of co-solvent use of PPA, see: **Reagents**, 10, 321.

REFERENCES:

1. D.P Curran and S.-C. Kuo, **J. Org. Chem.**, (1984), 49, 2063.

2. S.K. Rosen, **Tetrahedron Lett.**, (1983), 4351.

3. S. Oae and H. Togo, **Synthesis**, (1982), 152.

4. I. Agranat and Y.-S. Shih, **Synthesis**, (1974), 865.

5. B. Amit and A. Hassner, **Synthesis**, (1978), 932.

EXAMPLES:

PPA / N_2, 160°, 20 hrs (80%) ① [2]

PPA, 130°, 30 min (72%) ② [2]

PPA, Sulfolane, 95-100° (58%) ③ [5]

PPA, 140°, 10 hrs (25%) ④ [4]

PPA, 130°, 10 min (70%) ⑤ [5]

REAGENT: POLYPHOSPHORIC ESTER (**PPE**)
(POLYPHOSPHORIC ACID ETHYL ESTER, ETHYL POLYPHOSPHATE)

==

STRUCTURE:

⚠1

$$RO-\left[-\overset{\overset{\displaystyle O}{\|}}{\underset{\underset{\displaystyle OR}{|}}{P}}-O-\right]_n-R$$

Fieser's Reagents for Organic Synthesis: Vol., pages

1	892- 94	**4**	394- 95	**7**		**10**	
2	333- 34	**5**	539- 40	**8**		**11**	430
3	229- 31	**6**	474	**9**	376-77	**12**	

March's Advanced Organic Chemistry: Not indexed

MAJOR USES:

1. Condensation 2. Dehydration 3. **Friedel-Crafts Acylation**

PREPARATION:

1. The reagent can be prepared from P_2O_5 , Et_2O and $CHCl_3$. See: W. Pollmann and G. Schramm, **Biochem. Biophys. Acta,** (1964), 80, 1. For a different approach, see: M. Cava, M. Lakshmikantham and M. Mitcell, **J. Org. Chem.**, (1969), 34, 2665.

PRECAUTION:

NOTES:

1. This is only a general structure for PPE. The actual reagent will have both cyclic and straight-chain units; however, usually not a large number.

REFERENCES:

1. G.W. Rewcastle and W.A. Denny, **Synthesis**, (1985), 220.

2. S. Oae and H. Togo, **Synthesis**, (1982), 152

3. T. Imanoto, M. Kodera and M. Yokoyama, **Synthesis**, (1982), 134.

4. T. Imanoto, T. Takaoka and M. Yokoyama, **Synthesis**, (1983), 142.

5. J.L. Irvine and C. Piantadosi, **Synthesis**, (1972), 568.

EXAMPLES:

PPE. $CHCl_3$, Reflux (96%) ① [3]

PPE / $CHCl_3$, Reflux, 6 hrs (43%) ② [1]

PPE, MeOH, 55°, 15 hrs (97%) ③ [1]

PPE, $CHCl_3$; PPE (90%) ④ [2]

PPE, $CHCl_3$, Reflux, 5 hrs (92%) ⑤ [1]

REAGENT: POTASSIUM t-BUTOXIDE

STRUCTURE:

MW = 112.22
MP = 256- 58°C (Dec)

$$KO-C(CH_3)_3$$

Fieser's Reagents for Organic Synthesis: Vol., pages

1	911- 27	**4**	399- 405	**7**	296- 98	**10**	323
2	336- 44	**5**	544- 53	**8**	407- 08	**11**	432
3	233- 34	**6**	477- 79	**9**	380- 81	**12**	401

March's Advanced Organic Chemistry: 339, 567

MAJOR USES: ⚠1

1. Strong base 2. **Oppenauer oxidation** 3. **Wolff-Kishner reduction**

PREPARATION:

1. Can be prepared by the reaction of t-butyl alcohol and potassium under nitrogen (See: W. Johnson and W. Schneider, **Org. Synth.**, (1963), Coll. Vol. 4, 132).

2. Commercially available.

PRECAUTION:

1. Highly moisture sensitive. 2. The reactions should be carried out under an inert atmosphere. 3. The solid is flammable.

NOTES:

1. For a review, see: D. Pearson and C,. Buehler, **Chem. Rev.**, (1974), 74, 45.

REFERENCES:

1. H. Ahlbrecht and W. Raab, **Synthesis**, (1980), 320.

2. H. Verkruysse, H. Bos, L. deNorten and L. Brandsma, **Rec. Trav. Chem**, (1981), 100, 244.

3. J.C. Gilbert and B.E. Wiechman, **J. Org. Chem.**, (1986), 51, 258.

4. J. Durman and S. Warren, **Tetrahedron Lett.**, (1985), 2895.

5. P. Bravo, G. Resnati and F. Viani, **Tetrahedron Lett.**, (1985), 2913.

EXAMPLES:

$CH_3-C(CN)(CH_3)-N(CH_3)_2$ —[K^+t-BuO$^-$ / $MeOCMe_3$]→ $CH_2=C(CH_3)NMe_2$ (64%) ① [1]

$HC\equiv CCH_2NMe_2$ —[$K^{+-}OCMe_3$-$HOCMe_3$ / HMPT, 23°]→ $MeC\equiv CNMe_2$ + $H_2C=C=CHNMe_2$ (Excess of 85%, 4:96) ② [1]

(4-methylcyclohex-3-enone + (2-methylcyclopent-1-enyl)methanol, OH) —[K^+t-BuO$^-$ / $(EtO)_2P(O)CHN_2$]→ (50%, Z:E=60:40) ③

PhS, CO_2Et —[1.) t-BuOK, THF; 2.) $PhCH_2Br$]→ Ph, PhS, CO_2Et (76%) ④ [1]

O, OH, $^{\ominus}BF_4$, Ph, H, $\overset{\oplus}{S}$, $4CH_3Ph$, CH_3 —[t-BuOK]→ O, O, Ph, H (96%) ⑤ [1]

REAGENT: POTASSIUM HYDRIDE

==

STRUCTURE:

MW = 40.11 KH

Fieser's Reagents for Organic Synthesis: Vol., pages

1	935	**4**	409	**7**	301- 02	**10**	327- 28
2	346	**5**	557	**8**	412- 15	**11**	435- 37
3		**6**	482- 83	**9**	386- 87	**12**	407- 11

March's Advanced Organic Chemistry: Not Indexed

MAJOR USES:

1. Base

PREPARATION:

1. Commercially available as a dispersion in mineral oil.

PRECAUTION:

1. Reacts slowly with moist air, and will react violently with water.

2. Should be used under an inert atmosphere.

NOTES:

1. **Oxy-Cope rearrangement**. This reaction is often anion-accelerated.

2. The **Brook Rearrangement** (See: A.G. Brook, **Acc. Chem. Res.**, (1974), 7, 77) involves removal of the trimethylsilyl group via a rearrangement of the Me_3Si-group from carbon to oxygen before final removal.

3. A **Claisen condensation**.

REFERENCES:

1. a. M. Kahn, **Tetrahedron Lett.**, (1980), 4547.
   b. S. Levine and R. McDaniel, Jr., **J. Org. Chem.**, (1981), 46, 2471.

2. P. Wender and S. Sieburth, **Tetrahedron Lett.**, (1981), 2199.

3. S. Wilson, M. Hague and R. Misra, **J. Org. Chem.**, (1982), 47, 747.

4. R. L. Shone, J.R. Deason and M. Miyano, **J. Org. Chem.**, (1986), 51, 268.

5. S. Knapp and D.V. Patel, **Tetrahedron Lett.**, (1982), 3539.

EXAMPLES:

CH_2=CH OH — KH, THF → (83%) ① △1

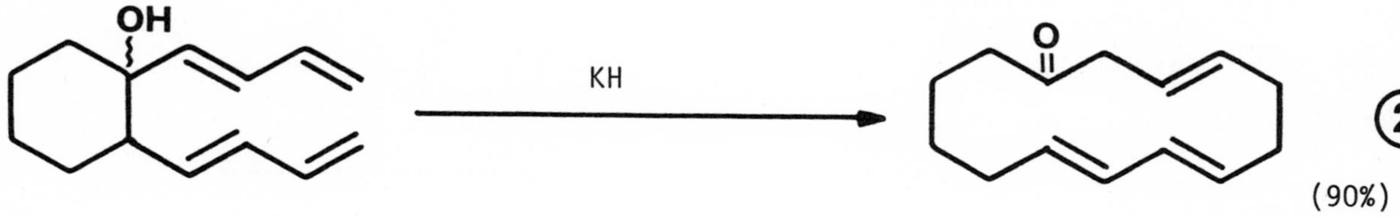

OH — KH → O (90%) ② △1

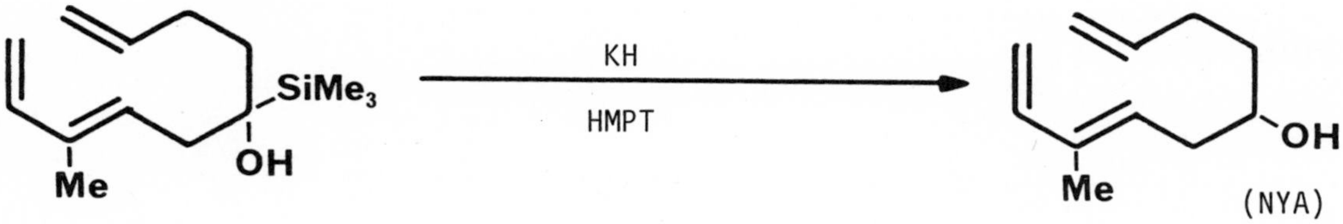

SiMe$_3$, OH, Me — KH / HMPT → OH, Me (NYA) ③ △2

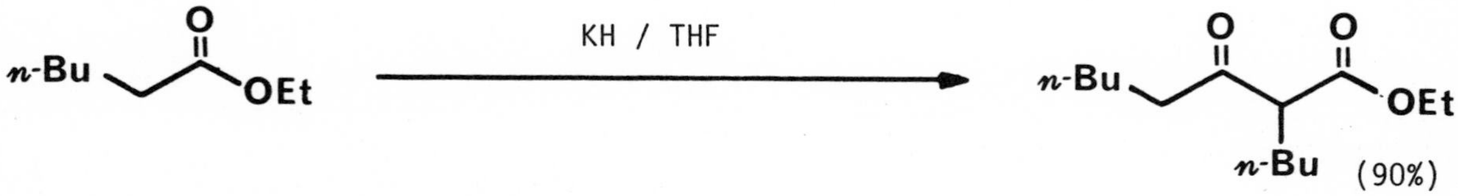

n-Bu, O, OEt — KH / THF → n-Bu, O, O, OEt, n-Bu (90%) ④ △3

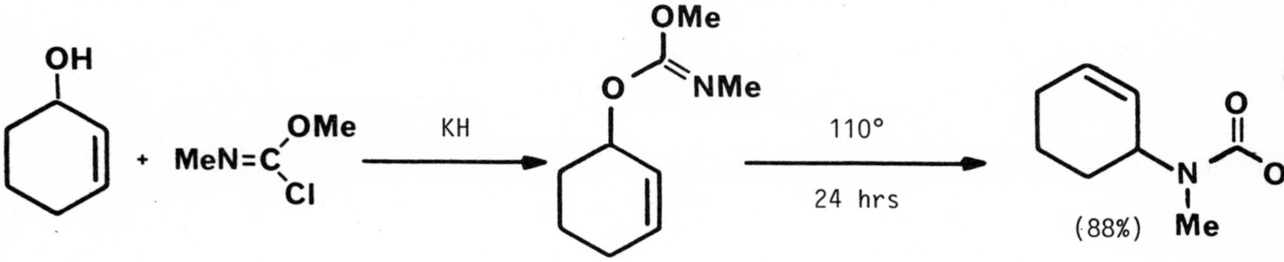

OH + MeN=C(OMe)Cl — KH → O, OMe, NMe — 110° / 24 hrs → O, N, OMe, Me (88%) ⑤

REAGENT: POTASSIUM NITROSODISULFONATE (FREMY'S SALT)

STRUCTURE:

MW = 268.34 $(KSO_3)_2NO$

<u>**Fieser's Reagents for Organic Synthesis**</u>: **Vol., pages**

1	940 42	**4**	411	**7**		**10**	329- 30
2	347- 48	**5**	562	**8**		**11**	
3	238	**6**		**9**		**12**	

March's <u>Advanced Organic Chemistry</u>: 1061

MAJOR USES: ⚠1

1. Oxidizing agent

PREPARATION:

1. Can be prepared by the oxidation of sodium hydroxylamine disulfonate with permanganate .

2. Commercially available.

PRECAUTION:

NOTES:

1. For a Review, see: H. Zimmer and D. Lankin, **Chem. Rev.**, (1971), <u>71</u>, 229.

REFERENCES:

1. A. Kozikowski, K. Sugiyama and P. Springer, **J. Org. Chem.**, (1981), <u>46</u>, 2426.

2. M.P. Vazquez Tato, L. Castedo and R. Riguera, **Chem. Lett.**, (1985), 623.

3. J. Morey, A. Dzielenziak and J.M. Saa, **Chem. Lett.**, (1985), 263.

4. L. Castedo, R. Riguera and M.J. Rodriguez, **Tetrahedron**, (1982), <u>38</u>, 1569.

5. P.A. Wherli and B. Schaer, **Synthesis**, (1974), 288.

EXAMPLES:

(1) 5-hydroxy-4-acetyl-2-methyl-2-carbomethoxy-benzoxazine $\xrightarrow{(KSO_3)_2NO\cdot}$ quinone (84%)

OH, $COCH_3$, N, CH_3, O, CO_2CH_3 → O, $COCH_3$, N, CH_3, O, CO_2CH_3, O (84%)

(2) $PhCH(OH)CH(Me)-NHMe$ $\xrightarrow[H_2NOH\cdot HCl\ /\ Na_2CO_3\ /\ Pyridine]{\cdot ON(SO_3K)_2}$ $PhCH(OH)CH(Me)-N(Me)-N{=}O$ (85%)

(3) piperonyl alcohol (OH) $\xrightarrow[2\text{-Octenol}\ /\ C_6H_6\ /\ \text{Adogen-464}]{K_2CO_3\ /\ \cdot ON(SO_3K)_2}$ piperonal (CHO) (94%)

(4) $PhCH_2NH_2$ $\xrightarrow[Na_2CO_3,\ RT,\ 2\ hrs]{\cdot ON(SO_3K)_2}$ [$PhCH{=}\overset{\oplus}{N}H_2$ $\xrightarrow{PhCH_2NH_2}$ $PhCH(HNCH_2Ph)(NH_2)$] ⟶ $PhCH{=}NCH_2Ph$ (88-97%)

(5) tetrahydroisoquinoline (NH) $\xrightarrow[Na_2CO_3,\ RT,\ 2\ hrs]{\cdot ON(SO_3K)_2}$ dihydroisoquinoline (N) ⟶ isoquinoline (N) (74%)

REAGENT: POTASSIUM PERMANGANATE

STRUCTURE:

$KMnO_4$

MW = 158.04
MP = decomposes at about 240°

Fieser's Reagents for Organic Synthesis: Vol., pages

1	942- 52	**4**	412- 13	**7**		**10**	330
2	348	**5**	562- 63	**8**	416- 17	**11**	440- 41
3		**6**		**9**	388- 91	**12**	413

March's Advanced Organic Chemistry: Not indexed

MAJOR USES:

Strong oxidizing agent

PREPARATION:

Commercially available in high purity

PRECAUTION:

As with all strong oxidizing agents, this reagent should be used with care. It is an irritant.

NOTES:

1. This is an example of oxidative aromatization.

2. See **Nef Reaction**.

3. Dilute solutions of $KMnO_4$ can be used to hydroxylate (cis addition from the less-hindered side) alkenes. In this example the secondary alcohol is further oxidized to a ketone.

REFERENCES:

1. A. Poulose and R. Croteau, **Chem. Commun.**, (1979), 243.

2. J. Clark and D. Cork, **Chem. Commun.**, (1982), 635.

3. N. Kornblum, A. Erickson, W. Kelly ad B. Henggeler, **J. Org. Chem.**, (1982), 47, 4534.

4. M.G. Constantino, P.M. Donate and N. Petragnani, **J. Org. Chem.**, (1986), 51, 253.

5. D. Chambers. W.A. Denny, J.S. Buckleton and G.R. Clark, **J. Org. Chem.**, (1985), 50, 4736.

EXAMPLES:

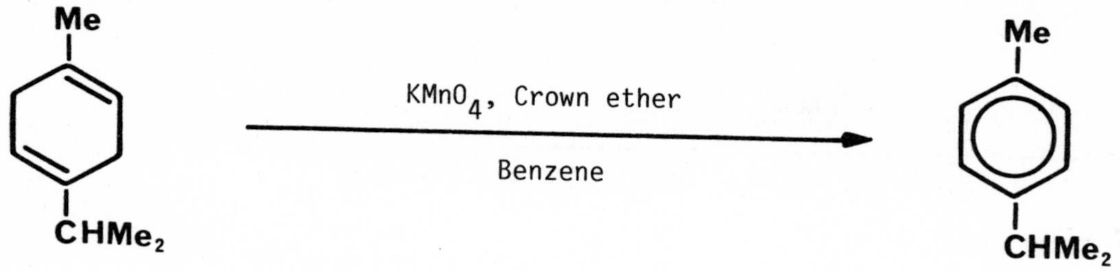

Me / CHMe$_2$ (1,4-cyclohexadiene) $\xrightarrow[\text{Benzene}]{KMnO_4,\ \text{Crown ether}}$ Me / CHMe$_2$ (benzene) ① △1 (100%)

$PhC(=O)CH(Ph)CH(H)CH(Me)NO_2$ $\xrightarrow[SiO_2]{KMnO_4}$ $PhC(=O)CH(Ph)CH(H)C(=O)Me$ ② △2 (91%)

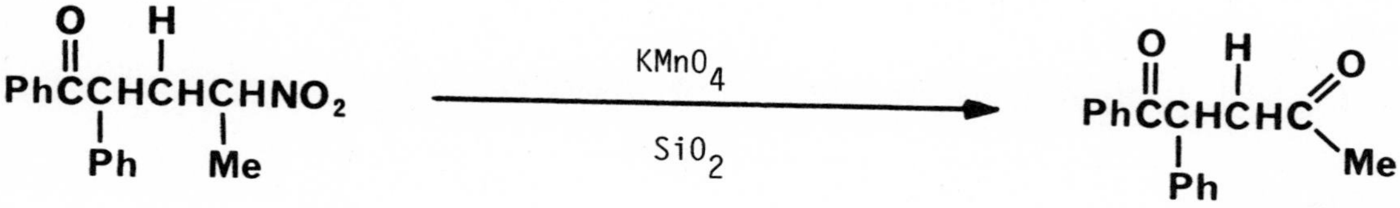

Br–C$_6$H$_4$–CH$_2$NO$_2$ $\xrightarrow[\text{t-BuOH, Pentane}]{KMnO_4\ /\ DMF}$ Br–C$_6$H$_4$–CHO ③ △2 (90%)

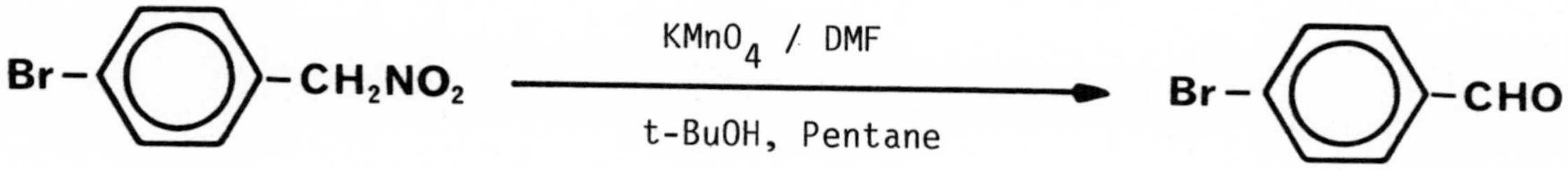

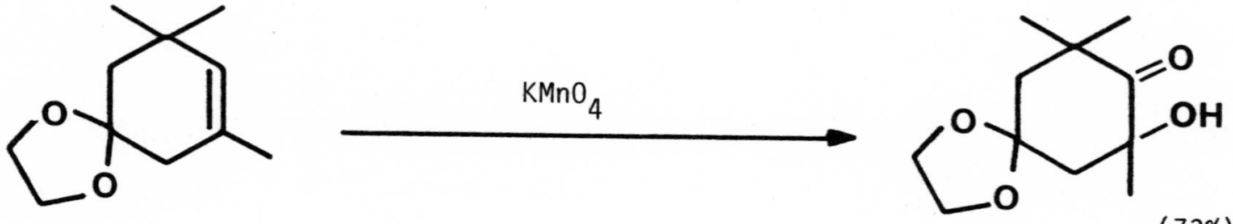

$\xrightarrow{KMnO_4}$ O, OH ④ △3 (73%)

HN—N, Me, Me, Me $\xrightarrow[HCl]{KMnO_4,\ H_2O}$ HN—N, HO_2C, CO_2H, CO_2H ⑤ (32%)

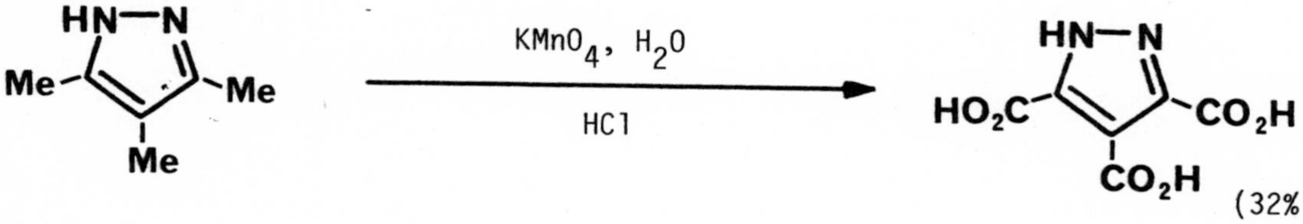

REAGENT: PYRIDINIUM CHLORIDE (PYRIDINE HYDROCHLORIDE)

STRUCTURE:

MW = 115.56
MP = 144°C

Fieser's Reagents for Organic Synthesis: Vol., pages

1	964- 66	**4**	415- 18	**7**	308	**10**	333
2	352- 53	**5**	566- 67	**8**	424- 25	**11**	449- 50
3	239- 40	**6**	497- 98	**9**		**12**	

March's Advanced Organic Chemistry: 797, 910

MAJOR USES:

1. Weak acid 2. Hydrolysis reactions 3. Isomerizations

PREPARATION:

1. The reagent is prepared from HCl and pyridine (M.D. Taylor and L.R. Grant, (1955), **J. Chem. Ed.**, (1955), 32, 39.)

2. Also commercially available.

PRECAUTION:

1. Irritant, hygroscopic

NOTES:

REFERENCES:

1. I.G.C. Coutts, M. Edwards and D.J. Richards, **Synthesis**, (1981), 487.

2. B.A. Keay and R. Rodrigo, **J. Am. Chem. Soc.**, (1982), 104, 4725.

3. L. Pellacani, P.A. Tardella and M.A. Loreto, **J. Org. Chem.**, (1976), 41, 1282.

4. S.M. Rosen and J.A. Moore, **J. Org. Chem.**, (1972), 37, 3770.

5. I. Vlattas, L. Della Vecchia and J.J. Fitt, **J. Org. Chem.**, (1973), 38, 3749, 4412.

EXAMPLES:

C_5H_5NHCl, Pyridine, Δ

(80%) ① [1]

PyH^+Cl^-

(84%) ② [3]

PyH^+Cl^-

Dry Pyridine, Reflux 20 hrs

CH_2Cl

(76%) ③ [3]

PyH^+Cl^- / CH_2Cl_2

(70%) ④ [3]

Me
$H_2NOCHSCH_2Ph$

Anhydrous PyH^+Cl^-

Pyridine, RT, 15 hrs

$NOCHSCH_2Ph$

$(CH_2)_6CO_2Me$

(100%) ⑤ [1]

REAGENT: PYRIDINIUM CHLOROCHROMATE (**PCC**) ⚠1

STRUCTURE:

Pyridinium ($C_5H_5NH^+$) $^{\ominus}ClCrO_3$

MW = 215.56
MP = 205° C

Fieser's Reagents for Organic Synthesis: Vol., pages

1		**4**		**7**	308- 09	**10**	334- 35
2		**5**		**8**	425- 27	**11**	450- 52
3		**6**	498- 99	**9**	397- 99	**12**	

March's Advanced Organic Chemistry: 1063, 1085

MAJOR USES: ⚠2

1. Versatile oxidizing agent for oxidation of alcohols.

2. Also oxidizes allylic methylene groups to a carbonyl.

PREPARATION:

1. The reagent is prepared from reaction of a molar equivalent of pyridine added to 6M HCl solution of CrO_3.

2. Commercially available.

PRECAUTION:

1. Strong oxidizing agent. 2. Carcinogen suspect.

NOTES:

1. Review: See: G. Piancatelli, A. Scettri and M.D'Auria, **Synthesis**, (1982), 245.

2. A stronger oxidizing agent than **PDC**.

REFERENCES:

1. P. Lansbury and T. Nickson, **Tetrahedron Lett.**, (1982), 2627.

2. A. Cisneros, S. Fernandez and J. hernandez, **Synth. Commun.**, (1982), 833.

3. G. Piancatelli, A. Scettri anbd M. D'Auria, **Tetrahedron**, (1980), 36, 661.

4. A. Gopalan and P. Magnus, **J. Am. Chem. Soc.**, (1980), 102, 1756.

5. T. Fujisawa, T. Itoh, M. Nakai and T. Sato, **Tetrahedron Letters**, (1985), 771.

EXAMPLES:

1.) B_2H_6

2.) PCC

① (55%)

PCC

$C(CH_2)_4CPh$

(70%) ②

PCC / CH_2Cl_2

$C_{12}H_{25}$

$C_{12}H_{25}$ (90%) ③

HO

PCC

OHC

④ (74%)

$C_{11}H_{23}$

OH

PCC, CH_2Cl_2

RT

$C_{11}H_{23}$

⑤ (88%)

REAGENT: PYRIDINIUM DICHROMATE (**PDC**)

==

STRUCTURE:

$[C_5H_5NH^{+}]_2\ Cr_2O_7^{-2}$

MW = 376.21
MP = 152- 53° C

Fieser's Reagents for Organic Synthesis: Vol., pages

1	**4**	**7**	**10** 335
2	**5**	**8**	**11** 453
3	**6**	**9** 399	**12**

March's Advanced Organic Chemistry: 1057

MAJOR USES:

1. A mild and "neutral" oxidizing agent. Useful for compounds having acid-sensitive functional groups.

2. More "neutral" than **PCC**.

PREPARATION:

1. The reagent is prepared by dissolving CrO_3 in a minimum amount of water, adding pyridine and recovering the precipitate.

PRECAUTION:

1. A suspected carcinogen.

NOTES:

1. The secondary alcohol is oxidized to the ketone; followed by bond isomerization. This sequence is then followed by allylic oxidation.

REFERENCES:

1. J. Zalikowski, K. Gilbert and W. Borden, **J. Org. Chem.**, (1980), 45, 346.

2. R. Antonioletti, M. D'Auria, A. DeMico, G. Piancatelli and A. Scettri, **Tetrahedron** , (1983), 1765.

3. M. D'Auria, A. DeMico, F. D'Onofrio and A. Scettri, **Synthesis**, (1985), 988.

4. A.V. Rama Rao, E. Rajarathnam Reddy, G.V.M. Shama, P. Yadagiro and J.S. Yadav, **Tetrahedron Lett.**, (1985), 465.

5. B. Maurer and A. Hauser, **Tetrahedron Lett.**, (1984), 1061.

EXAMPLES:

$-CH_2OH$, HO — PDC, CH_2Cl_2, 2 hrs → CH_2, O, OH — PDC, 16 hrs → =O, O (75%) ①

OAc — PDC, 4 Å Molecular sieves, Dry CH_2Cl_2, N_2 → OAc, O (65%) ②

C_8H_{17}, HO — PDC / DMF, MeOH, 15 hrs, 70° → O, O (75%) ③ [1]

PhCO(O)$(CH_2)_7CH_2OH$ — 1.) PDC, CH_2Cl_2 2.) $AgNO_3$, KOH, H_2O → PhCO(O)$(CH_2)_7CO_2H$ (70%) ④

HO, O — PDC, CH_2Cl_2, 25° → O, O (75%) ⑤

REAGENT: PYRIDINIUM HYDROBROMIDE PERBROMIDE ⚠1

STRUCTURE:

MW = 319.87
MP = 132- 34° (dec)

$C_5H_5N^{+}H \; Br_3^{-}$

<u>Fieser's Reagents for Organic Synthesis</u>: **Vol., pages**

1	967- 70	**4**		**7**		**10**	
2		**5**	568	**8**		**11**	
3		**6**	499- 500	**9**	399	**12**	

March's <u>**Advanced Organic Chemistry**</u>: 728

MAJOR USES:

1. A useful brominating agent
2. Dehydrogenations

PREPARATION:

1. This reagent is prepared by mixing pyridine with HBr. After cooling, bromine is added. The product is collected and crystallized from HOAc (See: **Reagents**, Vol. 1, 967).

2. Commercially available.

PRECAUTION:

1. Corrosive, lachrymator

NOTES:

1. Also known as Pyridinium bromide perbromide

REFERENCES:

1. N. Confalone, G. Pizzolato, E.G. Baggiolini, D. Lollar and M.R. Uskokovic, **J. Am. Chem. Soc.**, (1975), <u>97</u>, 5936.

2. U. Husstedt and H.J. Schafer, **Synthesis**, (1979), 964.

3. U. Husstedt and H.J. Schafer, **Synthesis**, (1979), 966.

4. J.M. Riordan and C.H. Slammer, **J. Org. Chem.**, (1974), <u>39</u>, 654.

5. W.L. Meyer, G.B. Clemans and R.A. Manning, **J. Org. Chem.**, (1975), <u>40</u>, 3686.

EXAMPLES:

$C_5H_5NHBr_3$, MeOH (60%) ①

$PyH^+Br_3^-$, $CHCl_3$, 25°, 4 hrs (99%) ②

$C_5H_5N^+HBr_3^-$, THF, -60°, 1-3 hrs (75%, 99% selective) ③

$Py{\cdot}HBr_3$, Acetic anhydride, Pyridine, 80°, then to RT (62%) ④

PyH^+Br^-, HOAc, 20°, ½ hr (93%) ⑤

REAGENT: QUININE

STRUCTURE:

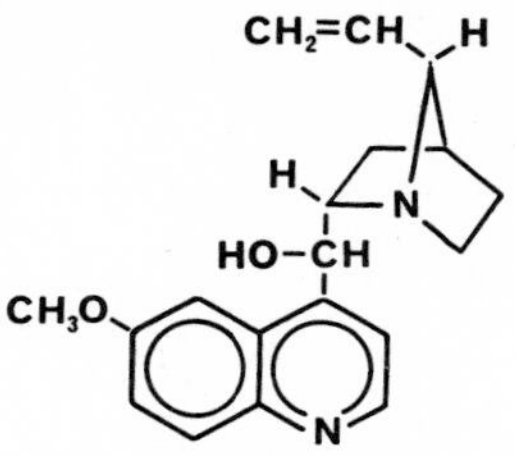

MW = 324.44
MP = 173- 75°C

$CH_2{=}CH$, H, H, N, HO–CH, CH_3O, N

Fieser's Reagents for Organic Synthesis: Vol., pages

1		**4**		**7**	311	**10**	338
2		**5**		**8**	430 -31	**11**	456
3		**6**	501	**9**	403	**12**	

March's Advanced Organic Chemistry: Not Indexed

MAJOR USES:

1. A chiral auxiliary for asymmetric reactions

PREPARATION:

1. Commercially available

2. A copolymer can be made by copolymerization of cinchona alkaloids with acrylonitrile, using **AIBN** as initiator (Ref. 1).

PRECAUTION:

NOTES:

REFERENCES:

1. N. Kobayashi and K. Iwai, **J. Am. Chem. Soc.**, (1978), 100, 7071.

2. H. Plum and H. Wynberg, **Tetrahedron Lett.**, (1979), 1251.

3. M. Shibasaki, A. Nishida and S. Ikegami, **Chem. Commun.**, (1982), 1324.

4. A.A. Smaardijk, S. Noorda, F. van Bolhuis and H. Wynberg, **Tetrahedron Lett.**, (1985), 493.

5. N. Kobayashi and K. Iwai, **J. Org. Chem.**, (1981), 46, 1823.

EXAMPLES:

① △1

Indanone-2-carboxylate (H, $COOCH_3$) + $CH_2=CHCOCH_3$ —Quinine copolymer→ product with $CH_2CH_2COCH_3$, $COOCH_3$ (98%, 30%ee)

② △1

CH_3-substituted benzeneselenol (SeH) + cyclohexenone —CH_3-Ph, Cinchonidine, 20°→ 3-(Se-4MePh)cyclohexanone (Excess of 95%)

③ △1

4-(SO_2Ph)azetidin-2-one (NH) —PhSH, C_6H_6, Cinchonidine, 35°→ 4-(SPh)azetidin-2-one (96%, 54%ee)

④

2-Chlorobenzaldehyde (Cl, CHO) + H-P(=O)(OMe)-OMe —Quinine, Toluene→ 2-Cl-C_6H_4-CH(OH)-P(=O)(OMe)-OMe (NYA)

⑤

$PhCH=CHNO_2$ + $HSCH_2CO_2H$ —Quinine, Toluene, N_2, RT→ $PhCH(SCH_2CO_2H)CH_2NO_2$ (86%, 58%ee)

REAGENT: RANEY NICKEL

STRUCTURE:

Ni_2H

Fieser's Reagents for Organic Synthesis: Vol., pages

1	723- 31	**4**		**7**	312	**10**	339- 340
2	293- 94	**5**	570- 71	**8**	433	**11**	457- 58
3		**6**	502	**9**	405- 6	**12**	422

March's Advanced Organic Chemistry: 509, 511, 652, 692, 795, 1097, 1100.

MAJOR USES: ⚠1

1. Reducing agent
2. Catalytic reduction
3. Dehydrogenation
4. Desulfurization
5. Decarbonylation
6. Dehalogenation

PREPARATION:

1. A variety of procedures are available for the preparation. See **Reagents,** Vol.1, 723. Commercially available as a 50% slurry in water at pH 10.

PRECAUTION:

1. Flammable. Ignites on contact with air. A suspected carcinogen.

NOTES:

1. The activated reagent slowly loses activity during storage.

2. **MOM** = Methoxymethyl

REFERENCES:

1. J. Forsek, **Tetrahedron Lett.**, (1980), 1071.

2. a. D. Curran, **J. Am. Chem. Soc.**, (1982), 104, 4024.

   b. A. Kozikowski and M. Adamczyk, **Tetrahedron Lett.**, (1982), 3123.

3. M.T. Reetz ad H. Muller-Starke, **Tetrahedron Lett.**, (1984), 3301.

4. C. Iwata, H. Kubota, M. Yamada, Y. Takemoto, S. Uchida, T. Tanaka and T. Imanishi, **Tetrahedron Lett.**, (1984), 3339.

5. T.H. Chan and C.V.C. Prasad, **J. Org. Chem.**, (1986), 51, 3012.

EXAMPLES:

Raney Ni

(97%)

① [1]

H_2, Ni, MeOH, H_2O

(88%)

② [1]

Raney Ni

(71%)

③ [4]

H_2, Raney Ni

EtOH, RT

(97%)

④ [1] △2

Raney Ni

Absolute EtOH

(82%)

⑤ [4]

REAGENT: RHODIUM (II) ACETATE

==

STRUCTURE:

MW = 441.99

$[(CH_3CO_2)_2Rh]_2$

Exists as the dimer.

Fieser's Reagents for Organic Synthesis: Vol., pages

1		**4**		**7**	571- 72	**10**	340- 42
2		**5**	571- 72	**8**	434- 35	**11**	458- 60
3		**6**		**9**	406- 08	**12**	423- 26

March's Advanced Organic Chemistry: Not indexed

MAJOR USES:

1. Catalyst for reactions of diazo compounds. 2. Cyclizations

---

PREPARATION:

1. For preparations of a variety of rhodium reagents, see: P. Legzdins, R.W. Mitchel, G.L. Rempel, J.D. Ruddick and G. Wilkinson, **J. Chem. Soc.**, A, (1970), 3322.

2. Commercially available

PRECAUTION:

NOTES:

REFERENCES:

1. P. Dowd, P. Garner, R. Shappert, H. Irngartinger and A. Goldman, **J. Org. Chem.**, (1982), 47, 4240, and earlier papers cited.

2. D.F. Taber and E.H. Petty, **J. Org. Chem.**, (1982), 47, 4808.

3. N. Ikota, N. Takamura, S.D. Young and B. Ganem, **Tetrahedron Lett.**, (1981), 4163.

4. M.G. Martin and B. Ganem, **Tetrahedron Lett.**, (1984), 251.

5. P.G.M. Wuts, M.L. Obrzut and P.A. Thompson, **Tetrahedron Lett.**, (1984), 4051.

EXAMPLES:

$AcOCH_2C{\equiv}CCH_2OAc$ + $N_2CHCO_2CMe_3$ —$Rh_2(OAc)_4$→ AcO, OAc, CO_2CMe_3 (60%) ① [1]

—$Rh_2(OAc)_4$, CH_2Cl_2→ O, CO_2Me, $CH{=}CH_2$ (48%) ② [1]

$Me_2CHCH_2C(=N_2)CO_2Me$ —$Rh_2(OAc)_4$→ Me_2C, CO_2Me, C=C, H, H (99%) ③ [1]

—$Rh_2(OAc)_4$ / Dimethyl diazomalonate, Toluene→ (80%) ④ [1]

$PhCH_2O$, HO —$Rh_2(OAc)_4$, PPh_3, 24 atm, 100°, 6 hrs→ [$PhCH_2O$, O, HO] —PCC→ $PhCH_2O$, O, O (86%) ⑤ [2]

REAGENT: RUTHENIUM TETROXIDE

==

STRUCTURE:

MW = 165.07
MP = 25.4°C
d = 3.29

RuO_4

Fieser's Reagents for Organic Synthesis: Vol., pages

1	986- 89	**4**	420- 21	**7**	315	**10**	343
2	357- 59	**5**		**8**	438	**11**	462-63
3	243 44	**6**	504- 6	**9**		**12**	

March's Advanced Organic Chemistry: 1051, 1057, 1071, 1080, 1088

MAJOR USES: 1

An oxidizing agent, similar in strength to OsO_4, but not as easy to work with.

PREPARATION:

The reagent can be prepared by oxidation of $RuCl_3$: see L. Berkowitz and P. Rylander, **J. Am. Chem. Soc.**, (1958), 80, 6682; F. Dean and J. Knight, **J. Chem. Soc.**, (1962), 4745. An in situ preparation from RuO_2 and $NaIO_4$ is available: H. Veale, J. Levin and D. Swern, **Tetrahedron Lett.**, (1978), 503. The reagent is also commercially available.

PRECAUTION:

The reagent is irritating to the eyes and respiratory system.

NOTES:

See **Reagents**, 1, 956; and 11, 462 for limitations on solvent use.

REFERENCES:

1. P. Carlsen, T. Katsuki, V. Martin and K.B. Sharpless, **J. Org. Chem.**, (1981), 46, 3936.

2. P.F. Shuda, J.L. Phillips and T.M. Morgan, **J. Org. Chem.**, (1986),51, 2742.

3. C. Cainelli, M. Contento, D. Giacomini and M. Panunzio, **Tetrahedron Lett.**, (1985), 937.

4. S. Torii, T. Inokuchi and K. Konco, **J. Org. Chem.**, (1985), 50, 4980.

5. S. Torii, T. Inokuchi and T. Sugivea,**J. Org. Chem.**, (1986), 51, 155.

EXAMPLES:

1.) $NaIO_4$, $RuCl_2\cdot(H_2)_n$, CCl_4, MeCN
2.) CH_2N_2

① (76%)

$RuO_2\cdot H_2O$ / H_2O / $NaIO_4$
CH_3CN, CCl_4

② (98%)

RuO_2, $NaIO_4$
Acetone, H_2O

③ (40%)

RuO_4, $NaIO_4$

(16%) + (80%) ④

Electrooxidation / $RuO_2\cdot 2H_2O$
CCl_4

⑤ (79%)

REAGENT: SELENIUM DIOXIDE (SELENIUM OXIDE)

==

STRUCTURE:

MW = 110.96
MP = 315°C

SeO_2

Fieser's Reagents for Organic Synthesis: **Vol., pages**

1	992- 1000	**4**	422- 24	**7**		**10**	345- 46
2	360- 62	**5**	575- 76	**8**	439- 40	**11**	
3	245- 47	**6**	509- 10	**9**	409- 10	**12**	

March's Advanced Organic Chemistry: 627, 930, 1053, 1077, 1088.

MAJOR USES: △1

1. Oxidizing agent

PREPARATION:

1. Commercially available

PRECAUTION:

1. Corrosive and highly toxic.

NOTES:

1. Use of this reagent often leaves a red colloidal form of Se. This can be removed with DMF. See: S. Milstein and E. Coats, **Aldrichemica Acta**, (1978), 11, 10.

2. By this procedure the build-up of colloidal Se is avoided.

REFERENCES:

1. H. Nagaoka, G. Schmid, H. Ito and Y. Kishi, **Tetrahedron Lett.**, (1981), 899.

2. R.W. Curley, Jr. and C.J. Ticoras, **J. Org. Chem.**, (1986), 51, 256.

3. T. Mandai, M. Takeshita, M. Kawada and J. Otera, **Chem. Lett.**, (1984), 1259.

4. F. Ogura, T. Obubo, K. Ariyoshi and H. Yamaguchi, **Chem. Lett.**, (1983), 1833.

5. N. Miyoshi, T. Yamamoto, BN. Kambe, S. Murai and N. Sonoda, **Tetrahedron Lett.**, (1982), 4813.

EXAMPLES:

Me, OMe, MeO, COEt substituted tetralone $\xrightarrow[70°]{SeO_2,\ HOAc}$ Me, OMe, MeO, COEt substituted 1,2-naphthoquinone (53%) ①

$Me_2C{=}CH{-}COOEt$ $\xrightarrow[\text{Reflux}]{SeO_2,\ EtOH}$ $(HOCH_2)(Me)C{=}CH{-}COOEt$ (90%) ②

$CH_2{=}C(OMe)-CH(Me)(SPh)$ + $C_6H_{13}Br$ $\xrightarrow[\text{HMPA, THF}]{\text{n-BuLi}}$ $CH_2{=}C(OMe)-C(Me)(SPh)-C_6H_{13}$ $\xrightarrow[H_2O_2,\ \text{RT, 3 hrs}]{SeO_2}$ $CH_2{=}C(Me)-C(=O)-C_6H_{13}$ (84%) ③

$PhCH_2OH$ $\xrightarrow[SeO_2\ \text{(catalyst), Dioxane, } N_2\text{, 1 day}]{\text{Excess Bis(para-methoxyphenyl)selenoxide}}$ $PhCHO$ (92%) ④ △2

Cyclohexanone + PhSeSePh $\xrightarrow[H_2SO_4\ \text{(cat), 10°, 10 hrs}]{SeO_2\ /\ CH_2Cl_2}$ 2-(phenylseleno)cyclohexanone (SePh) (84%) ⑤

REAGENT: SILICA/ SILICA GEL (SILICON DIOXIDE / SILICIC ACID)

STRUCTURE:

	SILICA	SILICA GEL
	SiO_2	$H_2SiO_3 \cdot n\,H_2O$
MW =	60.08	78.10

Fieser's Reagents for Organic Synthesis: Vol., pages

1		**4**		**7**		**10**	346- 47
2		**5**		**8**		**11**	466
3		**6**	510	**9**	410	**12**	431- 32

March's Advanced Organic Chemistry: 331, 787, 988

MAJOR USES:

1. Chromatography 2. Catalysts (acid or base)

PREPARATION:

1. Silica gel is precipitated silica. The material is ground to varying mesh sizes.

2. Commercially available in a variety of mesh sizes and activities.

PRECAUTION:

1. Prolonged inhalation can lead to silicosis (fibrosis of the lung).

NOTES:

REFERENCES:

1. P. Bartlett, A. Blakeney, M. Kimura and W. Watson, **J. Am. Chem. Soc.**, (1980), 102, 1383.

2. S. Tamagaki, K. Suzuki and W. Tagaki, **Chem. Lett.**, (1982), 1237.

3. K.S. Kim, Y.H. Song, B.H. Lee, and C.S. Hahn, **J. Org. Chem.**, (1986), 51, 404.

4. T. Mandai, T. Moriyama, Y. Nakayama, K. Sugino, M. Kawada, and J. Otera, **Tetrahedron Lett.**, (1984), 5913.

5. S. Tsuboi, H. Fujita, K. Muranaka, K. Seko and A. Takeda, **Chem. Lett.**, (1982), 1909.

EXAMPLES:

SiO_2

(84%, 3:2) ①

$SiO_2(H_2O)$

(95%) ②

$FeCl_3$-SiO_2

$CHCl_3$, CH_3COCH_3

(85%) ③

Silica Gel, Benzene

Reflux, 12-24 hrs

(755) ④

2-6 eq Silica Gel

Xylene, Reflux, 12 hrs

(79%) ⑤

REAGENT: SILVER (I) OXIDE

STRUCTURE:

MW = 231.74
MP = ca 200° C
d = 7.143

Ag_2O

Fieser's Reagents for Organic Synthesis: Vol., pages

1	1011	**4**	430- 1	**7**	321- 22	**10**	350- 52
2	368	**5**	583- 85	**8**	442- 43	**11**	468- 69
3	252- 54	**6**	515- 18	**9**		**12**	

March's Advanced Organic Chemistry: 629, 906, 974, 1061, 1079, 1091.

MAJOR USES:

1. Oxidizing agent

PREPARATION:

Commercially available.

PRECAUTION:

1. The reagent is sensitive to light.

NOTES:

REFERENCES:

1. G.A. Kraus and M.J. Tashner, **J. Org. Chem.**, (1980), 45, 1174, 1175.

2. A.B. Pepperman, **J. Org. Chem.**, (1981), 46, 5039.

3. H. Kikughi, K. Kogure and M. Toyoda, **Chem. Lett.**, (1984), 341.

4. J.A. Marshall and P.G.M. Wuts, **J. Org. Chem.**, (1977), 42, 1794.

5. H. Ulrich and D.V. Rao, **J. Org. Chem.**, (1977), 42, 3444.

EXAMPLES:

HO, CO_2Me, OH + CH_2, Me, $CH(CH_2)_2OBzl$ —Ag_2O, C_6H_6→ O, Me, $(CH_2)_2OBzl$, CO_2Me, O (95%) ①

CHO —Ag_2O, NaOH, H_2O→ COOH, O (72%) ②

OMe, CH=CHMe —Ag_2O, I_2 / 5:1 Dioxane / H_2O, RT, 4 hrs→ OMe, MeCHCHO (95%) ③

$CH(OAc)_2$ —Ag_2O / MeOH / NaOH, RT, 3 hrs→ COOH (82%) ④

OH, O, O, OH —Ag_2O / Na_2SO_4 / Ethyl acetate→ O, O, O, O (60%) ⑤

REAGENT: SILVER (II) OXIDE

STRUCTURE:

MW = 123.87
MP = > 100° (decomp)
d = 7.483

AgO

Fieser's Reagents for Organic Synthesis: Vol., pages

1		**4**	430- 1	**7**	322	**10**	352- 54
2	369	**5**		**8**		**11**	469
3		**6**	518	**9**	412- 3	**12**	

March's Advanced Organic Chemistry: 1074, 1083

MAJOR USES:

Oxidizing agent

PREPARATION:

This reagent is generally prepared by the oxidation of silver nitrate in alkaline solution with potassium peroxydisulfate. See: R.N. Hammer and J. Kleinberg, **Inorganic Synthesis**, (1953), 4, 12. The reagent is also commercially available.

PRECAUTION:

The reagent is an oxidizing agent and an irritant.

NOTES:

REFERENCES:

1. C. Escobar, F. Farina, R. Martinez-Utrilla, and M.C. Paredes, **J. Chem. Res.**, (1980), 156.

2. G.A. Kraus and K. Neuenschwander, **Synth. Commun.**, (1980), 9.

3. A.S. Kence, M.-P. Gesson and T.P. Demuth, **Tetrahedron Lett.**, (1981), 1667.

4. M.P. Sibi, J.W. Dankwardt and V. Snieckus, **J. Org. Chem.**, (1986), 51, 271.

5. I.H. Sanchez, S. Mendoza, M. Calderon, M.I. Larraza and H.J. Flores, **J. Org. Chem.**, (1985), 50, 5077.

EXAMPLES:

MeO OH

MeO OAc

AgO, H^+

MeO O

MeO O (70%) ①

CH_2=CH(CH_2)$_3$ OCH_3

AgO, Dioxane

O OH

CH_2=CH(CH_2)$_3$ O (92%) ②

OMe O

MeO OMe

AgO, $MeCO_2Me$, HNO_3

O

MeO O (NYA) ③

MeO OH

MeO

AgO, HNO_3

MeO O

MeO O (31%) ④

O O OMe

AgO, HNO_3, Dioxane

OH

O O (56%) ⑤

REAGENT: SILVER PERCHLORATE

STRUCTURE: ⚠1

$AgClO_4$

MW = 207.32
MP = 486°C (Dec)
d = 2.806

Fieser's Reagents for Organic Synthesis: Vol., pages

1		**4**	432- 35	**7**	322-23	**10**	354- 55
2	369- 7	**5**	585- 87	**8**		**11**	469- 70
3		**6**	518- 19	**9**	413- 14	**12**	

March's Advanced Organic Chemistry: Not Indexed

MAJOR USES:

1. Rearrangements   2. Lewis acid   3. Lactonizations

PREPARATION:

1. Commercially available as the hydrate. Can be prepared by the reaction of perchloric acid on silver carbonate.

PRECAUTION:

1. Oxidizing agent. Irritates skin and mucous membranes.

NOTES:

1. The monohydrate decomposes at 43°.

REFERENCES:

1. A. Padwa, T.J. Blacklock and R. Loza, **J. Am. Chem. Soc.**, (1981), 103, 2402.

2. J.S. Nimitz and R.H. Wollenberg, **Tetrahedron Lett.**, (1978), 3523.

3. T. Wakamatsu, K. Hashimoto, M. Ogura and Y. Ban, **Synth. Commun.**, (1978), 8, 319.

4. R.T. Reddy and U.R. Nayak, **Synth. Commun.**, (1986), 16, 973.

5. H. Saimoto, T. Hiyama and H. Nozaki, **J. Am. Chem. Soc.**, (1981), 103, 4975.

EXAMPLES:

① [2]

Ph(CH₃)-cyclopropene with gem-dimethyl → $AgClO_4$, MeOH → $PhC(=O)C(CH_3)=C(CH_3)_2$ (55%) + $PhCH=C(CH_3)C(CH_3)_2OCH_3$ (45%)

② [3]

Me / NH_2 pyrimidine–$SC(=O)(CH_2)_{13}OH$ → $AgClO_4$, Benzene, 80° → macrolactone (91%)

③ [2]

2-hydroxy-2-[$(CH_2)_6CO_2CH_3$]cyclopentanone → $AgClO_4$, C_6H_6, 23° → 2-[$=CH(CH_2)_5CO_2CH_3$]cyclopentanone + 2-[$(CH_2)_6CO_2CH_3$]cyclopent-2-enone (66%, 1:2)

④ [2]

Br / H, Br / H → $AgClO_4$-50% aq acetone → Br / H, HO / H (64%)

⑤ [2]

Ph / HO $CHC\equiv CCMe_2OH$ → $AgClO_4$, Pyridine / Acetic anhydride → Ph, O, OAc, Me, Me dihydrofuran (84%)

REAGENT: SILVER TETRAFLUOROBORATE

STRUCTURE:

MW = 194.68
MP = 200°C (dec)

$AgBF_4$

Fieser's Reagents for Organic Synthesis: Vol., pages

1	1015- 18	**4**	428- 29	**7**		**10**	
2	365- 66	**5**	587- 88	**8**	443- 44	**11**	471
3	250- 51	**6**	519- 20	**9**	414	**12**	434

March's Advanced Organic Chemistry: Not Indexed

MAJOR USES: ⚠1

1. Catalyst for reactions displacing halogens 2. Method for removing **TMS** groups. 3. Lewis acid

PREPARATION:

1. The reagent can be prepared from the reaction of BF_3 and AgF in benzene or nitromethane [K. Heyns and H. Paulsen, **Angew Chem.**, (1960), 72, 349.]

2. See **Reagents**, 1, 1015 for comments on an explosion during the preparation.

3. Commercially available.

PRECAUTION:

1. Corrosive and moisture sensitive.

NOTES:

1. For a review of silver salts in organic synthesis, see: **Aldrichimica Acta,** (1981), 14, 63.

REFERENCES:

1. A.J. Fry and Y. Migron, **Tetrahedron Lett.**, (1979), 3357.

2. J. Grimaldi and A. Cormoms, **Tetrahedron Lett.**, (1985), 825.

3. P.C. Ting and P.A. Bartlett, **J. Am. Chem. Soc.**, (1984), 106, 2668.

4. S. Shatzmiller, E. Shalom and E. Bahar, **Chem. Commun.**, (1984), 1522.

5. Y. Kawanami, T. Katsuki and M. Yamaguchi, **Tetrahedron Lett.**, (1983), 5131.

EXAMPLES:

$CH_3C(=O)C(Br)(CH_3)_2$ —[$AgBF_4$]→ $CH_3C(=O)C(F)(CH_3)_2$ (87%) ① [1]

$H(Me)C{=}C{=}C(H)C(Me)_2CH{=}NOH$ —[$AgBF_4$ / $CHCl_3$]→ Me, Me, Me substituted O–N seven-membered ring (70%) ② [3]

Br-substituted 2,2-dimethyltetrahydropyran with *i*-Pr —[$AgBF_4$ / aq Acetone, 21°, 2 hrs]→ $Me_2C(OH)$-substituted tetrahydrofuran with *i*-Pr (88%) ③ [1]

Cyclohexene + $MeON{=}C(Me){-}CH_2Br$ —[1.) $AgBF_4$, Cl_2CHCH_3 25°, 18 hrs; 2.) 10% KCN]→ N-OMe octahydroindole with CN, Me (76%) ④ [1]

$Me_2CHC(=O)SEt$ + $Me_3SiC{\equiv}CC_6H_{13}$ —[Anhydrous $AgBF_4$ / CH_2Cl_2; N_2, RT]→ $Me_2CHC(=O)C{\equiv}CC_6H_{13}$ (83%) ⑤ [2]

REAGENT: SINGLET OXYGEN

==

STRUCTURE: △1

MW =
BP =

$^{1}O_2$

Fieser's Reagents for Organic Synthesis: Vol., pages

1		**4**	362- 63	**7**	261- 69	**10**	294- 95
2		**5**	486- 91	**8**	367- 74	**11**	385- 87
3		**6**	431- 36	**9**	338- 41	**12**	363- 65

March's Advanced Organic Chemistry: 634, 738, 787

MAJOR USES: △2

1. Oxidations, photooxidations 2. Hydroperoxidations 3.Endoperoxidations

---

PREPARATION:

1. Photosensitization of ground state oxygen

2. Hypochlorite oxidation of hydrogen peroxide

3. Reaction of ozone and triphenylphosphite

PRECAUTION:

NOTES:

1. Singlet oxygen is simply oxygen in the singlet state , rather than the ground state triplet.

2. For a review see: H.H. Wasserman and J. Ives, **Tetrahedron**, (1981), 1825.

REFERENCES:

1. H.H. Wasserman and J. Pickett, **J. Am. Chem. Soc.**, (1982), 104, 4695.

2. H. Ensley, R. Carr, R. Martin and T. Pierce, **J. Am. Chem. Soc.**, (1980), 102, 2836.

3. T. Akasaka, R. Sato, Y. Miyama and W. Ando, **Tetrahedron Lett.**, (1985), 843.

4. C.W. Jefford and G. Barghietto, **Tetrahedron Lett.**, (1977), 4531.

5. M.L. Graziano, M.R. Iesce, A. Carotenuto and R. Scarpati, **Synthesis**, (1977), 572.

EXAMPLES:

OH / CH_3 (p-cresol) $\xrightarrow[(n\text{-}Bu)_4NF]{^1O_2}$ O / H_3C OOH (87%) ①

O / CMe_2 / Me $\xrightarrow{^1O_2}$ HO / O–O / Me / Me / Me (75%) ②

=NN=PPh_3 $\xrightarrow[CH_2Cl_2,\ \text{Methyl blue}]{^1O_2}$ =O + $Ph_3P=O$ (68%) ③

$Me_2CH-C(=PPh_3)-CO_2Me$ $\xrightarrow[\text{meso-tetraphenylporphine (as a sensitizer)}]{^1O_2\ /\ C_6H_6}$ $Me_2CHC(=O)CO_2Me$ (98%) ④

Me / N / EtO / O / Ph $\xrightarrow[CHCl_3,\ \text{Methylene blue}]{^1O_2\ /\ \text{Triethylenediamine}}$ Me / EtO_2C / N / O / O / Ph (98%) ⑤

REAGENT: SODIUM AMALGAM

STRUCTURE:

$$Na(Hg) \text{ or } Na_nHg_n$$

The "structure" and thus the molecular weight depends on the sodium content of the amalgam.

Fieser's Reagents for Organic Synthesis: Vol., pages

1	1030- 33	**4**		**7**	326- 27	**10**	
2	373	**5**		**8**		**11**	473- 75
3	259	**6**		**9**	416- 17	**12**	439

March's Advanced Organic Chemistry: 394, 926, 1110

MAJOR USES:

1. Reducing agent
2. Desulfurization

PREPARATION:

1. Generally prepared by the addition of mercury to sodium under an inert atmosphere (See: **Reagents**, 1, 1030)

PRECAUTION:

1. The sodium amalgam is decomposed by water and should be kept well sealed.

NOTES:

REFERENCES:

1. O. DeLucchi and G. Modena, **Chem. Commun.**, (1982), 914.

2. H.-J. Bestmann, K. Kumar and W. Schaper, **Angew. Chem. Int. Ed.**, (1983), 22, 167.

3. Y. Ikegami, S. Kubota and H. Watanabe, **Bull. Chem. Soc. Japan**, (1979), 52, 1563.

4. P.J. Kocienski and J. Tideswell, **Synth. Commun.**, (1979), 411.

5. P.A. Bartlett, F.R. Green (III) and E.H. Rose, **J. Am. Chem. Soc.**, (1978), 100, 4852, 4858.

EXAMPLES:

NaH$_2$PO$_4$, Na(Hg), MeOH
RT, 8 hrs
(69%) ① [2]

2% Na(Hg)
THF, 5°, 17 hrs
MeC≡CPh (80%) ② [2]

3% Na(Hg) / Acetonitrile
Dark, Ice bath, 1-2 hrs
(85%) ③ [1]

5.6% Na(Hg)
MeOH / THF
(76%) ④ [2]

Na(Hg)
THF / DMSO
0°
(77%) ⑤ [2]

REAGENT: SODIUM BIS(2-METHOXYETHOXY)ALUMINUM HYDRIDE (**SMEAH**)

STRUCTURE: 1

$[(CH_3OCH_2CH_2O)_2AlH_2]Na$

MW = 202.16

Fieser's Reagents for Organic Synthesis: Vol., pages

1		**4**	441-2	**7**	327-29	**10**	357
2		**5**	596	**8**	448-9	**11**	476-77
3	260-61	**6**	528-29	**9**	418-20	**12**	440

March's Advanced Organic Chemistry: 1102

MAJOR USES: 2 3

1. A versatile reducing agent
2. Alkylations
3. Dehydrogenation
4. Reductive methylation
5. Reductive cleavage

PREPARATION:

For preparative procedures, see: **Reagents**, 3, 260. The reagent is also commercially available in toluene under the name, "**Red-Al™**".

PRECAUTION:
Corrosive and a skin irritant.

NOTES:

1. Known as **Red-Al** and **Vitride**

2. For Reviews, see: J. Malek and M. Cerny, **Synthesis**, (1972), 217; J. Vit, **Eastman Organic Chemical Bulletin**, (1970), 42, 3.

3. For use as a modified reagent, see: R. Kanazawa and T. Tokoroyama, **Synthesis**, (1976), 526.

REFERENCES:

1. N. Cohen, R.J. Lopresti and G. Saucy, **J. Am. Chem. Soc.**, (1979), 101, 6710.

2. E.J. Corey and J.G. Smith, **J. Am. Chem. Soc.**, (1979), 101, 1038.

3. A.M. Mubarak and D.M. Brown, **Tetrahedron Lett.**, (1980), 2455.

4. C. Iwata, H. Kuboto, M. Yamada, Y. Takemoto, S. Uchida, T. Tanaka and T. Imanishi, **Tetrahedron Lett.**, (1984), 3339.

5. A.I. Meyers, B.A. Lefker, K.T. Wanner and R.A. Aitken, **J. Org. Chem.**, (1986), 51, 1936.

EXAMPLES:

①
SMEAH
Xylene, 140°
(83%)

②
1.) NaH
2.) SMEAH
(60%)

③
SMEAH
(Quant., 12:1)

④
SMEAH
Ether, -78°
(85%)

⑤
1.) LDA-MeI
2.) Red-Al
3.) H^+
(47%)

REAGENT: SODIUM BOROHYDRIDE

STRUCTURE:

$NaBH_4$

MW = 37.83
MP = 400°C

Fieser's Reagents for Organic Synthesis: Vol., pages

1	1049- 55	**4**	443- 44	**7**	329- 31	**10**	357- 59
2	377- 78	**5**	597- 601	**8**	449- 51	**11**	477- 79
3	262- 64	**6**	530- 39	**9**	420 -21	**12**	441- 44

March's Advanced Organic Chemistry: See Index for extensive listing

MAJOR USES: 1

1. Reducing agent

PREPARATION:

1. Commercially available in high purity.

PRECAUTION:

1. Flammable

NOTES:

1. This reducing agent is less reactive than lithium aluminum hydride, and unlike the latter (which must have reactions carried out in aprotic solvents) $NaBH_4$ reductions are carried out in protic solvents.

REFERENCES:

1. H. Wyss, M. Vogeli and R. Sheffold, **Helv.**, (1981), 64, 775.

2. G. Gribble, W. Kelly and M. Sibi, **Synthesis**, (1982), 143.

3. R. Walchli and M. Hess, **Helv.**, (1980), 65, 2299.

4. S. Raucher and P. Klein, **J. Org. Chem.**, (1986), 51, 123.

5. I. Murata, **J. Org. Chem.**, (1986), 51, 251.

EXAMPLES:

$NaBH_4$-Al_2O_3, Ether

(53%, 11:1)

①

$NaBH_4$, CF_3COOH

(92%)

②

1.) NH_3

2.) $NaBH_4$, HOAc

(41%)

③

$NaBH_4$

(100%)

④

$NaBH_4$

(85%)

⑤

REAGENT: SODIUM CYANOBOROHYDRIDE ⚠1

==

STRUCTURE:

MW = 62.84 $NaBH_3CN$
MP = >242°C (dec)

Fieser's Reagents for Organic Synthesis: Vol., pages

1	**4** 448- 51	**7** 334- 35	**10** 360- 61
2	**5** 607- 09	**8** 454	**11** 481- 83
3	**6** 537- 38	**9** 424- 26	**12** 445- 46

March's Advanced Organic Chemistry: 391, 394, 799, 815, 1097, 1098

MAJOR USES:

1. A selective reducing agent

PREPARATION:

1. Commercially available.

PRECAUTION:

1. Very flammable and highly toxic. Use with caution.

NOTES:

1. For a review on this reagent, see: R. Hutchins and N. Natale, **Org. Prep. Proc. Int.**, (1979), 11, 201.

REFERENCES:

1. T. Jones, J. Franko. M. Blum and H. Fales, **Tetrahedron Letters**, (1980), 789.

2. P. Garegg, H. Hultberg and S. Oscarson, **J. Chem. Soc., Perkin I**, (1982), 2395.

3. M. Hanaoka, M. Iwasaki and C. Mukai, **Tetrahedron Letters**, (1985), 917.

4. P.H. Morgan and A.H. Beckett, **Tetrahedron**, (1975), 31, 2595.

5. R.O. Hutchins and D. Kandasamy, **J. Org. Chem.**, (1975), 40, 2530.

EXAMPLES:

①

$CH_3\overset{O}{\overset{||}{C}}CH_2CH_2\overset{O}{\overset{||}{C}}CH_3$ → 1.) $NaCNBH_3$, NH_4OAc, KOH, MeOH; 2.) $NaBH_4$ → (cis- and trans-2,5-dimethylpyrrolidine)

(62%, 1:1)

②

$NaBH_3CN$, HCl / THF

(79%)

③

$NaCNBH_3$, THF / Tosic acid

(81%)

④

MeO, MeO–C_6H_3–$CH_2\overset{Me}{C}H{=}NOH$ → $NaCNBH_3$ / MeOH, HCl → MeO, MeO–C_6H_3–$CH_2\overset{Me}{C}HNHOH$

(90%)

⑤

$NaBH_3CN$ / Acidic HMPT

(57%)

REAGENT: SODIUM DITHIONITE (SODIUM HYDROSULFITE)

STRUCTURE:

$Na_2S_2O_4$

MW = 174.11
MP = > 300° (dec.)

Fieser's Reagents for Organic Synthesis: Vol., pages

1	1081- 83	**4**		**7**	336	**10**	363- 64
2		**5**	615	**8**	456- 58	**11**	485- 86
3		**6**		**9**	426 27	**12**	

March's Advanced Organic Chemistry: 391, 810, 1110

MAJOR USES:

1. A versatile reducing agent 2. Catalyst

PREPARATION:

1. Commercially available

PRECAUTION:

1. Moisture sensitive, flammable.

2. Should be used in solutions of pH > 7 to prevent decomposition.

NOTES:

REFERENCES:

1. K. Boddy, P. Boniface, R. Cambie, P. Graw, D. Larsen, H. McDonald, P. Rutledge and P. Woodgate, **Tetrahedron Lett.**, (1982), 4407.

2. F. Camps, J. Coll and M. Riba, **Chem. Commun.**, (1979), 1080.

3. J.T. Edward and R.H. Sheffler, **J. Org. Chem.**, (1985), 50, 4855.

4. S.-K. Chung and Q.-Y. Hu, **Synth. Commun.**, (1982), 261.

5. F. Camps, J. Coll, A. Guerrero, J. Guitart and M. Riba, **Chem. Letters**, (1982), 715.

EXAMPLES:

① 1,4-bis(allyloxy)anthraquinone → 1,4-dihydroxy-2,3-diallylanthraquinone

Reagents: $Na_2S_2O_4$, NaOH, DMF; H_2O, Heat

OCH₂CH=CH₂ / OCH₂CH=CH₂ → OH, CH₂CH=CH₂, CH₂CH=CH₂, OH (72%)

② 2-methylcyclohexanone → cis- + trans-2-methylcyclohexanol

Reagent: $Na_2S_2O_4$

(76%, 33:67)

③ 2-methylpyridine + pyridine → (product with S⁻, N⁺)

Reagent: $Na_2S_2O_4$

(40%)

④ $C_6H_5COCH_2Br$ → $C_6H_5COCH_3$

Reagents: $Na_2S_2O_4$; H_2O, N_2, RT, 24 hrs

(90%)

⑤ C_5H_{11}CH=CHCH=CHCO_2Et → C_5H_{11}CH₂CH=CHCH₂CO_2Et

Reagents: $Na_2S_2O_4$ / $NaHCO_3$; Pyridine, H_2O, 80°, 1 hr

(75%, E:Z=47:53)

REAGENT: SODIUM HYDRIDE

STRUCTURE:

NaH

MW = 24.00
MP = 800°C (Dec)

Fieser's Reagents for Organic Synthesis: Vol., pages

1	1075- 81	**4**	452- 55	**7**		**10**	
2	382- 83	**5**	610- 14	**8**	458- 59	**11**	
3		**6**	541- 42	**9**	427	**12**	447- 49

March's Advanced Organic Chemistry: 511, 1052, 1062, 1106

MAJOR USES:

1. Base

PREPARATION:

1. Commercially available; often used as an oil dispersion.

PRECAUTION:

1. Moisture sensitive; can react violently with water.

2. Flammable.

NOTES:

1. **MOM** = Methoxymethyl

REFERENCES:

1. H. Liu and H. Lai, **Tetrahedron Letters**, (1979), 1193.

2. C. Iwata, H. Kubota, M. Yamada, Y. Takemoto, S. Uchida, T. Tanaka and T. Imanishi, **Tetrahedron Letters**, (1984), 3339.

3. W. Verboom, H.J. Berga, W.A. Trompenaars and D.N. Reinhoudt, **Tetrahedron Letters**, (1985), 685.

4. M.E. Kuehne and P.J. Seaton, **J. Org. Chem.**, (1985), 50, 4790.

5. S. Knapp, G. Sankar Lal and D. Sahai, **J. Org. Chem.**, (1986), 51, 380.

EXAMPLES:

NaH, DME, EtSH

20°

(91%) ①

1.) NaH, DME, Reflux

2.) SMEAH, DME, Reflux

(50%) ② ⚠1

NaH, Toluene

100-110°

(79%) ③

1.) NaH, DMF, N_2

2.) MeI

(86%) ④

NaH, MeNCS, MeI

(100%) ⑤

REAGENT: SODIUM HYPOCHLORITE

==

STRUCTURE:

MW = 74.44 NaOCl
MP = 18°C

<u>**Fieser's Reagents for Organic Synthesis**</u>: **Vol., pages**

1	1084- 87	**4**	456	**7**	337- 38	**10**	365
2		**5**	617	**8**	461- 63	**11**	487- 88
3		**6**	543	**9**	430	**12**	

March's <u>**Advanced Organic Chemistry**</u>: 574, 785, 934, 1057, 1071, 1083

MAJOR USES:

1. Versatile oxidizing agent 2. Chlorination

PREPARATION:

1. Can be prepared from NaOH and Cl_2 (F. Mallory, **Org. Synth.**, (1963), <u>Coll. Vol. 4</u>, 74. The reagent is commercially available as **Chlorox**, a solution with 5% chlorine minimum.

PRECAUTION:

1. Crystals are decomposed by CO_2.

2. Very explosive in anhydrous state.

3. Avoid prolonged inhalation or contact with skin.

NOTES:

REFERENCES:

1. R. Stevens, K. Chapman, C. Stubbs, W. Tam and K. Albizati, **Tetrahedron Letters**, (1982), 4647.

2. R.D. Larsen and F.E. Roberts, **Synth. Commun.**, (1986), 899.

3. M.P. Cooke, **J. Org. Chem.**, (1983), <u>48</u>, 744.

4. D. Scholz, **Ann.**, (1984), 264.

5. A. Nohara, K. Ukawa and Y. Sanno, **Tetrahedron**, (1974), <u>30</u>, 3563.

EXAMPLES:

$CH_3C(=O)(CH_2)_4CHO$ —[NaOCl, CH_3OH, CH_3COOH]→ $CH_3C(=O)(CH_2)_4COOCH_3$ (90%) ① [1]

6-EtO-benzothiazole-2-SH —[NaOCl / aq NH_3]→ 6-EtO-benzothiazole-2-SNH_2 (90%) ② [2]

$PhCH_2CH_2C(=O)C(CO_2Et)=PPh_3$ —[NaOCl / NaOH, THF]→ $PhCH_2CH_2CO_2H$ (94%) ③

2-($SO_2CH_2C_5H_{11}$)cycloheptanone (=O) —[NaOCl / NaOH, EtOH, 0°]→ $HO_2C(CH_2)_5CCl(SO_2CH_2C_5H_{11})$... (Cl, CO_2H, $SO_2CH_2C_5H_{11}$) (97%) ④ [1]

3-CHO-chromone (4-oxo) —[aq NaOCl, Acetic acid]→ 3-Cl-chromone (86%) ⑤ [1] [2]

REAGENT: SULFUR DIOXIDE (SULFUROUS ACID)

STRUCTURE: ⚠1

MW = 64.06
BP = -10°C

SO_2 or H_2SO_3

Fieser's Reagents for Organic Synthesis: Vol., pages

1	1122	**4**	469	**7**	346	**10**	
2	392	**5**	633	**8**	464- 66	**11**	495
3		**6**	558	**9**	440- 41	**12**	

March's Advanced Organic Chemistry: 551, 637, 648, 987, 939

MAJOR USES:

1. Catalyst for rearrangements 2. Cycloaddition 3. General acid catalyst

PREPARATION:

1. Commercially available as SO_2 or as water solution (H_2SO_3).

PRECAUTION:

1. Corrosive and extremely irritating to eyes and respiratory system.

NOTES:

1. Often used in aqueous solution as sulfurous acid.

REFERENCES:

1. D. DeLucchi and V. Lucchini, **Chem. Commun.**, (1982), 1105.

2. J. Dalling, J. Gall and D. MacNicol, **Tetrahedron Lett.**, (1979), 4789.

3. T. Tanabe and T. Nagai, **Bull. Chem. Soc. Jpn.**, (1978), 51, 1459.

4. D. Masilamani and M.M. Rogic, **J. Am. Chem. Soc.**, (1978), 100, 4634.

5. B. Burczyk and Z. Kortylewicz, **Synthesis**, (1982), 831.

EXAMPLES:

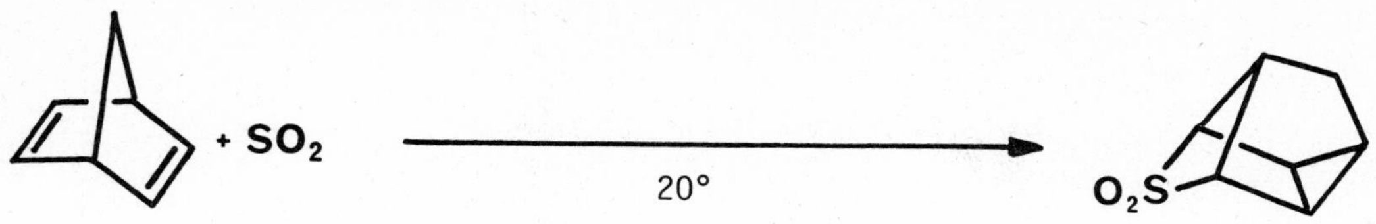

$+ SO_2$ —— 20° ——→ O_2S (99%) ① [2]

SO_2, 150° ——→ SO_2 ② [1] ("moderate") [2]

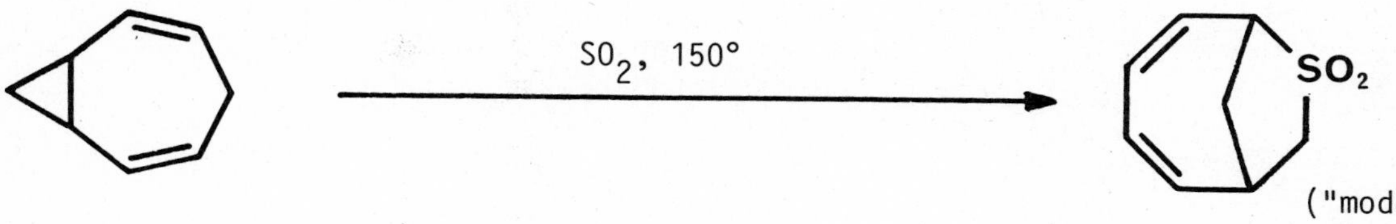

$Ph_2CN_2 + SO_2 + EtOH$ —— 20° 1 hr ——→ Ph_2CHSO_3Et (87%) ③

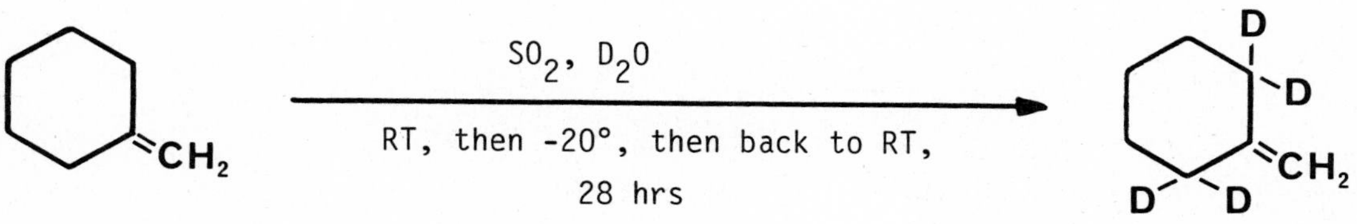

$=CH_2$ —— SO_2, D_2O; RT, then -20°, then back to RT, 28 hrs ——→ D, D, D, D, $=CH_2$ (100%) ④ [3]

$Et(Me)C{=}O$ + n-$C_5H_{11}SH$ —— Gaseous SO_2; C_6H_6 ——→ $Et(Me)C(S$-n-$C_5H_{11})_2$ (69%) ⑤ [3]

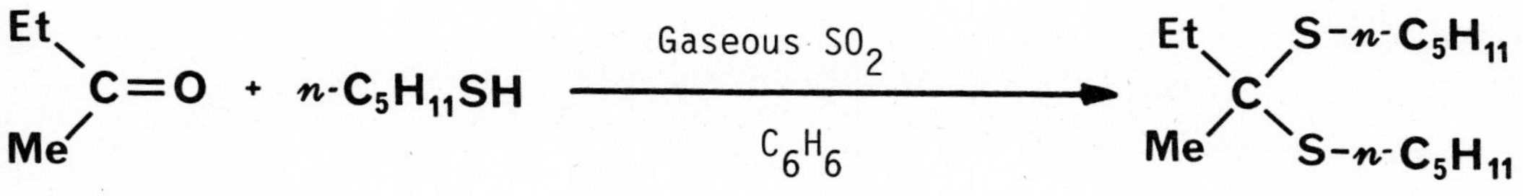

REAGENT: SULFURYL CHLORIDE

STRUCTURE:

MW = 134.97
BP = 68- 70°C
d = 1.680

SO_2Cl_2

<u>Fieser's Reagents for Organic Synthesis</u>: **Vol., pages**

1	1128- 31	**4**	474- 75	**7**	349- 50	**10**	375- 76
2	394- 95	**5**	641	**8**		**11**	
3	276	**6**	561	**9**		**12**	

March's <u>**Advanced Organic Chemistry**</u>: 529, 532, 550, 621, 725.

MAJOR USES:

1. Chlorinating agent

PREPARATION:

1. Can be prepared from sulfur dioxide and chlorine with activated charcoal; however, is commercially available in high purity.

PRECAUTION:

1. A lachrymator.

2. Vapors and liquid corrosive and harmful to mucous membranes.

NOTES:

1. Chlorination followed by thermal dehydrochlorination.

REFERENCES:

1. J. Press, C. Hofmann and S. Safir, **J. Org. Chem.**, (1979), <u>44</u>, 3292.

2. A.L. Schroll and G. Barany, **J. Org. Chem.**, (1986), <u>51</u>, 1866.

3. D.T.W. Chu, **J. Org. Chem.**, (1983), <u>48</u>, 3571.

4. C.C. Fortes, H.C. Fortes and D.C.R.G. Gonsalves, **Chem. Commun.**, (1982), 857.

5. W.H. Gilligan and S.L. Stafford, **Synthesis**, (1979), 600.

EXAMPLES:

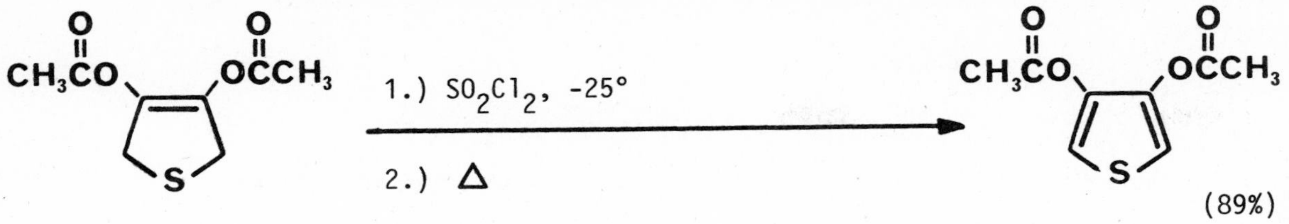

1. $CH_3COC(O)$ / $OC(O)CH_3$ dihydrothiophene → 1.) SO_2Cl_2, -25°; 2.) Δ → 3,4-bis(acetoxy)thiophene (89%) ① [1]

2. i-PrOC(S)-S-C(S)Oi-Pr → XS SO_2Cl_2, $CaCO_3$; Pet ether → ClC(O)-S-S-SC(O)Cl (80%) ②

3. SPh acetal → SO_2Cl_2; CH_2Cl_2, 0°, 3 min → CHO (72%) ③

4. $C_8H_{17}CH_2SPh$ → SO_2Cl_2; CCl_4, Pyr, -5°, 2 hrs → [$C_8H_{18}CCl_2SPh$] → MeOH; H_2O, -5°, 30 min → $C_8H_{17}CO_2Me$ ④

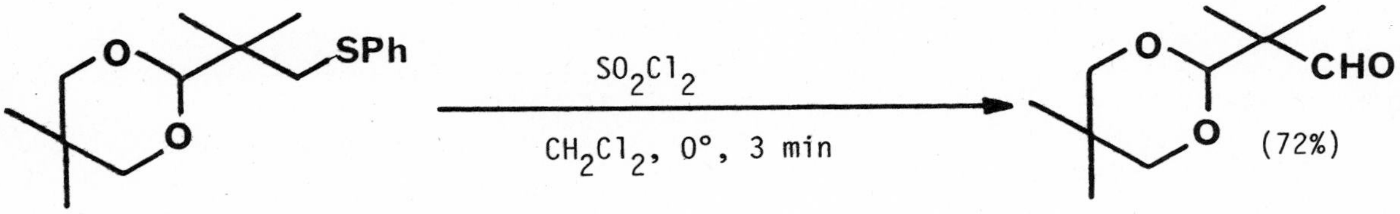

5. $(NO_2)_3CCH_2OC(O)SEt$ → XS SO_2Cl_2; CH_2Cl_2, Reflux, 6 hrs → $(NO_2)_3CCH_2OC(O)Cl$ (97%) ⑤

REAGENT: TETRA-n-BUTYLAMMONIUM FLUORIDE (**TBAF**)

==

STRUCTURE: $[CH_3(CH_2)_3]_4NF$

MW = 261.47

Fieser's Reagents for Organic Synthesis: Vol., pages

1		**4**	477- 78	**7**	353- 54	**10**	378- 81
2		**5**	645	**8**	467-68	**11**	499- 500
3		**6**		**9**	444- 46	**12**	458- 462

March's Advanced Organic Chemistry: Not indexed

MAJOR USES:

1. Removal of silyl groups 2. Cyclizations 3. Catalyst

PREPARATION:

1. The reagent can be prepared from tetrabutyl ammonium hydroxide and HF: E.J. Corey and A. Venkateswarlu, **J. Am. Chem. Soc.**, (1972), 94, 6190.

2. Commercially available.

PRECAUTION:

Hygroscopic.

NOTES:

1. This cycloreversion reaction is 10^6 times faster with the reagent than without it.

REFERENCES:

1. Y. Ito, M. Nakatsuka and T. Saegusa, **J. Am. Chem. Soc.**, (1980), 102, 863.

2. D.B. Grotjahn and N.H. Anderson, **Chem. Commun.**, (1981), 306.

3. O. Papies and W. Grimme, **Tetrahedron Lett.**, (1980), 2799.

4. D. Seebach, A.K. Beck, T. Mukhopadhyay and E. Thomas, **Helv.**, (1982), 65, 1101.

5. E.W. Collington , H. Finch and I.J. Smith, **Tetrahedron Lett.**, (1985), 681.

EXAMPLES:

$C_6H_4(CH_2SiMe_3)(CH_2\overset{\oplus}{N}Me_3Cl^{\ominus})$ + $CH_2{=}CHCN$ —[$(Bu)_4N^+F^-$, CH_2Cl_2]→ 2-cyano-1,2,3,4-tetrahydronaphthalene (82%) ① [1]

2-$SiMe_3$-2-$[(CH_2)_3CH{=}CHCMe(=O)]$-1,3-dithiane —[n-Bu_4NF]→ spiro dithiane-cyclopentane with $CH_2CMe(=O)$ (64%) ② [1]

MeO_2C, MeO_2C, $OSiMe_3$ tricyclic adduct —[n-Bu_4NF, 60°]→ MeO_2C, MeO_2C, $OSiMe_3$ benzene + 1,4-cyclohexadiene (NYA) ③ [1] △1

PhCHO + $Me_3C{-}Si(Me)_2{-}O{-}N(\to O){=}C(H)C_5H_{11}$ —[n-Bu_4NF, THF; -78°, then 10°]→ $Ph{-}CH(OSiMe_2CMe_3){-}CH(NO_2){-}C_5H_{11}$ (50-85%, erythro in excess of 95%) ④ [1]

t-$BuMe_2Si{-}CH_2CH_2{-}C_6H_4{-}O{-}t$-$BuSiMe_2$ —[n-Bu_4NF, THF]→ t-$BuMe_2Si{-}CH_2CH_2{-}C_6H_4{-}OH$ (96%) ⑤

REAGENT: TETRAKIS(TRIPHENYLPHOSPHINE)PALLADIUM (0)

==

STRUCTURE: △1

MW = 1155.58

MP = 100- 105°C (Dec)

$[(C_6H_5)_3P]_4Pd$

Fieser's Reagents for Organic Synthesis: Vol., pages

1		**4**		**7**	357- 58	**10**	384- 91
2		**5**		**8**	472- 76	**11**	503- 14
3		**6**	571- 73	**9**	451- 58	**12**	468- 75

March's Advanced Organic Chemistry: 932

MAJOR USES:

1. Alkylation
2. Oxidation

PREPARATION:

1. The reagent can be prepared from palladium dichloride.

2. Commercially available.

PRECAUTION:

1. The reagent is light-sensitive and should be used under an inert atmosphere.

NOTES:

1. The reagent can be supported on silica gel or polystyrene. See: B.M. Trost and E. Klinam, **J. Am. Chem. Soc.**, (1978), 100, 7779.

REFERENCES:

1. M. Suzuki, Y. Oda and R. Noyori, **Tetrahedron Lett.**, (1981), 4413.

2. L.V. Dunkerton, A.J. Serino, **J. org. Chem.**, (1982), 47, 2812.

3. E. Negishi, H. Matsushita, S. Chatterjee and R.A. John, **J. Org. Chem.**, (1982), 47, 3188.

4. H. Watanabe, M. Saito, N. Sutov and Y. Nagai, **Chem. Commun.**, (1981), 617.

5. S.E. Bystrom, R. Aslanian and J.E. Backvall, **Tetrahedron Lett.**,(1985), 1749.

EXAMPLES:

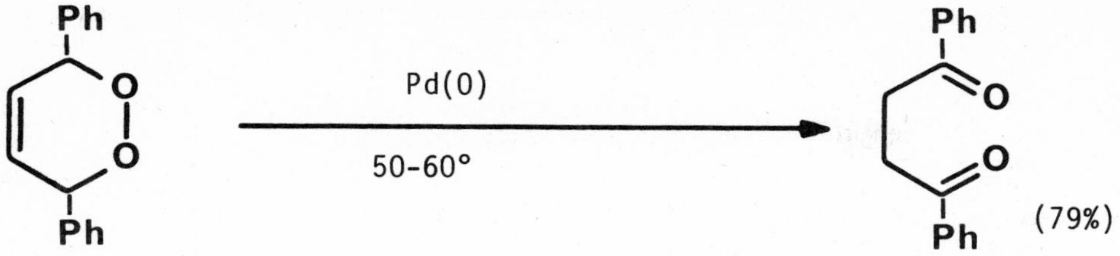

Ph
O
O
Ph
Pd(0)
50-60°
Ph
O
O
Ph
(79%)

(1)

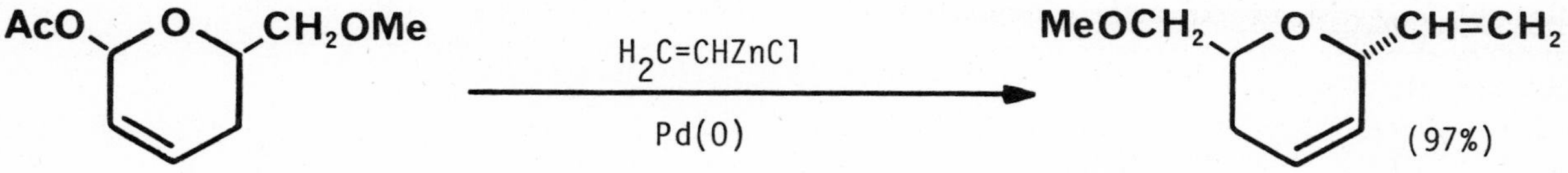

AcO
O
CH₂OMe
H₂C=CHZnCl
Pd(0)
MeOCH₂
O
CH=CH₂
(97%)

(2)

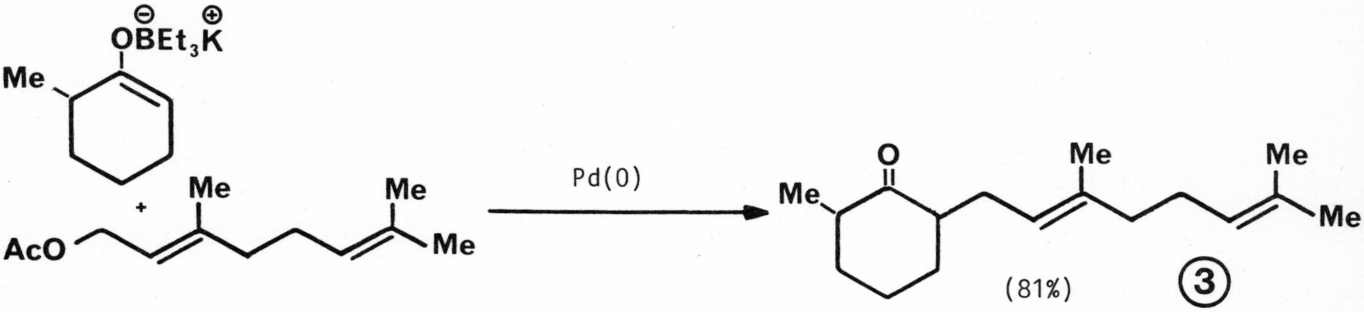

OBEt₃K
Me
+
AcO
Me
Me
Me
Pd(0)
O
Me
Me
Me
Me
(81%)

(3)

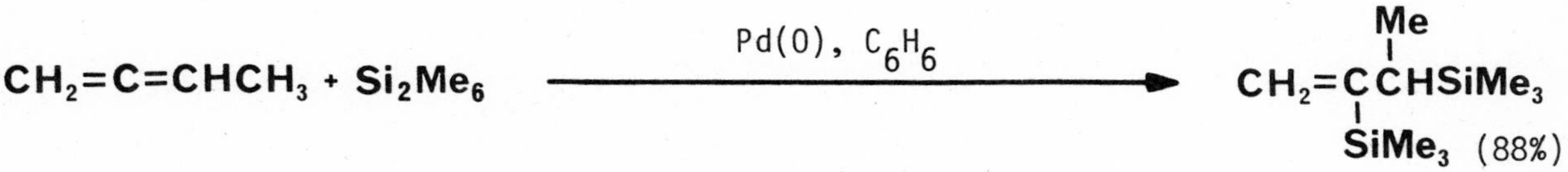

CH₂=C=CHCH₃ + Si₂Me₆
Pd(0), C₆H₆
Me
CH₂=CCHSiMe₃
SiMe₃
(88%)

(4)

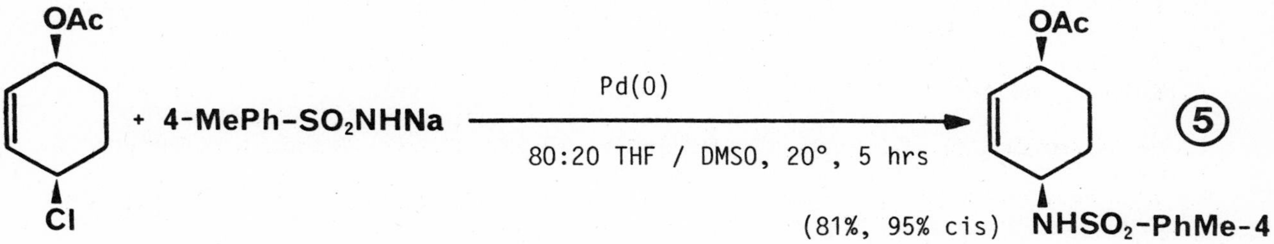

OAc
Cl
+ 4-MePh-SO₂NHNa
Pd(0)
80:20 THF / DMSO, 20°, 5 hrs
OAc
(81%, 95% cis)
NHSO₂-PhMe-4

(5)

REAGENT: THALLIUM (III) NITRATE, **[TTN]** THALLIC NITRATE

==

STRUCTURE:

MW = 266.38

MP = 206°

$Tl(NO_3)_3$

Fieser's Reagents for Organic Synthesis: Vol., pages

1	**4** 492- 97	**7** 362- 65	**10** 395- 96
2	**5** 656- 57	**8** 476- 78	**11**
3	**6** 578- 79	**9** 460 -62	**12**

March's Advanced Organic Chemistry: 784, 1051, 1072, 1085, 1089, 1120

MAJOR USES:

1. Versatile oxidizing agent

PREPARATION:

1. The reagent can be prepared from thallium (III)oxide and nitric acid [**Reagents**, 4, 492.], and is commercially available in very high purity.

PRECAUTION:

1. Extremely toxic. Use with caution.

2. Hygroscopic.

NOTES:

REFERENCES:

1. D. Crouse, M. Wheeler, M. Goemann, P. Tobin, S. Basu and D. Wheeler, **J. Org. Chem.**, (1981), 46, 1814.

2. A. McKillop, D.W. Young, M. Edwards and R.P. Hug, **J. Org. Chem.**, (1978), 43, 3773.

3. E. Mincione and F. Lanciano, **Tetrahedron Lett.**, (1980), 1149.

4. M. Meyer-Dayan, B. Bodo, C. Deschamps-Vallet and D. Molho, **tetrahedron Letters**, (1978), 3359.

5. A.J. Irwin and J.B. Jones, **J. Org. Chem.**, (1977), 42, 2176.

EXAMPLES:

OH / HO (1,5-dihydroxynaphthalene) —TTN→ O / O / HO (5-hydroxy-1,4-naphthoquinone) ① (64%)

O ‖ $PhCCH_3$ —$Tl(NO_3)_3$, Acetonitirle→ O ‖ $PhCCH_2ONO_2$ ② (97%)

(cyclohexene oxide) O —TTN, Hexane, RT, 3 hrs→ OH / ONO_2 ③ (76%)

$ClO_4^{\ominus}$ $\overset{\oplus}{O}$ Ph —TTN, MeOH, RT, 5 hrs→ O / O / Ph ④ (65%)

O (bicyclic ketone) —TTN / MeOH, MeOH, 20°, 8 hrs→ COOMe ⑤ (85%)

REAGENT: THALLIUM (III) TRIFLUOROACETATE (**TTFA**)
(THALLIC TRIFLUOROACETATE)

==

STRUCTURE:

MW = 543.42 $(CF_3CO_2)_3Tl$

Fieser's Reagents for Organic Synthesis: Vol., pages

1		**4**	498- 501	**7**	365	**10**	397
2		**5**	658- 59	**8**	478- 81	**11**	515- 16
3	286- 69	**6**		**9**	462- 64	**12**	481- 83

March's Advanced Organic Chemistry: 484, 547

MAJOR USES:

1. Biphenyl synthesis
2. Aromatic alkylation
3. Oxidative coupling

PREPARATION:

1. Can be prepared from thallium (III) oxide and TFA [see: A. McKillop, J.S. Fowler, M.J. Zelesko, J.D. Hunt, E.C. Taylor and G. McGillivray, **Tetrahedron Lett.**, (1969), 2423.]

2. Commercially available.

PRECAUTION:

1. Very toxic. [For information regarding antidote to thallium poisoning, see: H. Heydlauf, **Evr. J. Pharmacol.**, (1969), 6, 340.]

2. Hygroscopic.

NOTES:

REFERENCES:

1. M. Ochiai, M. Arimoto and E. Fujita, **Tetrahedron Lett.**, (1981), 4491.

2. S. Nishiyama and S. Yamamura, **Chem. Letters**, (1981), 1511.

3. M. Ochiai, M. Arimoti and E. Fujita, **Tetrahedron Lett.**, (1981), 4491.

4. E.C. Taylor, J.G. Anorade, G.J.H. Rall and A. McKillop, **Tetrahedron Lett.**, (1978), 3623.

5. A. McKillop, A.G. Turrell, D.W. Young and E.C. Taylor, **J. Am. Chem. Soc.**, (1980), 102, 6504.

EXAMPLES:

$+ Me_3SiCH_2CH=CH_2$ —TTFA→ (46%) ① [2]

TTFA / TFA; BF_3, EtOH, 25° (36%) ② [2]

TTFA; CH_2Cl_2, 0°, 5 hrs (56%) ③ [2]

$CH=CHCO_2H$ OMe OMe —TTFA, $BF_3 \cdot Et_2O$; TFA, CH_2Cl_2→ Ph-3,4(OMe)$_2$ H H 3,4-(MeO)$_2$Ph (47%) ④ [2]

2 (OMe, Me, Me) —TTFA / TFA; RT→ MeO- -OMe (Me, Me, Me, Me) (60%) ⑤ [1] [3]

REAGENT: THIONYL CHLORIDE (SULFUROUS OXYCHLORIDE)

STRUCTURE:

MW = 118.97 $SOCl_2$
BP = 79°
d = 1.631

Fieser's Reagents for Organic Synthesis: Vol., pages

1	1158- 63	**4**	503- 05	**7**	366- 67	**10**	399
2	412	**5**	663- 67	**8**	481	**11**	
3	290	**6**	585	**9**	465	**12**	

March's Advanced Organic Chemistry: Many references

MAJOR USES:

1. Chlorination 2. Acids to acid chlorides 3. Dehydrations

4. Cyclizations 5. Dehydrogenation

PREPARATION:

Available commercially. For purification, see: **Reagents**, Vol. 1, 1158.

PRECAUTION:

Corrosive and a powerful lachrymator. Irritating. Should be handled with care. Avoid vapors. The reagent will attack rubber, so should be handled in all-glass equipment. By-products of reagent under most reaction conditions are SO_2 and HCl, so provisions should be made to handle these gases.

NOTES:

REFERENCES:

1. Y. Uchida and S. Kozuka, **Chem. Commun.**, (1981), 510.

2. J. Barluenga, J. Lopez-Ortiz and V. Groton, **Chem. Commun.** (1979), 891.

3. J. Garcia, R. Greenhouse, J.M. Muchowski and J.A. Ruiz, **Tetrahedron Lett.**, (1985), 1827.

4. W.L. Brown and A.G. Fallis, **Tetrahedron Lett.**, (1985), 607.

5. D. Chambers, W.A. Denny, J.S. Buckleton and G.R. Clark, **J. Org. Chem.**, (1985), 50, 4736.

EXAMPLES:

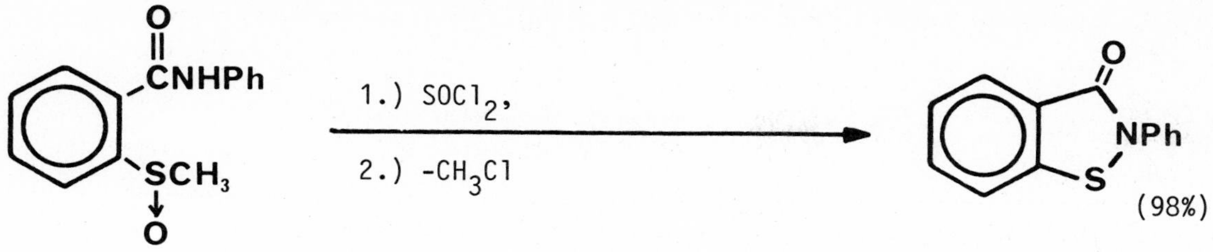

CNHPh
SCH3
1.) SOCl2,
2.) -CH3Cl
NPh
S
(98%)

① [1]

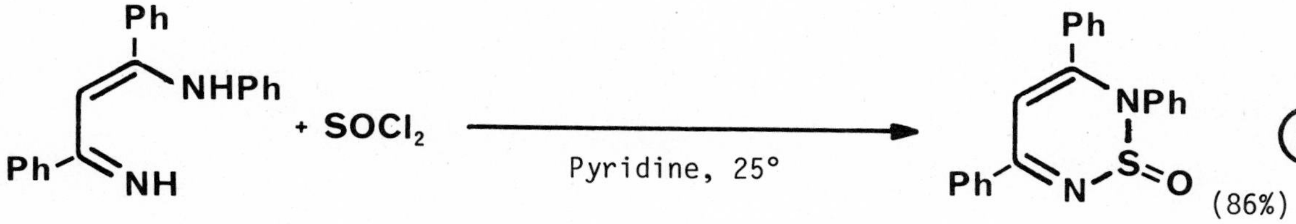

Ph
NHPh
Ph
NH
+ SOCl2
Pyridine, 25°
Ph
NPh
S=O
Ph
N
(86%)

②

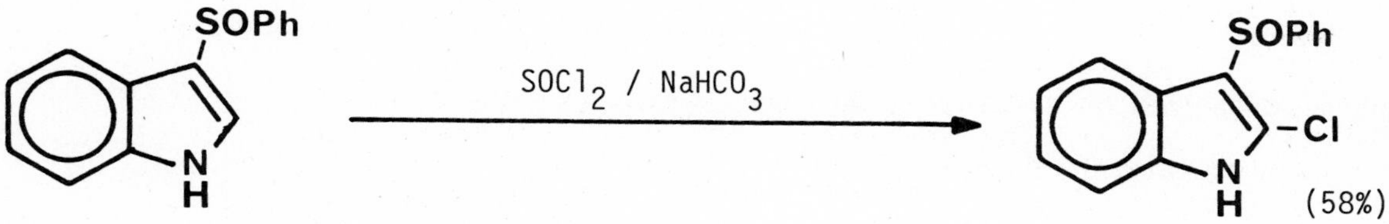

SOPh
N
H
SOCl2 / NaHCO3
SOPh
-Cl
N
H
(58%)

③ [1]

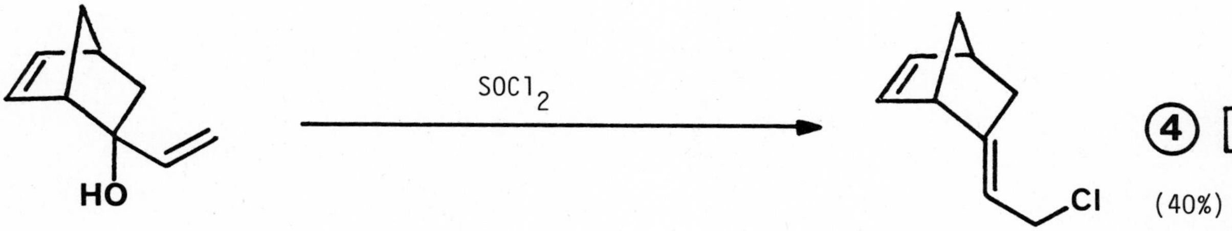

HO
SOCl2
Cl
(40%)

④ [1]

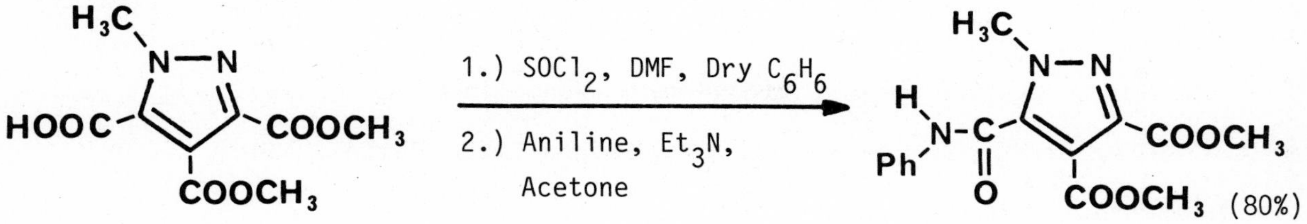

H3C
N—N
HOOC
COOCH3
COOCH3
1.) SOCl2, DMF, Dry C6H6
2.) Aniline, Et3N,
Acetone
H3C
N—N
H
N-C
Ph
O
COOCH3
COOCH3
(80%)

⑤ [2]

REAGENT: TIN (IV) CHLORIDE

==

STRUCTURE:

MW = 260.50
BP = 114.1°C
d = 2.226

$SnCl_4$

<u>**Fieser's Reagents for Organic Synthesis**</u>**: Vol., pages**

1	1111- 13	**4**		**7**	342- 45	**10**	370- 73
2		**5**	627- 31	**8**		**11**	522- 24
3	269	**6**	553-54	**9**	436-38	**12**	486- 90

March's <u>**Advanced Organic Chemistry**</u>: Not Indexed

MAJOR USES:

1. Lewis acid 2. Catalyst

PREPARATION:

1. Commercially available in very high purity.

PRECAUTION:

1. Corrosive as well as moisture-sensitive.

2. Irritates eyes and mucous membranes.

NOTES:

REFERENCES:

1. M. Reetz, I. Chatzhosifidis and K. Schwellnus, **Angew. Chem. Int. Ed.**, (1981), <u>20</u>, 687.

2. D. Lindler, J. Doherty, Shoham and R. Woodward, **Tetrahedron Lett.**, (1982), 5111.

3. D. Liotta, M. Saindane and C. Barnum, **J. Am. Chem. Soc.**, (1981), <u>103</u>, 3224.

4. M.T. Reetz and H. Muller-Starke, **Tetrahedron Lett.**, (1984), 3301.

5. I.K. Stamos, **Tetrahedron Lett.**, (1985), 477.

EXAMPLES:

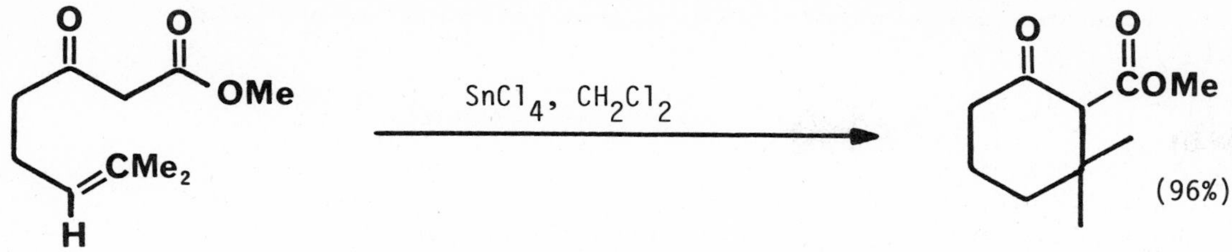

$SnCl_4$, CH_2Cl_2 (96%) ① [1]

$SnCl_4$, CH_3NO_2, 25° (85%) ② [2]

$SnCl_4$, CH_3CN, 25° (85%) ③ [2]

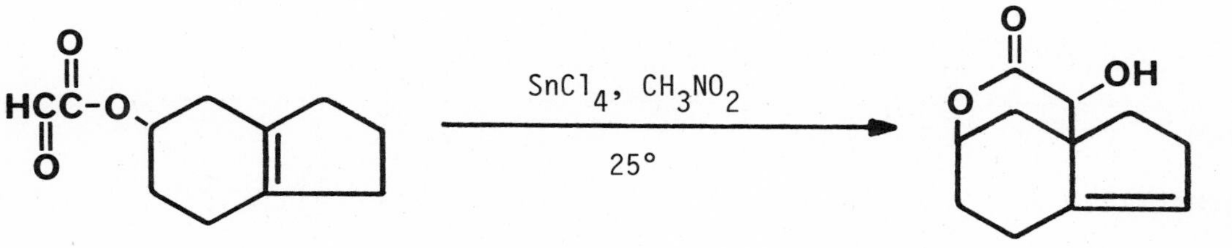

$SnCl_4$, CH_2Cl_2, 0°, 4 hrs (93%) ④ [1]

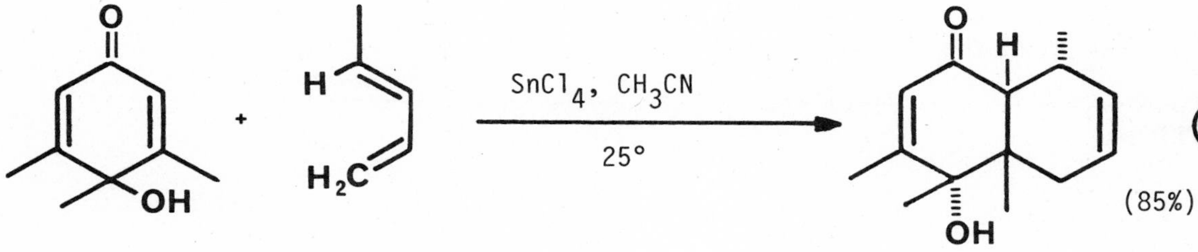

$SnCl_4$, CH_2Cl_2, 30 min, 0° (92%) ⑤ [1]

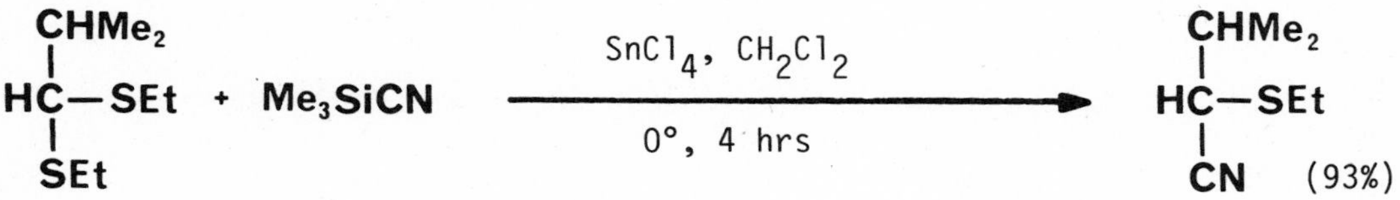

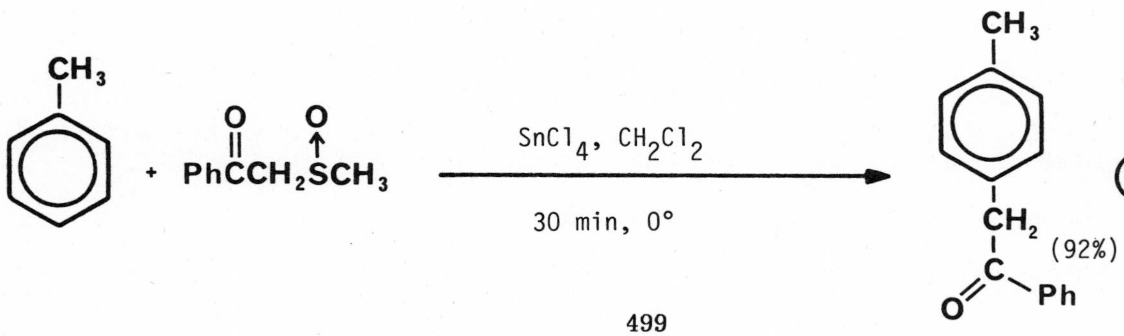

REAGENT: TITANIUM (III) CHLORIDE (TITANIUM TRICHLORIDE)

STRUCTURE:

MW = 154.26
MP = 440° (dec)
d = 2.640

$TiCl_3$

Fieser's Reagents for Organic Synthesis: Vol., pages

1		**4**	506-08	**7**	369	**10**	400
2	415	**5**	669-71	**8**	482-83	**11**	529
3		**6**		**9**	467	**12**	492

March's Advanced Organic Chemistry: 391, 407, 784, 787, 919, 921, 925, 1103, 1108, 1111.

MAJOR USES:

1. Dehalogenation
2. Alkene synthesis (**McMurry Reaction**)
3. Reductive coupling
4. Reductions

PREPARATION:

1. Commercially available

PRECAUTION:

1. Pyrophoric
2. Corrosive

NOTES:

REFERENCES:

1. P. Mattingly and M. Miller, **J. Org. Chem.**, (1980), 45, 410.

2. A. Citterio and E. Vismara, **Synthesis**, (1980), 291.

3. B. Stanovnik, M. Tisler, M. Kocevar, B. Koren, M. Bester and V. Kermavner, **Synthesis**, (1979), 194.

4. Y. Watanabe, T. Numata and S. Oae, **Synthesis**, (1981), 204.

5. J.-M. Surzar and L. Stella, **Tetrahedron Lett.**, (1974), 2191.

EXAMPLES:

$PhC(=O)NHOH \xrightarrow[H_2O,\ NaOAc]{TiCl_3,\ CH_3OH} PhC(=O)NH_2\ (74\%) + PhC(=O)H\ (26\%)$ ① [4]

$PhN_2^{\oplus}Cl^{\ominus} + CH_2{=}CHC(=O)Me \xrightarrow[Acetone]{TiCl_3} PhCH_2CH_2C(=O)Me\ (72\%)$ ② [3]

4-diazo-3,5-dimethylpyrazole $\xrightarrow[EtOH]{15\%\ aq\ TiCl_3}$ 3,5-dimethyl-4H-pyrazole (70%) ③ [4]

$Ph_2S \xrightarrow[MeOH\ /\ H_2O,\ RT,\ 5\ min]{TiCl_3,\ H_2O_2} Ph{-}S(=O){-}Ph\ (100\%)$ ④

N-chloro-N-allyl-4-pentenylamine $\xrightarrow[HOAc\ /\ H_2O,\ -10°]{15\%\ TiCl_3\ /\ N_2}$ 2-(chloromethyl)pyrrolizidine (66%) ⑤

REAGENT: TITANIUM (IV) CHLORIDE (TITANIUM TETRACHLORIDE)

STRUCTURE:

MW = 189.71
BP = 136.4°
MP = -24°
d = 1.730

$TiCl_4$

Fieser's Reagents for Organic Synthesis: Vol., pages

1	1169- 71	**4**	507- 08	**7**	370- 72	**10**	401- 03
2	414- 15	**5**	671- 72	**8**	483- 86	**11**	529- 33
3	291	**6**	590- 96	**9**	468- 70	**12**	494- 502

March's Advanced Organic Chemistry: 372,410, 682, 724, 788, 871, 933, 981, 1002, 1092, 1111

MAJOR USES:

1. Lewis acid 2. Reductive coupling reactions 3. Rearrangements

---

PREPARATION:

Commercially available

PRECAUTION:

Reacts readily with water. Can be harmful to the eyes, mucous membranes and respiratory system.

NOTES:

Reduction of $TiCL_4$ provides a reagent useful for the reductive coupling of carbonyl compounds to pinacols.

REFERENCES:

1. A. Hosami and H. Sakurai, **J. Am. Chem. Soc.**, (1977), 99, 1673.

2. M. Tius, **Tetrahedron Lett.**, (1981), 3335.

3. M. Saidi, **Heterocycles**, (1982), 19, 1473.

4. H. Nishiyama and K. Itoh, **J. Org. Chem.**, (1982), 47, 2496.

5. A.G. Holba, V. Premasager, B.C Barot and E.J. Eisenbraun, **Tetrahedron Lett.**, (1985), 571.

EXAMPLES:

$PhCH{=}CHC(O)CH_3$ + $(CH_3)_3SiCH_2CH{=}CH_2$ —[$TiCl_4$, CH_2Cl_2, -40°]→ $PhCH(CH_2CH{=}CH_2)CH_2C(O)CH_3$ (80-87%) ① [3]

(p-tolyl)C(OH)(CH₂CH=CHSiMe₃)CH₂CH(OMe)₂ —[$TiCl_4$]→ 4-methylbiphenyl (59%) ② [3]

2-naphthyl-$OCH_2CH{=}CH_2$ —[$TiCl_4$, CH_2Cl_2]→ 2-methyl-dihydronaphthofuran (84%) ③ [1]

$Me_3SiCH_2CH{=}CH_2$ + 2-OCH_2CH_2OMe-chroman —[$TiCl_4$]→ 2-$CH_2CH{=}CH_2$-chroman (90%) ④

2 (1-tetralone) —[1.) $TiCl_4$, -10°; 2.) Et_3N]→ 2-(3,4-dihydronaphthalen-1-yl)-1-tetralone (70%) ⑤ [1]

REAGENT: p-TOLUENESULFONIC ACID (monohydrate) TOSIC ACID

==

STRUCTURE:

$H_3C-C_6H_4-SO_3H$

MW = 172.20 hydrate (190.22)
MP = 106- 107° 103- 106°

Fieser's Reagents for Organic Synthesis: Vol., pages

1	1172- 78	**4**	508- 10	**7**	374- 75	**10**	
2		**5**	673- 75	**8**	488- 89	**11**	535
3		**6**		**9**	471	**12**	507

March's Advanced Organic Chemistry: 920, 965

MAJOR USES:

1. Acid catalyst

PREPARATION:

1. The reagent can be made by the sulfonation of toluene, but is commercially available in high purity.

PRECAUTION:

1. Hygroscopic and corrosive. The reagent is irritating to mucus membranes and to skin.

NOTES:

1. **Bz** = **Bzl** = Benzyl = $Ph-CH_2-$

REFERENCES:

1. B. Barot and H. Pinnick, **J. Org. Chem.**, (1981), 46, 2981.

2. M. Pirrung, **J. Am. Chem. Soc.**, (1979), 101, 7130.

3. T. Durst, E.C. Kozma and J.L. Charlton, **J. Org. Chem.**, (1985), 50, 4829.

4. S. Ogawa and T. Takagaki, **J. Org. Chem.**, (1985), 50, 5075.

5. J.J. Eisch and R. Sanchez, **J. Org. Chem.**, (1986), 51, 1848.

EXAMPLES:

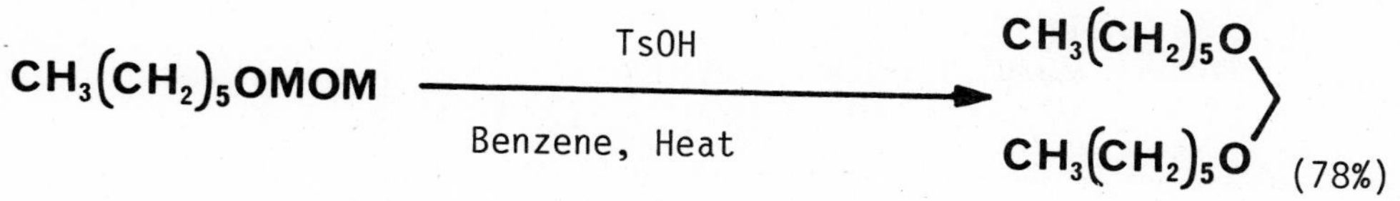

CH3(CH2)5OMOM
TsOH
Benzene, Heat
CH3(CH2)5O
CH3(CH2)5O
(78%)

①

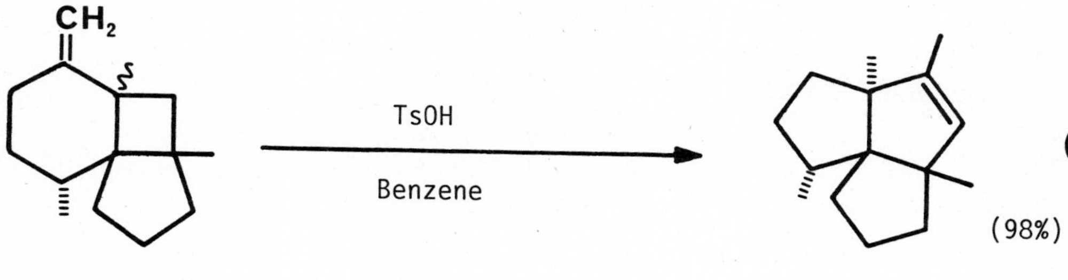

CH2
TsOH
Benzene
(98%)

②

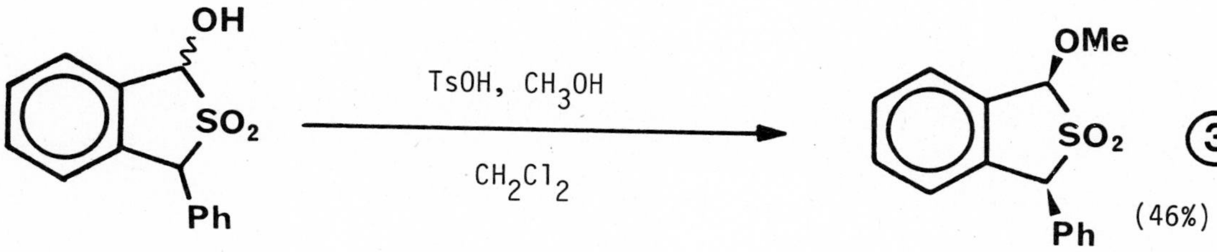

OH
SO2
Ph
TsOH, CH3OH
CH2Cl2
OMe
SO2
Ph
(46%)

③

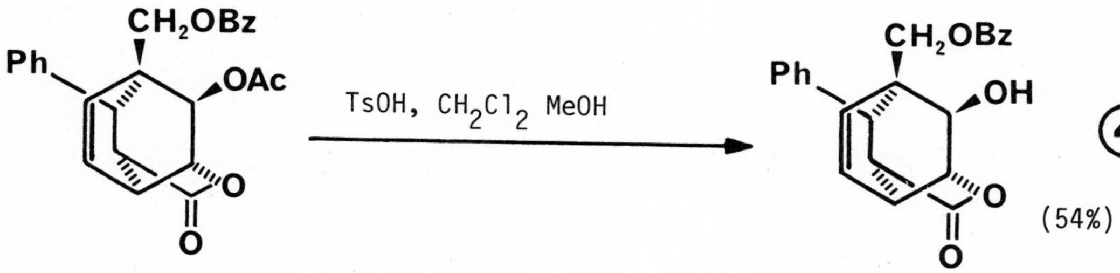

CH2OBz
Ph
OAc
O
O
TsOH, CH2Cl2 MeOH
CH2OBz
Ph
OH
O
O
(54%)

④ ⚠

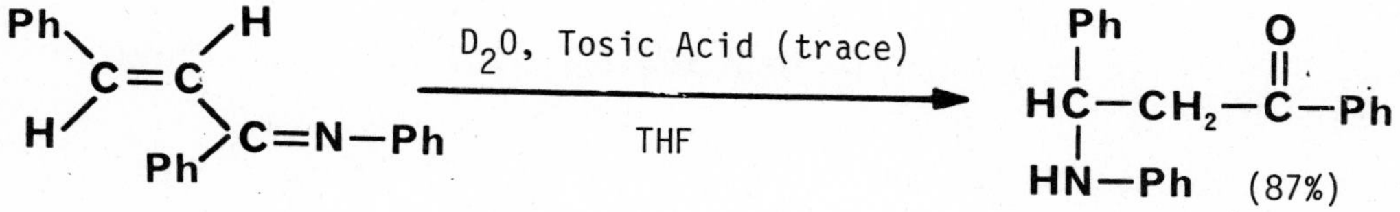

Ph
H
C=C
H
Ph
C=N—Ph
D2O, Tosic Acid (trace)
THF
Ph
O
HC—CH2—C—Ph
HN—Ph
(87%)

⑤

REAGENT: p-TOLUENESULFONYL AZIDE (TOSYL AZIDE)

STRUCTURE:

MW = 197.23
MP = 22°C

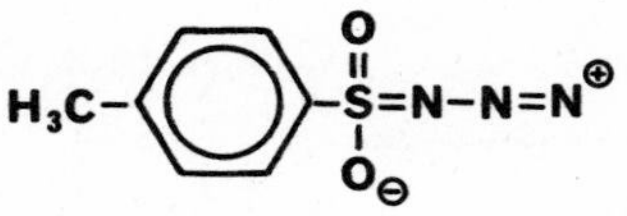

$H_3C-C_6H_4-S(=O)(-O^{\ominus})=N-N=N^{\oplus}$

Fieser's Reagents for Organic Synthesis: Vol., pages

1	1178- 79	**4**	510	**7**		**10**	
2	415- 17	**5**	675	**8**		**11**	535- 36
3	291- 92	**6**	597	**9**	472	**12**	

March's Advanced Organic Chemistry: 534, 535, 554, 573.

MAJOR USES: ⚠1

1. Synthesis of diazo compounds
2. Azide synthesis

PREPARATION:

1. The reagent can be prepared from tosyl chloride and sodium azide: W. von E. Doering and C. DePuy, **J. Am. Chem. Soc.**, (1953), 75, 5955.

PRECAUTION:

NOTES:

1. For review on synthesis of diazo compounds, see: **Angew. Chem. Int. Ed.**, (1967), 6, 733.

REFERENCES:

1. B. Stanovnik, M. Tisler, M. Kunaver, D. Gabrijelcic and M. Kocevar, **Tetrahedron Lett.**, (1978), 3059.

2. N.S. Narasimhan and R. Ammanamanchi, **Tetrahedron Lett.**,(1983), 4733.

3. I. Yamamoto, H. Tokanou, H. Uemura and H. Gotoh, **J. Chem. Soc.**Perkin 1, (1977), 1241.

4. A.L. Fridman, Y.S. Andreichikov, V.L. Gein and L.F. Gein, **J. Org. Chem. USSR**, (1976), 12, 457.

5. S.J. Weininger, S. Kohem, S. Mataka, G. Koga and J.-P. Anselme, **J. Org. Chem.**, (1974), 39, 1591.

EXAMPLES:

$$PhNHNH_2 + TsN_3 \xrightarrow[\text{Xylene, NaOH / } H_2O]{Et_4N^+Br^-} C_6H_6\ (40\%) + 2N_2 + TsNH_2 \quad (1)$$

$$C_6H_5OMe \xrightarrow[\text{Ether}]{\text{n-BuLi}} \text{2-LiC}_6H_4OMe \xrightarrow[\text{Ether}]{TsN_3} [\text{2-}N_3C_6H_4OMe] \xrightarrow[\text{Ni-Al}]{KOH} \text{2-}NH_2C_6H_4OMe\ (80\%) \quad (2)$$

$$\text{2-ClC}_6H_4\text{-N=C=N-C}_6H_4\text{Cl-2} \xrightarrow[CHCl_3\text{, Reflux, 25 hrs}]{TsN_3\text{, Cu (powder)}} \text{2-ClC}_6H_4\text{-NH-C(=O)-NH-C}_6H_4\text{Cl-2}\ (72\%) \quad (3)$$

$$\text{2-C}_{10}H_7SO_2CH_2C(=O)CO_2Et \xrightarrow[Et_2NH\text{, 10-15°, 1 hr}]{TsN_3} \text{2-C}_{10}H_7SO_2CHN_2\ (74\%) \quad (4)$$

$$PhCH(CO_2Et)_2 + TsN_3 \xrightarrow[\text{Dry glyme}]{NaH / N_2} PhC(N_3)(CO_2Et)_2\ (77\%) \quad (5)$$

REAGENT: p-TOLUENESULFONYL CHLORIDE (TOSYL CHLORIDE)

STRUCTURE:

$H_3C-C_6H_4-SO_2-Cl$

MW = 190.65
MP = 67- 69°C
BP = 146°C

<u>Fieser's Reagents for Organic Synthesis</u>: **Vol., pages**

1	1179- 84	**4**	510	**7**		**10**	
2		**5**	676- 77	**8**	489	**11**	536
3	292	**6**	598	**9**	472	**12**	

March's <u>Advanced Organic Chemistry</u>: 532, 934

MAJOR USES:

1. Making tosylate derivatives of alcohols 2. Dehydrations

3. Protecting group 4. Isomerizations and rearrangements

PREPARATION:

1. Commercially available. For pure compound, p-toluenesulfonic acid should be removed by crystallization.

PRECAUTION:

1. Corrosive and moisture sensitive.

NOTES:

1. After preferential formation of the primary tosylate, there occurs an internal displacement to form the epoxide.

REFERENCES:

1. T. Fujisawa, T. Itoh, M. Nakai and T. Sato, **Tetrahedron Lett.**, (1985), 771.

2. J.A. Marshall, B.S. DeHoff and D.G. Cleary, **J. Org. Chem.**, (1986), <u>51</u>, 1735.

3. T. Shoho and S. Kashimura, **J. Org. Chem.**, (1983), <u>48</u>, 1939.

4. T. Kolasa, **Synthesis**, (1983), 539.

5. T. Fujisawa, T. Sato, Y. Gotoh, M. Kawashima and T. Kawara, **Bull. Chem. Soc., Jpn.**, (1982), <u>55</u>, 3555.

EXAMPLES:

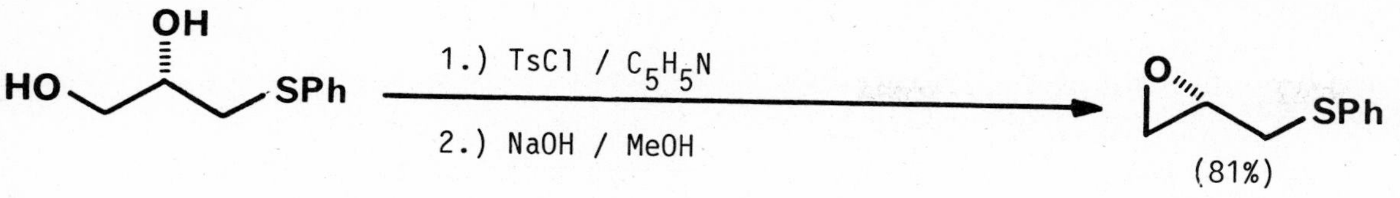

1.) $TsCl / C_5H_5N$

2.) $NaOH / MeOH$

(81%) ① [1] △1

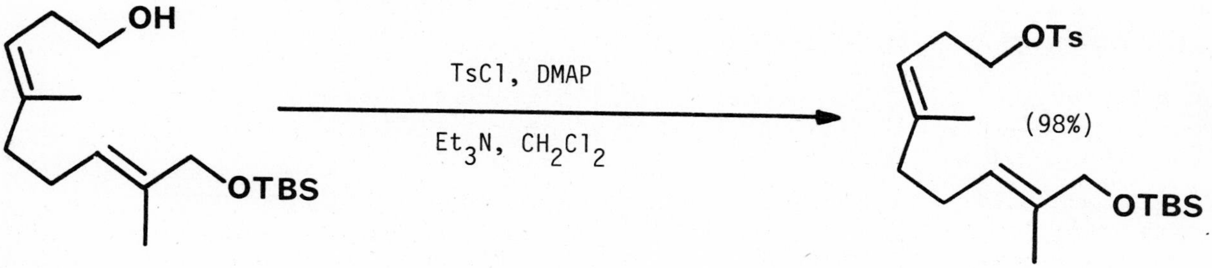

TsCl, DMAP

Et_3N, CH_2Cl_2

(98%) ② [1]

TsCl, Et_3N, CH_2Cl_2, 0°, 1 hr

$NaHCO_3$

CHO

(76%) ③ [1]

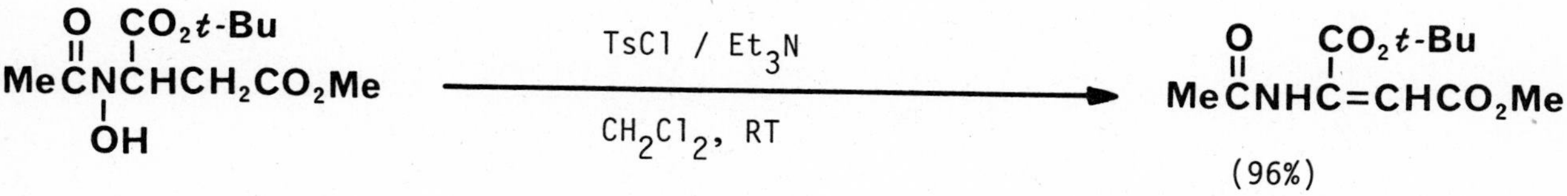

$MeC(O)N(OH)CH(CO_2t\text{-}Bu)CH_2CO_2Me$ → $MeC(O)NHC(CO_2t\text{-}Bu){=}CHCO_2Me$

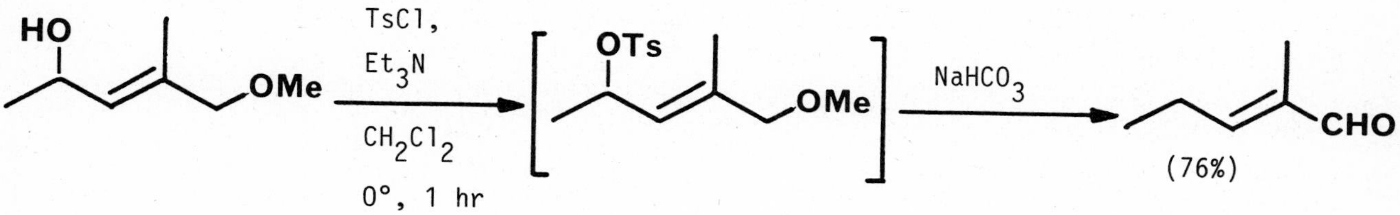

$TsCl / Et_3N$

CH_2Cl_2, RT

(96%) ④

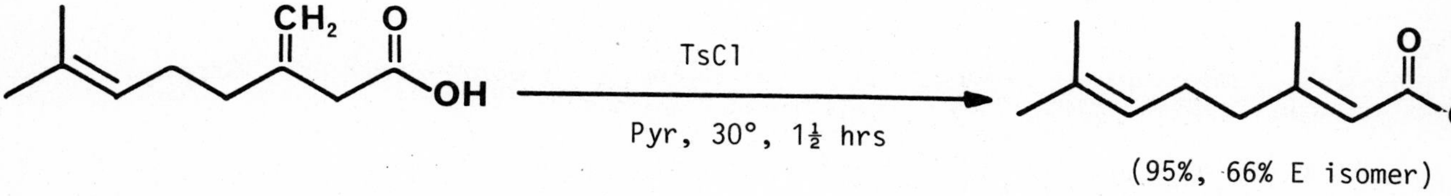

TsCl

Pyr, 30°, 1½ hrs

(95%, 66% E isomer) ⑤ [4]

REAGENT: P-TOLUENESULFONYL HYDRAZINE (TOSYL HYDRAZINE)

STRUCTURE:

MW = 186.24

MP = 104- 07°

$H_3C-C_6H_4-S(=O)_2-NHNH_2$

Fieser's Reagents for Organic Synthesis: Vol., pages

1	1185- 87	**4**	511- 12	**7**	375- 76	**10**	
2	417- 23	**5**	678- 81	**8**	489- 93	**11**	537
3	293	**6**	598- 600	**9**	472- 73	**12**	

March's Advanced Organic Chemistry: 1102

MAJOR USES: 1

1. Preparing tosylhydrazones

PREPARATION:

1. The reagent is prepared from tosyl chloride and hydrazine: See L. Friedman, R. Litle and W. Reichle, **Org. Synth.**, (1960), 40, 93.

PRECAUTION:

NOTES:

1. Tosylhydrazones are useful intermediates for a number of novel transformations. See, for example: **Bamford- Stevens Reaction.**

REFERENCES:

1. R.W. Gray and A.S. Dreiding, **Helv.**, (1977), 60, 1969.

2. R.O. Hutchins, C.A. Milewski and B.E. Maryanoff, **J. Am. Chem. Soc.**, (1973), 95, 3662.

3. A. deMeijere, D. Schallner and C. Weitemeyer, **Tetrahedron Lett.**, (1973), 3483.

4. C.D. Anderson, J.T. Sharp, H. RajSood and R. Stewart Strathdee, **Chem. Commun.**, (1975), 613.

5. K.M. Patel and W. Reusch, **Synth. Commun.**, (1975), 27.

EXAMPLES:

H_2N-NHTs

Acetic acid, Hexane, -70°, then RT

N-NHTs

(83%)

①

HO

$TsNHNH_2$

MeOH, Reflux, 3 hrs

TsNHN

$NaBH_4$

Reflux, 8 hrs

(73-76%)

②

$TsNHNH_2$

EtOH

=O

=NNHTs

(85%)

MeLi

Dry ether

(77%)

③ △1

Me, =CH_2, O, Me

$TsNHNH_2$

H^+ / EtOH

Me, N-Ts, N, Me

(77%)

④

OMe, O, Me, Me, Me

$TsNHNH_2$

EtOH

TsNHN, OMe, Me

(85%) Me Me

MeLi

OMe, Me

(95%) Me Me

⑤

REAGENT: TRI-n-BUTYLTIN HYDRIDE

==

STRUCTURE:

MW = 291.05 $[CH_3(CH_2)_3]_3SnH$
BP = 80°
d = 1.082

Fieser's Reagents for Organic Synthesis: Vol., pages

1	1192- 93	**4**	518- 20	**7**	379- 80	**10**	411- 13
2	424	**5**		**8**	497	**11**	545- 51
3	294	**6**	604	**9**	476	**12**	516- 23

March's Advanced Organic Chemistry: 390, 392, 396, 564, 646, 810, 919, 921, 923, 1108.

MAJOR USES:

1. Reducing agent 2. Desulfurization 3. Useful for initiating radical cyclizations

---

PREPARATION:

1. The reagent can be prepared form bis[tri-n-butyl]tin oxide by reduction with lithium aluminum hydride. See H. Kuivila, **Synthesis**, (1970), 499.

2. Commercially available.

PRECAUTION:

1. Moisture sensitive and an irritant.

NOTES:

1. **AIBN** = azobisisobutylnitrile

2. The I^- form the NaI most likely displaces the tosylate group, and then is reduced by the tin hydride reagent.

REFERENCES:

1. G. Stork and N. Baine, **J. Am. Chem. Soc.**, (1982), <u>104</u>, 2321.

2. M. Bachi and C. Hoornaert, **Tetrahedron Lett.**, (1981), 2693.

3. D.H. Barton, R. Motherwell and W. Motherwell, **J. Chem. Soc.**, <u>Perkin I</u>, (1981), 2363.

4. D.B. Gerth and B. Giese, **J. Org. Chem.**, (1986), <u>51</u>, 3726.

5. J.R. Schauder, J.N. Davis and A. Krief, **Tetrahedron Lett.**, (1983), 1657.

EXAMPLES:

n-Bu_3SnH, hν

(72%, cis:trans=1:1) ① [3]

n-Bu_3SnH, AIBN

(22-56%) + (24-42%) ② [3] △1

n-Bu_3SnH, AIBN, C_6H_6

(65%) ③ [1]

n-Bu_3SnH, NaI, AIBN, Glyme

(82%) ④ [1] △2

$C_{12}H_{25}$-CH—CH_2 (S) → n-Bu_3SnH, 110°, 1 hr → $C_{12}H_{25}CH=CH_2$ (81%) ⑤ [2]

REAGENT: TRIETHYLALUMINUM

STRUCTURE:

MW = 114.17 Et_3Al
BP = 128- 30° @ 50 mm
d = 0.835

Fieser's Reagents for Organic Synthesis: Vol., pages

1	1197- 98	**4**	526	**7**		**10**	
2	427	**5**	688- 89	**8**		**11**	
3	299	**6**		**9**		**12**	

March's Advanced Organic Chemistry: 828

MAJOR USES:

1. Alkylation
2. Reaction with amines to form amino complexes
3. Catalyst

PREPARATION:

1. Commercially available. Can be obtained in hexane or toluene solution.

PRECAUTION:

1. Pyrophoric. Reactions should be carried out dry and inert atmosphere.

NOTES:

1. The NaOH is added dropwise, with caution.

REFERENCES:

1. Y. Yamamoto, H. Yatagai and K. Maruyama, **J. Org. Chem.**, (1980), 45, 195.
2. G.H. Posner and S.R. Haines, **Tetrahedron Lett.**, (1985), 1823.
3. A. Yasuda, M. Takahashi and H. Takaya, **Tetrahedron Lett.**, (1981), 2413.
4. T. Kauffmann, L. Ban and D. Kuhlmann, **Ber.**, (1981), 114, 3684.
5. L.E. Overman and L.A. Flippin, **Tetrahedron Lett.**, (1981), 195.

EXAMPLES:

PhCH₂O–[furanose]–F, PhCH₂O, OCH₂Ph $\xrightarrow[0°,\ 10\ min]{Et_3Al,\ Toluene}$ PhCH₂O–[furanose]–Et, PhCH₂O, OCH₂Ph (86%) ① [1]

②

$PhSH \xrightarrow[C_6H_6,\ RT,\ 1\ hr]{Et_3Al}$ $[Et_2AlSPh]$ + (2-methyl-2-vinyloxirane) ⟶ PhS–CH₂–CH=C(CH₃)–CH₂–OH

(92%, 98% cis)

$Ph-C(NH_2)=NNH_2 \xrightarrow[C_6H_6\ /\ N_2,\ 5°]{Et_3Al}$ [Ph–C(N(H))⊖AlEt=N⊕–NHAlEt₂] $\xrightarrow[85°,\ H^+]{PhCN}$ $PhC(NH_2)=N-N=C(NH_2)Ph$ ③

(95%)

④ ⚠1

$(PhCH_2)_2NH$, CH_2Cl_2, N_2, Et_3Al ⟶ $[(PhCH_2)_2NAlEt_2]$

(cyclopentene oxide) $\xrightarrow{[(PhCH_2)_2NAlEt_2]}$ [cyclopentane–OAlEt₂, N(CH₂Ph)₂] $\xrightarrow{NaOH}$ cyclopentane–OH, N(CH₂Ph)₂

(78%)

n-BuPh + $CH_2=CH-CN$ $\xrightarrow[CH_2Cl_2\ /\ N_2]{Et_3Al}$ [*n*-BuPCH₂CHCN] $\xrightarrow{BuCHO}$ *n*-BuCHOH–C(=CH₂)–CN ⑤

(90%)

REAGENT: TRIETHYL ORTHOFORMATE (ETHYL ORTHOFORMATE)

===

STRUCTURE:

MW = 148.20

BP = 146°C

d = 0.089

$(EtO)_3CH$

Fieser's Reagents for Organic Synthesis: Vol., pages

1	1204- 09	**4**	527	**7**	385- 86	**10**	
2	430	**5**	690- 91	**8**		**11**	555
3		**6**	610- 11	**9**		**12**	

March's Advanced Organic Chemistry: 930

MAJOR USES:

1. Aldehyde synthesis 2. Esterification 3. Condensations

PREPARATION:

1. A simple procedure involves the addition of benzoyl chloride to abs. EtOH, formamide and ligroin (R. Ohme and E. Schmitz, **Ann.**, (1968), 716, 207).

2. Commercially available.

PRECAUTION:

1. Flammable and an irritant.

NOTES:

1. Yields were much higher using dimethylformamide dimethylacetal rather than triethylorthoformate.

REFERENCES:

1. P. Camps, J. Cardellach, J. Font, P. Ortuno and P. Ponsati, **Tetrahedron Lett.**, (1982), 2389.

2. A.H. Schmidt and M. Russ, **Ber.**, (1981), 114, 1099.

3. R.A. Swaringem, Jr., D.A. Yeuwell, J.C. Wisowaty, H.A. El-Sayad, E.L. Stewart and M.E. Darnofall, **J. Org. Chem.**, (1979), 44, 4825.

4. S.I. Goldberg and A.H. Lipkin, **J. Org. Chem.**, (1972), 37, 1823.

5. G.R. Weisman, V. Johnson and R.E. Fiala, **Tetrahedron Lett.**, (1980),

EXAMPLES:

$MeOCH_2$ lactone diol (HO, OH) $\xrightarrow{HC(OEt)_3}$ $MeOCH_2$ lactone orthoester (OEt) ① [3] (95%)

$Me_3SiCl + NaBr \xrightarrow{30°,\ 8\ min} [Me_3SiBr] + (EtO)_3CH \xrightarrow{Reflux,\ 2\ hrs} HCOEt$ (C=O) ② (92%)

3,4,5-(MeO)$_3$C$_6$H$_2$–$CH_2CH(CN)CHO$ $\xrightarrow{HC(OEt)_3}$ 3,4,5-(MeO)$_3$C$_6$H$_2$–$CH_2C(CN)(CHO)CH(OEt)_2$ ③ [3] (78%)

2 (2-pyridyl)CH_2CO_2Me + $HC(OEt)_3$ $\xrightarrow[Reflux,\ 8\ hrs]{Acetic\ anhydride}$ (2-pyridyl)-quinolizinone with CO_2Me ④ [3] (79%)

1,4,7-triazacyclodecane (HN, NH, NH) $\xrightarrow[HCl,\ Toluene,\ Heat]{HC(OEt)_3}$ tricyclic orthoamide (N, N, N) ⑤ [3] △1 (38%)

REAGENT: TRIETHYL PHOSPHITE

STRUCTURE:

MW = 166.16
BP = 156°C
d = 0.969

$(C_2H_5O)_3P$

Fieser's Reagents for Organic Synthesis: Vol., pages

1	1212- 16	**4**	529- 30	**7**	387	**10**	
2	432- 33	**5**	693	**8**	501	**11**	555
3	304	**6**	612	**9**		**12**	

March's Advanced Organic Chemistry: Not Indexed

MAJOR USES:

1. Deoxygenation 2. Reductions

PREPARATION:

1. The reagent can be prepared from PCL_3, EtOH ad N,N-diethylaniline; but is also commercially available.

2. For a purification, see: A. Ford- Moore and B. Perry (**Org. Synth.**, (1963), Coll. Vol. 4, 955.

PRECAUTION:

1. Moisture sensitive and skin irritant.

NOTES:

1. **Via:**

CH=NPh N–O–P(OEt)₃ → CH=NPh N S → C≡N S N Ph

REFERENCES:

1. R.S. Mali and V.J. Yadav, **Synthesis**, (1984), 862.

2. A. Koziara, K. Osowska- Pacewicka, S. Zawadski and A. Zwierzak, **Synthesis**, (1985), 202.

3. L. Hendriksen, **Synthesis**, (1982), 771.

4. V.M. Colburn, B. Iddon, H. Suschitzky and P.T. Gallagher, **Chem. Commun.**, (1978), 453.

5. R.J. Sundberg and R.H. Smith, **Tetrahedron Lett.**, (1971), 267.

EXAMPLES:

3,4-Dimethoxy-2-nitrocinnamate ethyl ester (MeO, MeO, NO_2, CO_2Et) → $(EtO)_3P$, 170°, 3 hrs → 5,6-dimethoxyindole-2-carboxylic acid ethyl ester (CO_2Et, NH) (63%) ① [2]

Bromocyclopentane → NaN_3, C_6H_6 → [azidocyclopentane (N_3) → $P(OEt)_3$, Below 30°, 8 hrs → $(OEt)_3P{=}N$-cyclopentane] → HCl → cyclopentylamine (NH_2) (80%) ② [2]

$2\ Et_2NH$ + CSe_2 → Tosyl-Cl, $CHCl_3$ → $Et_2NC(=Se)Se{-}SeC(=Se)NEt_2$ → $P(OEt)_3$, Toluene, 50°, 1 hr → $Et_2NC(=Se){-}Se{-}C(=Se)NEt_2$ (68%) ③ [2]

2-Nitro-3-thienyl $CH{=}NPh$ (NO_2, S) → $P(OEt)_3$ / t-Butylbenzene, 14 hrs → 1-phenylpyrrole-3-carbonitrile (CN, N–Ph) (55%) ④

Benzene + PhNO → $P(OEt)_3$, Trifluoroethanol → PhN (azepine) (34%) ⑤

REAGENT: TRIFLUOROACETIC ACID (**TFA**)

STRUCTURE:

CF_3COOH

MW = 114.02
BP = 72°
d = 1.480

Fieser's Reagents for Organic Synthesis: Vol., pages

Vol.	pages	Vol.	pages	Vol.	pages	Vol.	pages
1	1219- 21	**4**	530 -32	**7**	388- 89	**10**	418
2	433-34	**5**	695- 700	**8**	503	**11**	557- 59
3	305ˆ 08	**6**	613- 15	**9**	483	**12**	529

March's Advanced Organic Chemistry: 278, 286, 317

MAJOR USES:

1. Acid catalyst
2. Solvent

PREPARATION:

1. Commercially available.

PRECAUTION:

1. Toxic and corrosive.

NOTES:

REFERENCES:

1. K.C. Nicolaou and R. Zipkin, **Angew. Chem. Int. Ed.**, (1981), 20, 785.

2. L. Harwood, **Chem. Commun.**, (1982), 1120.

3. W. Dauben and A. Chollet, **Tetrahedron Lett.**, (1981), 1583.

4. L. Crombie, R.C.F. Jones and C.J. Pamer, **Tetrahedron Lett.**, (1985), 2933.

5. G.W.J. Fleet and P.W. Smith, **Tetrahedron Lett.**, (1985), 1469.

EXAMPLES:

TFA

①

(80%)

TFA

(32%) ②

(21%)

TFA, Benzene

Δ

(Quant.) ③

F_3CCO_2H

④

(53%)

1.) $CF_3COOH-H_2O$, RT, 1 hr

2.) $NaBH_4$, $EtOH-H_2O$

⑤

(81%)

REAGENT: TRIFLUOROMETHANESULFONIC ANHYDRIDE (TRIFLIC ANHYDRIDE)

STRUCTURE:

MW = 282.13
BP = 81- 83° / 745 mm.
d = 1.677

$(CF_3SO_2)_2O$

Fieser's Reagents for Organic Synthesis: Vol., pages

1		**4**		**7**	390	**10**	419- 20
2		**5**	702- 05	**8**		**11**	560 -61
3		**6**	618- 20	**9**		**12**	530

March's Advanced Organic Chemistry: 930

MAJOR USES:

1. Preparation of triflates
2. Ketene preparation

PREPARATION:

1. The reagent can be prepared from trifluoromethanesulfonic acid and phosphorous anhydride [T. Gramstad and R. Haszeldine, **J. Chem. Soc.**, (1957), 4069.]

2. Commercially available.

PRECAUTION:

1. Moisture sensitive and corrosive.

NOTES:

1. For a Review, see: **Aldrichimica Acta**, (1983), 16, 15.

2. This reaction requires the in situ generation of a keteneimminium salt.

REFERENCES:

1. J. Faumagne, J. Escudaro, S. Taleb-Sahraoui and L. Ghosez, **Angew. Chem. Int. Ed.**, (1981), 20, 879.

2. D. Hagiwara, K. Sawada, T. Ohnami, M. Aratani and M. Hashimoto, **Chem. Commun.**, (1982), 578.

3. G.W.J. Fleet and P.W. Smith, **Tetrahedron Lett.**, (1985), 1469.

4. X. Creary, **J. Org. Chem.**, (1985), 50, 5080.

5. C. Bladon and G. Kirby, **Chem. Commun.**, (1982), 1402.

EXAMPLES:

Cyclohexene + $CH_3CH_2\overset{O}{\overset{\|}{C}}N(CH_3)_2$ → Collidine, $(CF_3SO_2)_2O$, $CHCl_3$ → 7-methylbicyclo[4.2.0]octan-8-one (48%) ① [2] △2

H_2N-penicillanate (CO_2CH_2Ph) → $(CF_3SO_2)_2O$ / $CHCl_3$ → $(CF_3SO_2)_2N$-penicillanate (CO_2CH_2Ph) (Quant.) ② [1]

Furanoside (cyclic carbonate, $PhCH_2O$, OH, OMe) → 1.) $(CF_3SO_2)_2$, Pyr, CH_2Cl_2; 2.) NaN_3, DMF → azide ($PhCH_2O$, N_3, OMe) → MeOH / NaOMe → diol (HO, HO, $PhCH_2O$, N_3, OMe) (75%) ③

$PhCH_2MgCl$ → $(CF_3SO_2)_2O$ / Ether → $PhCH_2SO_2CF_3$ (72%) ④ [1]

$Me_3CN(OH)\overset{O}{\overset{\|}{C}}CH_2Ph$ → $(CF_3SO_2)_2O$ / CH_2Cl_2, -70° → Me_3CN aziridinone (H, Ph, O) (97%) ⑤

REAGENT: TRIMETHYL ORTHOFORMATE

==

STRUCTURE:

MW = 106.12 $(MeO)_3CH$
BP = 101- 102°C
d = 0.970

<u>**Fieser's Reagents for Organic Synthesis**</u>**: Vol., pages**

1		**4**	540	**7**		**10**	125
2		**5**	714	**8**	507- 08	**11**	568
3	313	**6**		**9**		**12**	

March's <u>**Advanced Organic Chemistry**</u>: 1120

MAJOR USES:

1. Trans ketalization 2. Formylations

PREPARATION:

Commercially available in high purity

PRECAUTION:

1. Flammable and an irritant.

NOTES:

1. **Nafion H** is a solid perfluorinated sulfonic acid resin.

REFERENCES:

1. S. Suzuki, A. Yanagiasawa and R. Noyori, **Tetrahedron Lett.**, (1982), 3595.

2. U. Ghatak, B. Sanyal, S. Ghosh, M. Raju and E. Wenkert, **J. Org. Chem.**, (1980), <u>45</u>, 1081.

3. P. Beslin, D. Lagain and J. Vialle, **J. Org. Chem.**, (1980), <u>45</u>, 2517.

4. E.C. Taylor and C.-S. Chiang, **Synthesis**, (1977), 467.

5. G.A. Olah, S.C. Narang, D. Meidar and G.F. Salem, **Synthesis**, (1981), 282.

EXAMPLES:

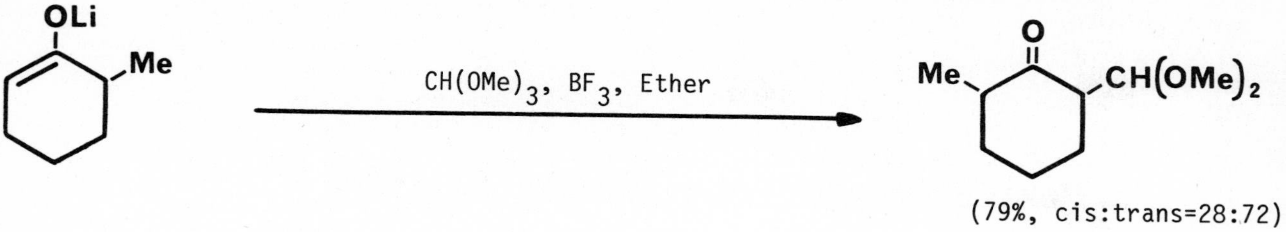

OLi, Me (cyclohexenolate) —[$CH(OMe)_3$, BF_3, Ether]→ Me, O, $CH(OMe)_2$ ① [2]

(79%, cis:trans=28:72)

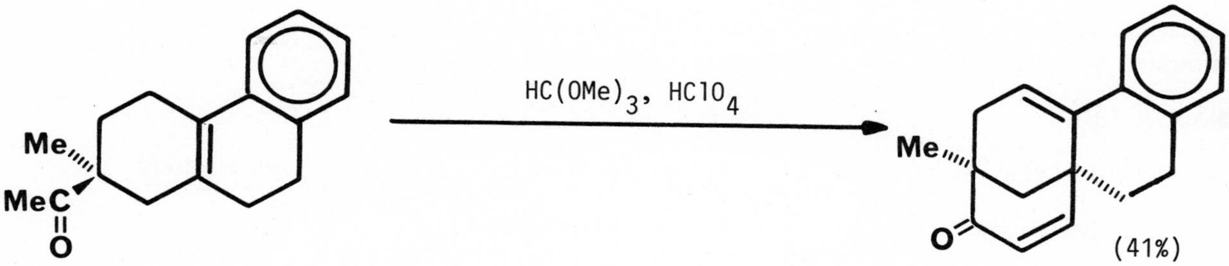

Me, MeC, O —[$HC(OMe)_3$, $HClO_4$]→ Me, O ②

(41%)

Me, Me, O, Me —[$HC(OMe)_3$, H_2S, $ZnCl_2$ / MeOH, Ice bath, 5 hrs]→ Me, Me, S, Me —[100° / 1 hr]→ Me, Me, S, Me ③

(90%)

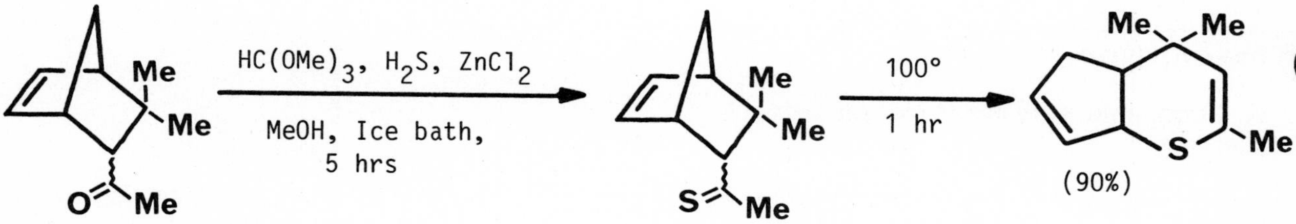

O (cyclohexanone) + $HC(OMe)_3$ —[Hexane / RT, 2 min]→ MeO, OMe ④ [1]

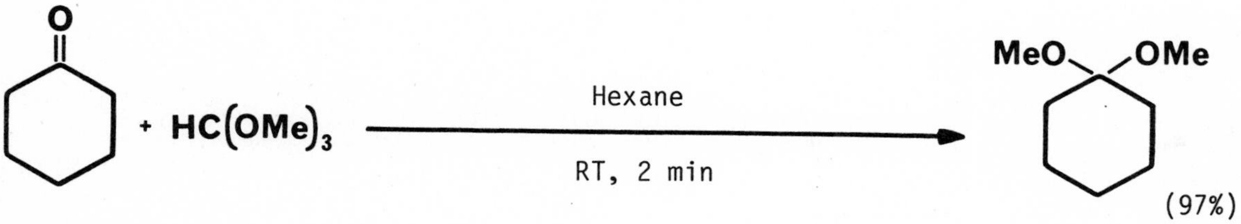

(97%)

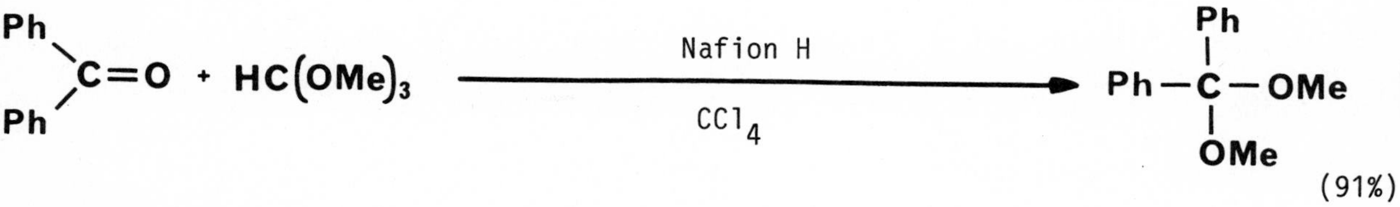

Ph, Ph, C=O + $HC(OMe)_3$ —[Nafion H / CCl_4]→ Ph, Ph—C—OMe, OMe ⑤ [1] △1

(91%)

REAGENT: TRIMETHYLPHOSPHITE

STRUCTURE:

MW = 124.08

BP = 111- 112° C

d = 1.052

$(MeO)_3P$

Fieser's Reagents for Organic Synthesis: Vol., pages

1	1233- 35	**4**	541- 42	**7**	393- 99	**10**	
2	439- 41	**5**	717	**8**		**11**	570
3	315- 16	**6**		**9**	490- 91	**12**	

March's Advanced Organic Chemistry: 919, 1034, 1066

MAJOR USES:

1. Molecular rearrangements
2. Enolizations
3. Phosphorylations
4. Deoxygenations
5. Desulfurizations

PREPARATION:

1. Commercially available

PRECAUTION:

1. Flammable and moisture sensitive

NOTES:

1. **Bz** = **Bzl** = benzyl

REFERENCES:

1. M. Hirayama, K. Gamoh and N. Ikekawa. **Tetrahedron Lett.**, (198), 4725.

2. A. Suarato, P. Lombardi, C. Galliani and G. Franceschi, **Tetrahedron Lett.**, (1978), 4059.

3. M. Kosugi, T. Ohya and N. Inamoto, **Bull. Chem. Soc. Jpn.**, (1983), 56, 3539.

4. K. Akiba, Y. Negishi and N. Inamoto, **Synthesis**, (1979), 55.

5. S.M. Katzman and J. Moffat, **J. Org. Chem.**, (1972), 37, 1842.

EXAMPLES:

$P(OMe)_3$, MeOH-THF

25°

(52%) ① [1]

②

$Bu_3Sn-SiMe_3$ + $BrCH_2C(O)Ph$

$PdCl_2 + 2\ P(OMe)_3$

Benzene, △, 5 hrs

$PhC(OSiMe_3)=CH_2$ (81%) + $PhC(O)Me$ (8%)

③

+ BzCl + $P(OMe)_3$

NaI

Dry acetonitrile

NBz

$(MeO)_2P=O$

(75%) △1

Ph Ph

N N⊕ O⊖

$P(OMe)_3$

△, Reflux

2 Ph−C≡N (87%) ④

NC S NC S =S

Excess $P(OMe)_3$

Toluene, Reflux, 18 hrs

NC S S CN

NC S S CN

(70%) ⑤

REAGENT: TRIPHENYLPHOSPHINE

==

STRUCTURE:

Ph_3P

MW = 262.29
MP = 79- 81°

Fieser's Reagents for Organic Synthesis: Vol., pages

1	1238- 47	**4**	548- 50	**7**	403- 04	**10**	
2	443- 45	**5**		**8**		**11**	588
3	317- 20	**6**	643- 44	**9**		**12**	550

March's Advanced Organic Chemistry: Not Indexed

MAJOR USES: ⚠1

1. **Wittig Reaction** 2. Deoxygenation 3. Desulfurization

PREPARATION:

1. Commercially available in high purity. Can be also obtained as a polymer supported reagent.

PRECAUTION:

1. An irritant.

NOTES:

1. $Ph_3P + CX_4$ (X = Br or Cl) is a useful reagent for converting primary alcohols to the corresponding organohalogen compound.

2. **THP** = Tetrahydropyranyloxy

REFERENCES:

1. B.P. Gunn, **Tetrahedron Lett.**, (1985), 2869.

2. H.R. Buser, P.M. Guerin, M. Toth, G. Szocs, A. Schmid, W. Francke and H. Arn, **Tetrahedron Lett.**, (1985), 403.

3. H. Wamhoff, W. Schupp, A. Kirfel and G. Will, **J. Org. Chem.**, (1986), 51, 149.

4. P. Lambert, M. Vaultier and R. Carrie, **Chem. Commun.**, (1982), 1224.

5. E.J. Corey, B. Samuelson and F.A. Luzzio, **J. Am. Chem. Soc.**, (1984), 106, 3682.

EXAMPLES:

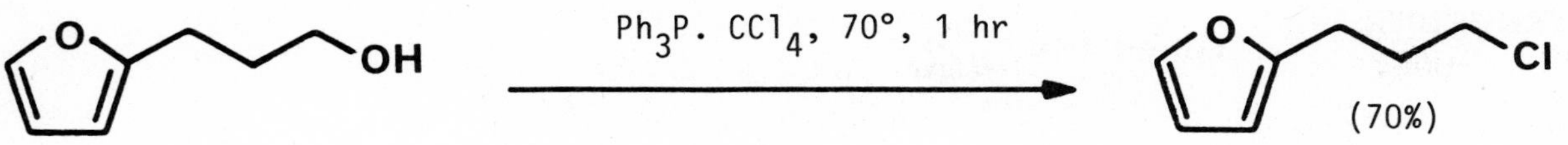

① △1

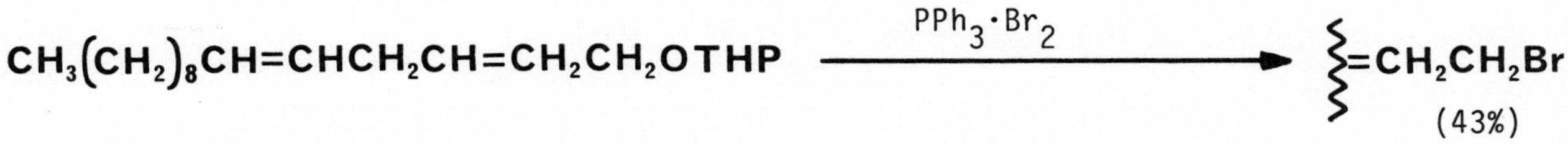

$Ph_3P.\ CCl_4$, 70°, 1 hr

Cl (70%)

②

$CH_3(CH_2)_8CH=CHCH_2CH=CH_2CH_2OTHP$ —$PPh_3 \cdot Br_2$→ $=CH_2CH_2Br$ (43%) △2

③

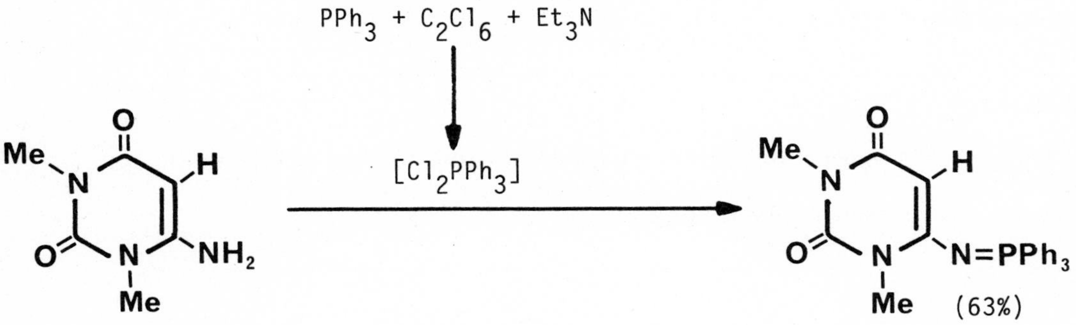

$PPh_3 + C_2Cl_6 + Et_3N$

$[Cl_2PPh_3]$

Me N H O O N NH_2 Me → Me N H O O N $N=PPh_3$ Me (63%)

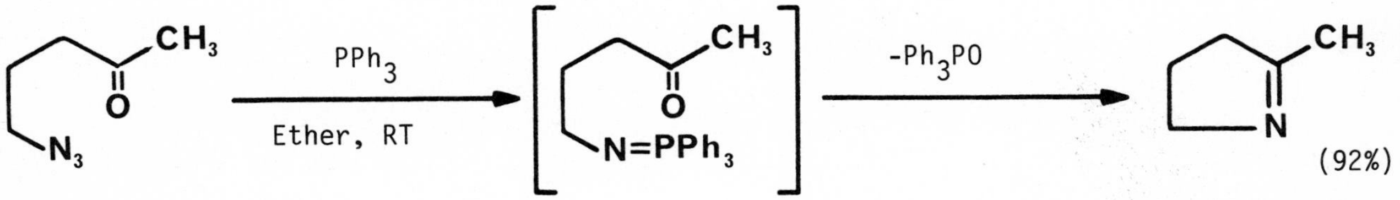

④

CH_3 O N_3 —PPh_3, Ether, RT→ [CH_3 O $N=PPh_3$] —$-Ph_3PO$→ CH_3 N (92%)

⑤

$C_8H_{17}N_3 + PPh_3$ —CH_2Cl_2, N_2, RT, 5 hrs→ [$C_8H_{17}N=PPh_3$] —O_3, CH_2Cl_2, -78°, 10 min→ $C_8H_{17}NO_2$ (70%)

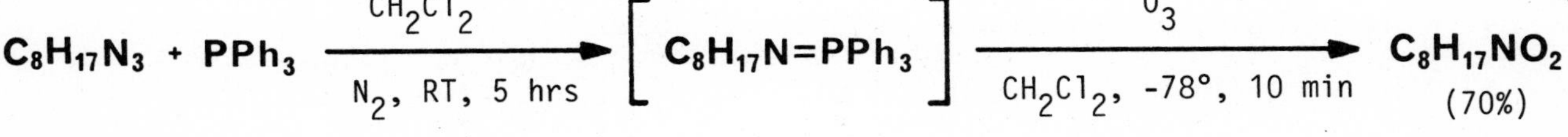

REAGENT: TRIPHENYLPHOSPHINE- DIETHYLAZODICARBOXYLATE (TPP-DEAD)

==

STRUCTURE:

$EtO_2CN{=}NCO_2Et/Ph_3P$

MW = 174.16 262.28
BP = 93- 95° @ 5 mm **MP** = 80.5°

Fieser's Reagents for Organic Synthesis: Vol., pages

1	245- 47	**4**	553- 55	**7**	404- 06	**10**	448- 49
2		**5**	727- 28	**8**	517	**11**	589
3		**6**	645	**9**	504- 06	**12**	552- 54

March's Advanced Organic Chemistry: 345, 934, 1092

MAJOR USES: 1 2

1. Condensations 2. Cyclizations 3. Alkylation 4. Dehydrogenation

5. **Mitzunobo Reaction**

PREPARATION:

1. **DEAD** is prepared from ethoxycarbonyl chloride and fuming nitric acid.

2. Both reagents are commercially available.

PRECAUTION:

NOTES:

1. Et_3P is becoming more preferred than TPP.

2. Review: O. Mitsunobo, **Synthesis**, (1981), 1.

REFERENCES:

1. B.H. Lee, A. Biswas and M.J. Miller, **J. Org. Chem.**, (1986), 51, 106.

2. P.-K. No and N. Davies, **J. Org. Chem.**, (1984), 49, 3027.

3. H. Nagasawa and O. Mitsunobu, **Bull. Chem. Soc. Japan**, (1981), 54, 2223.

4. W.A. Hoffman, **J. Org. Chem.**, (1982), 47, 5209.

5. P.G. Mattingly, J.F. Kerwin and M.J. Miller, **J. Am. Chem. Soc.**, (1979), 101, 3983.

EXAMPLES:

DEAD / TPP ① [1] (75%)

TPP / DEAD / $ZnCl_2$, THF ② (92%)

$PhC(O)NHC(NEt_2)=S + HOCH_2CH_2Ph$ → DEAD / TPP, THF / RT, 18 hrs → $PhC(O)N=C(NEt_2)-SCH_2CH_2Ph$ ③ (94%)

$2\ C_5H_{11}OH + CO_2$ → TTP / DEAD, Dry THF, 18 hrs → $(C_5H_{11}O)_2C=O$ ④ [1] (81%)

DEAD / TTP, THF, 50°, 6 hrs ⑤ [2] (80-90%)

REAGENT: WILKINSON'S CATALYST
CHLOROTRIS(TRIPHENYLPHOSPHINE)RHODIUM (I)
TRIS-(TRIPHENYLPHOSPHINE)RHODIUM(I) CHLORIDE 

==

STRUCTURE:

MW = 925.24
MP = 140°C (dec)

$[(C_6H_5)_3P]_3RhCl$

Fieser's Reagents for Organic Synthesis: Vol., pages

Vol.	pages	Vol.	pages	Vol.	pages	Vol.	pages
1	1252	**4**	559- 62	**7**	68	**10**	98- 99
2	348- 53	**5**	736- 40	**8**	109	**11**	130
3	325- 29	**6**	652- 53	**9**	113- 14	**12**	

March's Advanced Organic Chemistry: 510, 655, 692, 694-698, 918, 933

MAJOR USES:

1. Catalyst for cyclization and coupling reactions
2. Trans-halogenation

PREPARATION:

1. Can be prepared from the reaction of $RhCl_3 \cdot 3H_2O$ and triphenylphosphine

2. Commercially available.

PRECAUTION:

NOTES:

1. Also known as chlorotris(triphenylphosphine)rhodium(I).

2. For a Review, see: **Chem. Rev.**, (1973), 73, 21.

REFERENCES:

1. R. Grigg, R. Scott and P. Stevenson, **Tetrahedron Lett.**, (1982), 2691.

2. K. Sakai, Y. Ishiguro, K. Funakoshi, K. Vebo and H. Suemune, **Tetrahedron Lett.**, (1984), 961.

3. J.J. Talley and A.M. Colley, **J. Organomet. Chem.**, (1981), 215.

4. B.C. Laguzza and B. Ganem, **Tetrahedron Lett.**, (1981), 1483.

5. J.E. Lyons, **Chem. Commun.**, (1975), 418.

EXAMPLES:

$PhSH + HSnBu_3 \xrightarrow[\text{THF / } N_2]{ClRh[PPh_3]_3} PhS\text{-}Sn\text{-}Bu_3$ ③

(95%)

$CH_3I + CH_2Cl_2 \xrightarrow[\text{Argon, 100°, 2 hrs}]{ClRh[PPh_3]_3} CH_3Cl + CH_2ICl$ ⑤

(96%)

REAGENT: ZINC (DUST)

STRUCTURE:

Zn

MW = 65.37
MP = 419.5° C   **d** = 7.140

Fieser's Reagents for Organic Synthesis: Vol., pages

1	1276- 84	**4**	574- 77	**7**	426- 28	**10**	459
2	459- 62	**5**	753- 56	**8**	532	**11**	598
3	334- 37	**6**	672- 75	**9**		**12**	

March's Advanced Organic Chemistry: See Index

MAJOR USES:

1. Reduction   2. Dehalogenation   3. Dehydrohalogenation

4. **Reformatsky reaction**   5. Condensations   6. **Simmons- Smith reaction**

PREPARATION:

1. For preparation of a highly reactive zinc powder see; R. Rieke, P. Li, T. Burnsa and S. Uhm, **J. Org. Chem.**, (1981), 46, 4323.

PRECAUTION:

1. The fine powder is flammable.

2. Zinc dust may deteriorate (via air oxidation). For purification, see: **Reagents**, 1, 1276.

NOTES:

1. **TBS-** is a t-butyldimethylsilyl protecting group. It is formed from **TBSCl**.

REFERENCES:

1. T. Shono, H. Hamaguchi, I. Nishiguchi, N.M. Miyamoto and S. Fujita, **Chem. Lett.**, (1981), 1217.

2. B. Han and P. Boudjouk, **J. Org. Chem.**, (1982), 47, 751.

3. I. Murata, **J. Org. Chem.**, (1986), 51, 251.

4. F.E. Ziegler and A. Kneisley, **Tetrahedron Lett.**, (1985), 263.

5. L. Kolaczkowski and W. Reusch, **J. Org. Chem.**, (1985), 50, 4766.

EXAMPLES:

Zn, CH_3CN, Heat (74%) ① [5]

Zn (NYA) ② [2]

Zn, Acetic acid (92%) ③ [1]

Zn / aq HOAc (88%) ④ [3] △1

Zn, HOAc / H_2O (95%) ⑤ [1]

REAGENT: ZINC BROMIDE

STRUCTURE:

$ZnBr_2$

MW = 225.19
MP = 394°C
d = 4.201

Fieser's Reagents for Organic Synthesis: Vol., pages

1		**4**		**7**		**10**	461
2	463-64	**5**		**8**	535	**11**	600
3		**6**		**9**	520	**12**	

March's Advanced Organic Chemistry: 401

MAJOR USES:

1. Lewis acid
2. Useful for cleavage of MEM ethers

PREPARATION:

1. Commercially available in very high purity.

PRECAUTION:

1. Irritant and very hygroscopic.

NOTES:

REFERENCES:

1. H. Khan and I. Patterson, **Tetrahedron Lett.**, (1982), 2399.

2. E.J. Corey, J.-L. Gras and P. Ulrich, **tetrahedron Letters**, (1976), 809.

3. T. Morimoto, M. Aono and M. Sekiya, **Chem. Commun.**, (1984), 1055.

4. H.A. Kahn and I. Patterson, **Tetrahedron Lett.**, (1982), 5083.

5. A. Koshinen and M. Lounasmaa, **Tetrahedron Lett.**, (1983), 1951.

EXAMPLES:

$OSiMe_3$ / Me (1-trimethylsilyloxy-2-methylcyclohexene) + $ClCH(SPh)CH_2CH=CH_2$ → 1.) $ZnBr_2$; 2.) O_3, Me_2CO; 3.) Δ → 2-methyl-2-(CH=CH–CHO)cyclohexanone (45%) ① [1]

t-Bu–cyclohexyl–$OCH_2OCH_2CH_2OMe$ → $ZnBr_2$, CH_2Cl_2, 25° → t-Bu–cyclohexyl–OH (92%) ② [2]

$PhS-SiMe_3$ + $MeOCH_2N(SiMe_3)_2$ → $ZnBr_2$, N_2 / 60°, 3 hrs → $PhSCH_2N(SiMe_3)_2$ (98%) ③ [1]

$Me_3SiO(OSiMe_3)C=CHEt$ + $Me_3SiOCH_2CH_2CH(Cl)SPh$ → $ZnBr_2$, CH_2Cl_2, 20°, 45 min, then H^+Cl^- → 3-Et-4-SPh-tetrahydropyran-2-one (93%) ④ [1]

1-Me-2-CN-3-(CH_2Me)-1,2,5,6-tetrahydropyridine + $Me(OSiMe_3)C=CHCO_2Me$ → $ZnBr_2$, CH_2Cl_2, Dimethoxyethane, Argon, RT → MeN-tetrahydropyridine with CH_2Me and $CH(CO_2Me)C(=O)Me$ (80%) ⑤ [1]

REAGENT: ZINC CHLORIDE

==

STRUCTURE: ⚠1

$ZnCl_2$

MW = 136.28
MP = 293°C
d = 2.907

Fieser's Reagents for Organic Synthesis: Vol., pages

1	1289- 92	**4**		**7**	430	**10**	461- 62
2	464	**5**	763-64	**8**	536- 37	**11**	602- 04
3	338	**6**	676	**9**	522- 23	**12**	

March's Advanced Organic Chemistry: 495- 97

MAJOR USES:

1. Acetylation 2. Lewis acid catalyst 3. Alkylation

PREPARATION:

The reagent is commercially available in very high purity.

PRECAUTION:

The reagent is an irritant, and is toxic.

NOTES:

1. The anhydrous $ZnCl_2$ readily takes up water. Must be kept under anhydrous conditions.

2. The zinc complex of the enolate is used to direct the **aldol condensation**.

REFERENCES:

1. D. Widdowson, G. Wiebeuce and D. Williams, **Tetrahedron Lett.**, (1982), 4285.

2. R. Tirpak and M. Rathke, **J. Org. Chem.**, (1982), 47, 5099.

3. H. Klein and H. Mayr, **Angew. Chemie. Int. Ed.**, (1981), 20, 1027.

4. M. Renard and L. Hevesi, **Tetrahedron Lett.**, (1985), 1885.

5. J.-N. Denis, A.E. Green, A.A. Serra and M.-J. Luchi, **J. Org. Chem.**, (1986), 51, 46.

EXAMPLES:

γ-Butyrolactone + PhCHO —($LDA, ZnCl_2$)→ threo + erythro α-(hydroxy(phenyl)methyl)-γ-butyrolactones (83%, 30:70) ① 3 △2

$Me_2CHC(OSiMe_3){=}CMe_2$ + $MeC(O)Cl$ —($ZnCl_2$)→ $Me_2CHC(O)CMe_2(COMe)$ (94%) ② 1

$Me_2C{=}C(Me)CH(Cl)Me$ + $EtOCH{=}CMe_2$ —($ZnCl_2$)→ EtO-substituted hexamethylcyclopentene (73%) ③ 2

MeSe(H)C=CHCH$_2$OH —($MeSeH / ZnCl_2$; CH_2Cl_2, RT)→ MeSe(H)C=CHCH$_2$SeMe (82%, 100% E) ④ 2

Ph(H)C–O–C(H)CO_2Me (epoxide) —($Me_3SiN_3 / ZnCl_2$; THF / HOAc / HCl / CH_2Cl_2)→ N_3CH(M)CH(HO)C(O)OMe (90%) ⑤ 2

INDEX